Return or renew all Library Materials! The *Minimum* Fee for Book is $50.00.

on charging this material is responsible for to the library from which it was withdrawn ore the **Latest Date** stamped below.

underlining of books are reasons for discipli-
missal from the University.

WITHDRAWN
University of
Illinois Library
at Urbana-Champaign

Franklin J. Reiss

FARMLAND

By David A. Lins, Neil E. Harl
and Thomas L. Frey

Copyright © 1982 by Agri Business Publications, Division of Century Communications, Inc., 5520-G West Touhy Ave., Skokie, IL 60077. All rights reserved, including those to reproduce this book or parts thereof in any form. Printed in the United States of America. First Edition.

Library of Congress Card Number 81-71618

ISBN 0-930264-60-6 (hard cover edition)
ISBN 0-930264-57-6 (soft cover edition)

1281-1H-9S

PREFACE

This book is the result of discussions between the editors of AGRI FINANCE and the authors concerning the need for a book to clearly document the economic, legal and tax aspects of farm and ranch land ownership. The book was designed to cover issues related to the purchase, management, and disposition of farm and ranch land. People interested in these topics include not only farmers and ranchers, but also landlords, absentee owners, lenders, agribusiness firms and others who have a personal or financial interest in farm and ranch land. The diverse nature of the audience was kept in mind as the book was written.

At the outset we should clarify that this book is not devoted to farmland appraisals — although we do discuss appraisal techniques. Likewise, it is not a "get rich quick" guide to investment in land. Rather, the book deals with the economic, legal, and tax aspects of buying, selling and managing farm and ranch land.

Chapter 1 introduces you to the uses, concepts, and rights in land and water, while Chapter 2 identifies the characteristics of farm and ranch land transfers. Returns to land versus other investments are compared in Chapter 3. Economic and noneconomic reasons for owning land are also explored. Such information is essential for understanding the market in which land transactions occur. This lays the foundation for the discussion on investment, disinvestment and management decisions involving farm and ranch land.

Chapters 4 through 6 deal with evaluating land as an investment. Procedures for evaluating farmland purchases are reviewed. Methods of establishing the value of land between a "willing buyer and willing seller" are also discussed. Detailed examples of how to determine what the land is worth to you are presented.

Methods of financing purchases of farmland and procedures for measuring the cash flow feasibility of farmland purchases are discussed in Chapters 7 and 8. Estimates of how much various terms of financing influence land values are presented. Chapter 9 identifies alternative sources of funds for the purchase of land.

Chapters 10 and 11 deal with land acquisition. Discussion focuses on differing methods of buying and selling farm and ranch land and the legal aspects of land purchases.

Chapters 12 through 15 deal with farm and ranch management for the nonoperator owner. The focus on nonoperator owners is intentional because informational needs for farm production practices are extensive and far beyond the scope of this book. Consequently this section focuses on sources of management information, and different management arrangements. Liability considerations associated with land ownership are also identified.

Chapters 16 through 19 identify the tax aspects of land ownership and disposition. Separate chapters are devoted to property taxes and the income tax aspects of acquiring land. Disposition can occur by voluntary sale, forced sale, by gift, or by transfer at death. Legal procedures and tax consequences of these alternatives are identified.

The information contained in this book is based upon the state of the law as of November of 1981. Legal changes are inevitable and these changes should be taken into account as you use the materials contained herein.

TABLE OF CONTENTS

1 THE NATURE OF FARMLAND AND FARMLAND VALUES 1
Concepts and Definitions of Land
Rights of Ownership
Limitations Imposed by Government
Rights in Water
Interests in Real Estate
Valuation Concepts and Characteristics
Elements of Value
Land Characteristics That Create Value
Economic Characteristics
Summary

2 CHARACTERISTICS OF FARMLAND AND FARMLAND TRANSFERS 11
How Much Farm and Ranch Land Is There?
How Is the Land Used?
Who Owns the Land?
Who Buys Farmland?
Who Sells the Land?
Sizes and Prices of Parcels Sold
Financing Farmland Transfers
Real Estate Exchanges

3 WHY BUY OR SELL FARMLAND? 23
Personal Reasons for Buying Land
   Economic Incentives for Buying Land
   Farmland as a Hedge Against Inflation
   Returns on Investment
   Are Capital Gains on Farmland Justified?
Personal Reasons for Selling Land
   Economic Reasons for Selling Land
   Know What Motivates the Buyer/Seller

4 GATHERING DATA AND INFORMATION FOR EVALUATING VALUE OF REAL ESTATE 37

General Considerations
Climatic Factors
Economic
Soils and Land Useability
Land Classification
Land Use, Production, and Treatment History
Buildings and Improvements
Additional Considerations
Sources of Information and Assistance

5 DETERMINING THE FAIR MARKET VALUE OF LAND 57

Market Data Approach
Income Approach
Cost or Inventory Approach
Determining the Final Estimate of Fair Market Value
Summary

6 WHAT IS THE LAND WORTH TO YOU? 71

Time Value of Money
   Compounding
   Discounting
Net Present Value of Land
Examples of Net Present Value Calculations
Appendix

7 CASH FLOW FEASIBILITY OF FARMLAND PURCHASES 89

Estimating Cash Flows
Will the Land Pay for Itself?
Is There Adequate Cash Flow Available for Debt Servicing
Negotiating to Improve Cash Flow Feasibility
Appendix: Computing Payments on an Increasing Amortization Loan

8 HOW TERMS OF FINANCING INFLUENCE FARMLAND VALUES 111

Interest Rates
Downpayment
Length of Loan
Appendix: Computing the Annual Percentage Rate of Interest

9 FINANCING THE PURCHASE OF LAND 133

Terms Associated With Farm Real Estate Financing
Sources of Farm Real Estate Loan Funds
Federal Land Banks
Farmers Home Administration
Life Insurance Companies
Commercial Banks
Individuals
Sources of Outside Equity Capital

10 GETTING BUYERS AND SELLERS TOGETHER 147

Private Agreement
Auctions
Real Estate Brokers
Summary

11 THE LAND TRANSACTION 159

Preliminary Considerations
The Offer to Buy or Land Contract
Quality of Title
Types of Deeds
Financing the Transaction
The Closing of the Transaction

12 MANAGEMENT ARRANGEMENTS AND INFORMATION 183

The Farm Owner-Operator
Custom Hiring
Leasing the Farm
Professional Farm Managers
Where To Go For More Information

13 LIABILITY CONSIDERATIONS 197
A Litigious Era
Negligence
Special Situations
Insurance Coverage

14 FEDERAL AND STATE PROGRAMS AND REGULATIONS 205
Federal Programs
State Programs
Federal Regulations
State and Local Regulations

15 PROPERTY TAXES 215
Types of Property Taxes
Administration of the Property Tax

16 INCOME TAX ASPECTS OF ACQUIRING LAND 221
Income Tax Basis
Investment Tax Credit
Other Income Tax Deductions

17 DISPOSITION OF FARMLAND BY SALE OR EXCHANGE 243
Preparing For Sale
Reporting Income Tax Gains
Installment Sale
Private Annuity
Tax-Free Exchange
Income Tax Treatment of Entity Owning Land

18 TRANSFERRING LAND AT DEATH ... 261
Use Valuation of Land
Property Interests Subject to Federal Estate Tax
The Adjusted Gross Estate
Calculating the Taxable Estate and Tax Due
Federal Estate Tax Return

19 TRANSFERRING LAND BY GIFT **277**
Federal Gift Tax
How Valued
Changing Co-Ownership Patterns
Part Gift/Part Sale

APPENDIX A: INTEREST FACTOR TABLES **288**
APPENDIX B: SAMPLE OFFER TO BUY CONTRACT **296**
APPENDIX C: SAMPLE OF ABSTRACT OF TITLE **298**
APPENDIX D: SAMPLE DEED **300**
APPENDIX E: SAMPLE MORTGAGE **302**
APPENDIX F: SAMPLE INSTALLMENT CONTRACT **304**
APPENDIX G: SETTLEMENT SHEETS **306**
APPENDIX H: LEASE FORMS **308**

CHAPTER ONE
THE NATURE OF FARMLAND AND FARMLAND VALUES

This is a book about buying, selling, and managing farmland. It covers the very basic concepts and characteristics of land as well as the factors affecting value and how value differs from person to person. The impacts of financing and alternative financial strategies are explored. The legal concepts of buying, owning, and managing including details on initiating and closing real estate transactions are presented.

Management concerns from the viewpoint of the non-resident owner are covered. And finally, given the tremendously important tax implications associated with buying, owning, and disposing of real estate, we devote the last five chapters to tax related issues. In summary, the book deals with the economic and legal aspects of buying, selling, and managing farm and ranch land.

But what is land? What is real estate? What rights and limitations go with its ownership? What is value and what are the characteristics of value? How can it be estimated? The purpose of this chapter is to address these basic questions and thus build a base for the remainder of the book.

Concepts and Definitions of Land

There are many different views or concepts of land throughout the world. In some societies land is viewed as a diety or god. Under this concept, the uses of land are governed as much by moral judgments as by legal and economic decisions. Some societies, like China, view land as communal property. Under these conditions there is no market for land. Land is owned by society and it is neither bought nor sold.

In the United States we believe in a concept of individual rights in land. While Federal, state and local governments do own land, individuals also own land. This gives rise to a land market since individuals can buy or sell land. But even within the United States, there are several different concepts and definitions of land.

For example, the term *real estate* refers to the physical land and appurtenances, including structures affixed thereto. It is a physical definition that includes not only natural attachments such as trees and rocks, but also all the fixed improvements that are created and permanently attached to the land by man.

In contrast, the term *real property* refers to the interests, benefits, and rights inherent in the ownership of the physical real estate. It denotes the bundle of rights with which the ownership of real estate is endowed (the bundle of rights will be explained later). This broadens the narrower technical definition of real estate. But often (including this book) the term "real estate" is used as a composite term of the two definitions to include the tangible (physical) elements of real estate plus those intangible attributes which are rights of ownership. The term *land*, and *farmland* are also used interchangeably with *real estate*, again the intent in general discussion of land being to include both rights of ownership as well as the physical commodity itself.

The legal concept of real estate is a broader, more comprehensive theory which holds that land:

". . . includes not only the ground or soil, but everything which is attached to the earth, whether by course of nature, as trees and herbage, or by the hand of man, as houses and other buildings. It includes not only the surface of the earth, but everything under it and over it. Thus, in legal theory, a tract of land consists not only of the portion on the surface of the earth, but is an inverted pyramid having its tip or apex at the center of the earth, extending outward through the surface of the earth at the boundary lines of the tract, and continuing on upward to the heavens."[1]

The above description can be expressed with the diagram in Figure 1.1. The theory expressed by Kratovil does not hold completely. Through the Air Commerce Act of 1926 and the Civil Aeronautics Act of 1938, Congress declared that the United States has complete sovereignty in the air space over the nation. Legal interpretation, however, holds that land owners do have exclusive control of the space and air rights above the land, sufficient for enjoyment and use of the surface. Thus, a landowner cannot prevent commercial airlines from flying over the premises. But the landowner can erect a very tall building unless barred by zoning ordinances or prior conveyance of air rights to another. Air rights and mineral rights can be bought and sold separately from the surface.

There is also an *economic concept* of land. Land can be viewed as a factor of production, or as a consumption good in and of itself. Viewed as a factor of production, land is one of the inputs used to produce agricultural products. By combining the inputs of land, labor, capital and management the farm operator expects to generate income. An alternative view is land as a consumption good. Some people buy land because they like the view from the property, or because it affords them the opportunity to achieve the privacy they desire.

---

[1]Kratovil, Robert. Real Estate Law, 3rd Edition. Prentice-Hall, Inc., Englewood Cliffs, N.J., 1958.

Figure 1.1 Legal boundaries of land

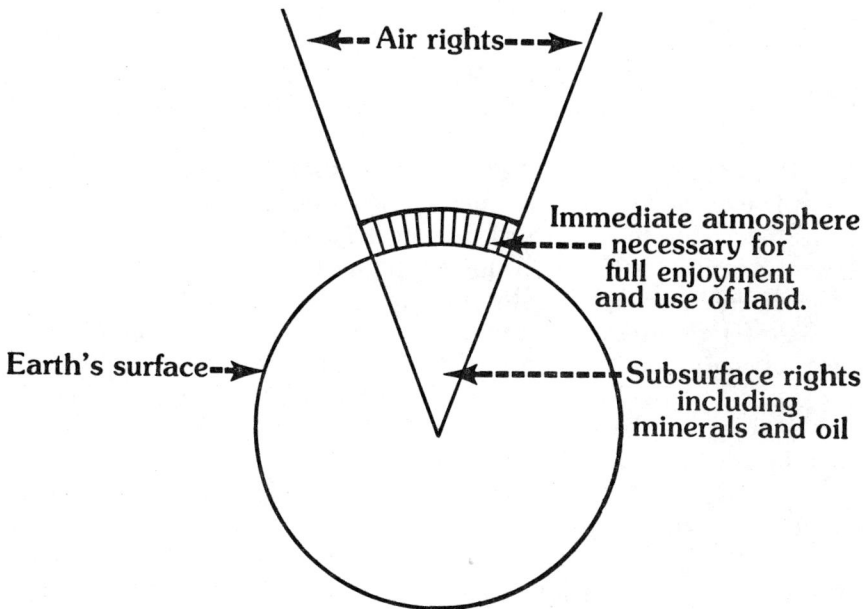

Rights of Ownership

The ownership of land in the United States has often been referred to as a "bundle of rights". These rights are sometimes compared to a bundle of sticks with each stick representing a separate right of ownership. These rights are:
- right of use
- right of disposition
- right of exclusion

Right of use—The owners of land have the right to use land in any way that they see fit, subject to the constraints imposed by law. Owners have the right to grow corn, to grow watermelons, to leave the land idle, to lease it to another, or to use it as a hunting preserve. However, previous owners may have imposed restrictions on one's ability to use land. For example, restrictive covenants may govern the use of land such as limitations on the size and type of building that may be constructed on the land.

Right of exclusion—The owners of land can exclude other people from their property. For example, land owners may prevent others from trespassing or hunting on their property, thereby exercising their right of exclusion.

Right of disposition—The owners of land can dispose of the land in any manner they see fit. They can sell it, will it to their heirs or donate it to a favorite charity. The owners of land can also grant easements, or subdivide the property.

Limitations Imposed by Government

Rights to land are not absolute—they are limited by four specific powers of government:
- Power of taxation
- Power of eminent domain
- Police power
- Escheat

Power of taxation—Real estate is subject to taxation by the government. While there is much discussion about the fairness of assessments, the fact remains that local governmental units, such as school districts, get most of their funding through the tax power of government. Ownership rights are subject to taxation.

Power of eminent domain—This is the right of government to take private property by condemnation procedures (eminent domain) for public benefit. Just compensation must be paid for property taken and the use must be for the public. Land acquisition for public roads represents a frequent use of this power.

Police power—Police power is a right of government to regulate property for promoting the public's safety, health, morals, and general welfare. Rights in property, its use and occupation may be restricted without any compensation whatsoever when the government deems such restrictions necessary. Zoning, building codes, and city and county planning are based upon this power.

Escheat—Escheat is not a limitation on current land ownership but rather provides for reversion of land to the state when an owner of land dies, leaving no heirs and without having disposed of the land by will.

The four powers summarized above are limitations imposed by government on land ownership. But keep in mind that for a specific piece of property further limitations may be imposed through the action of current or previous owners, i.e. there may be restrictive covenants or deed restrictions. For example, a seller may stipulate that the property is or is not to be used for certain purposes. Or maybe the mineral rights were retained by a previous owner. Both governmental and private limitations on property should be identified on real estate purchases.

Rights in Water

Closely related to rights in land are the rights in water which flows over, through, or under the earth's surface. Concerns over water rights are directly related to the available supply of water. For example, problems over water rights are far more common in the Western parts of the United States than in the Eastern states.

Much of the more humid regions of the United States are governed by the *riparian rights doctrine*. According to this doctrine, all landowners whose properties are bounded by or crossed by a river, stream, spring, or natural body of water have the right to that water for *reasonable use*. However, there must be an adequate supply of water remaining to meet the *natural* needs of other riparian owners. A similar set of rules governs the right to use ground water.

During settlement of the arid Western parts of the United States it became apparent that riparian rights were in conflict with the most beneficial uses of water. For example, when the Mormons settled the Great Salt Lake Valley they irrigated the land and acquired the water under an *appropriation doctrine*. Under this doctrine, both riparian and nonriparian landowners can file claims to use part of the water from streams or other bodies of water. The essence of the appropriation doctrine can be summarized as "first in time, first in right." Thus the first claimant has the highest priority while the last claimant has the lowest priority. However, all claimants are limited to the quantity of water which must be for "beneficial use".

Nine states, Arizona, New Mexico, Nevada, Utah, Colorado, Wyoming, Montana, Idaho, and Alaska have now completely accepted the appropriation doctrine over the riparian doctrine. Oregon, California, North Dakota, South Dakota, Nebraska, Kansas, Oklahoma, and Texas use the appropriation doctrine in combination with the riparian doctrine.

About one-quarter of the states have shifted to an *administrative* or *permit system* for allocating water. A state agency issues permits to use a specified quantity of water for a designated period (typically 10 to 25 years). Some uses, such as for domestic-purposes and livestock, do not require a permit.

Interests in Real Estate

The word "estate" in law means the degree, nature, extent and quality of interest or ownership that one has in land. This correctly implies that there can be variations in the extent of interest held by individuals in property. Estates can be classified into two major categories: *freehold estates,* and *leasehold estates*. Freehold estates are further classified into: (1) fee simple, (2) fee simple determinable or fee simple subject to a condition and (3) life estate and remainder. We turn to a description of each of these estates.

Fee Simple Estate—This is the highest type of legal interest in real estate. A holder with fee simple title is entitled to all the rights incident to the property, subject to the limitations described above. This estate is of unlimited duration in time and upon the death of the owner, it passes to the heirs or as provided for in the owner's will, or to the state if there are no heirs and no applicable will provision to govern property disposition. Three other terms are used to describe this estate, all meaning the same fee simple estate. They are *fee, fee simple,* and *fee simple absolute*.

Determinable Fee Estates—Here the ownership interests exist only "as long as" certain conditions exist. The estate is extinguished or ended upon the occurrence of a designated event—but the time of occurrence is uncertain, and there is no assurance that the event will ever happen. Title reverts back to the original grantor or the heirs or devisees under the will at the time the event occurs. An example, applicable in some rural areas, is the land deeded for school house use, but with the provision that if the land ceased being used for a school house, it would revert back to the grantor or the grantor's heirs. As with fee simple estates, determinable fee estates are of indeterminant length. A similar estate is the *"fee simple subject to a condition"*.

Life Estate and Remainder—A life estate denotes that ownership rights are vested in some designated person or entity but only during the lifetime of that person or some other specified person or persons. At the death of the specified person(s), the estate terminates and the right of possession passes to whomever was designated as *the remainderman* at the time the life estate was created.

A life tenant, the holder of the life estate, cannot encroach upon the rights of the remainderman. However, a life tenant is entitled to all the income and profits arising from the property during the term of ownership and may even sell or mortgage *the life interest*. However, acting alone a life tenant cannot sell and give good title or mortgage the *entire* ownership interest in the property. Together, the life tenant and remainderman may be able to sell or mortgage the land.

A special problem arises if a life tenant leases the land to a farm tenant. At the death of the life tenant, by the majority view the farm lease terminates at that time. The farm tenant may have the right, under state law, to harvest a crop already planted. But death of the life tenant late in the lease year, can interrupt the farm tenant's expectations for the next crop year. As a result, some states have adopted statutes continuing the lease at least until the date set for terminating farm tenancies in that state (often March 1).

If death of the life tenant occurs after the date set for giving notice to terminate the lease, state law may provide the lease continues for the next crop year. In states that do not have such statutes, and most do not, farm tenants leasing from a life tenant may wish to obtain the signatures of all remaindermen to the lease to prevent lease termination at death of the life tenant. The problem usually does not arise with life estates held in trust because the trustee typically has the right to bind both the life tenant and the remaindermen to farm leases.

The most typical life estate is for a given individual to have ownership rights during that person's own lifetime. For example, a spouse may die and leave a life interest in the farm to the surviving spouse, with the property passing to the children upon the death of the surviving spouse.

Leasehold Estates—A land owner can lease real estate to a tenant. The tenant's right to occupy the land for the duration of the lease is called a *leasehold estate*. When property is sold, the rights of existing tenants with enforceable rights to the land must be honored. For example, if someone has one year remaining on a five-year lease, the new owner must allow the *leasehold estate* to continue for that one year.

There are four categories of leasehold estates:

1) **Estate for years.** This is any estate which continues for a definite period of time, whether it be for a period of months or a period of years, with no continuation of the lease specified.

2) **Estate from year to year.** Under this arrangement the term is indefinite, since the lease states an initial period of time but allows for automatic continuation until such time as proper notice is given by one of the parties to terminate. This has been a common arrangement in farm leasing, where a tenant stays on for an additional year unless notification is given by the landlord or tenant by a certain date each year.

3) **Tenancy at will** is an estate created by possession. Rights for the tenant are granted by consent of the landlord. It is for an indefinite period and can be terminated with proper notice in accordance with the statutes of the state where the land is located. Death of either the tenant or landlord generally terminates this estate.

4) **Tenancy at sufferance** is an estate that arises from a tenant retaining possession of property after the tenant's rights have expired and without the consent of the landlord.

A buyer of real estate should be familiar with those estates that are relevant for the property being considered and should understand whether the seller has the full interests that are listed for sale. Such concerns are legal in nature and should be discussed with an attorney.

Valuation Concepts and Characteristics

What is value? Value means many things to many people. And yet any buyer or seller knows that, somehow, valuation of real estate is central to exchange of property. In this section valuation concepts and characteristics of value are put into perspective.

There are two major categories of value in real estate:

1) **Subjective value** is the value of property to a given individual—often termed utility value or value in use because it represents what a given owner-user is willing to pay. It is the value as perceived by a specific individual and may differ widely from one individual to the next.

2) **Objective value** is often called the *market value* or value in exchange. This value represents the price at which the property can be sold or exchanged at a given time, based on the "willing-buyer" and "willing-seller" concept. It represents a group judgment on value and clearly embodies demand and supply factors.

The objective or market value is a *value in exchange*. Appraisers are typically interested in trying to estimate this "market value", as shown by the following definitions of market value:

FARMLAND

"1. As defined by the courts, the highest price estimated in terms of money which a property will bring if exposed for sale in the open market, allowing a reasonable time to find a purchaser who buys with knowledge of all the uses to which it is adapted and for which it is capable of being used.
2. Frequently, it is referred to as the price at which a willing seller would sell and a willing buyer would buy, neither being under abnormal pressure."[1]

Fundamental to the concept of value is the theory of highest and best use. Highest and best use can be defined as that legal use which is most likely to produce the greatest net return to the land and buildings over a given period of time. The highest and best use concept is important because the objective value of income property (such as farm-land) is dependent on the amount of net income that can likely be produced in the future. Appraisers assume a typical buyer would use the property in the most profitable manner.

Readers should note that Chapters 4 through 8 deal with the two values described above—namely the market (objective) value and the subjective value (value to a given individual). Chapter 5 specifically relates to the process a professional appraiser would go through in estimating market value. Chapter 6, 7 and 8 deal with how individuals can determine the value of land to themselves. It may be uniquely different from, or close to, the market value. The important point is that prospective buyers and sellers of farmland appreciate and understand the different values and develop the ability to estimate such values themselves or learn to work with professionals who are skilled at such analysis.

Elements of Value

What is it that gives something value? The following four items are frequently referred to as the basic elements of value:

1. **Utility** is the power of a good to render a service or fill a need. The object must have the ability to arouse the desire for possession and then have the power to give satisfaction. Utility alone, however, is not sufficient to assure value. Otherwise, air and water would be valuable as measured in terms of dollars. Scarcity is missing.

2. **Scarcity** relates to supply and demand for the commodity. There must be some degree of scarcity for an item to have value. The air example above illustrates an item with the utmost utility value but in super abundance. Thus, no dollar value attaches. Utility and scarcity are both required, but still not sufficient by themselves to assure value. It takes effective demand.

3. **Effective Demand** is the desire of a purchaser who has sufficient resources to purchase the good or service. This is often termed *purchasing power*. The crucial economic concept implied by demand is that there is presence of both a need and the monetary power

---
[1]Source: American Institute of Real Estate Appraisers, *Appraisal Terminology and Handbook*, 5th Ed. Amer. Inst. of Real Estate Appraisers, 155 E. Superior St., Chicago, IL., 1967.

to fill that need. Many people *want* land and real estate, but only those with financial resources create a real demand for it. In real estate especially, purchasing power is a central focus.

4. **Transferability** is the fourth element of value. For an item to have value it must be possible to transfer the item, or at least title to the item, from one person to another. Otherwise no transaction could take place and there would be little value.

All other things constant, a relative change in one of the elements of value will alter the value of the commodity. One should evaluate carefully the future prospects related to these basic elements of value as a part of estimating value.

Land Characteristics that Create Value
Physical Characteristics.

1. **Immobility.** Land cannot be moved. Parts of a given tract of real estate such as topsoil and minerals can be separated and transferred to a different location, but the geographic location of the real estate is rigidly fixed. This leads to legal land descriptions being quite important, to allow descriptions for each tract of real property—that describes one and only one piece of real estate. Immobility leads to a market for real estate that is local in character. This is in contrast, for example, to a stock certificate representing ownership in a large corporation which is in the form of stock that may be readily marketable throughout the country.

2. **Indestructibility.** Land is durable and lasts through time. It is not a depleting resource like oil or other minerals. Conditions surrounding the land, such as the level of fertility, may change and alter its value but the land itself will still be present.

3. **Non-homogeneity.** No two parcels of land are exactly alike. Even though quite similar in physical characteristics, each has a geographic location that makes each tract unique. This characteristic ties back specifically to the scarcity and demand element of value. A given tract of land may offer desirable features that are not available elsewhere. With supply sharply curtailed, demand helps generate value for the land.

Economic Characteristics

1. **Scarcity.** Value exists only to the extent that some commodity is not an unlimited free good. There is a finite quantity of land in the world. But keep in mind that with the fixed supply of total land, the use can be altered. For example, more land is brought into cultivation each year. So broad generalizations about scarcity related to land could be misleading.

Consider the possibility that food can some day be produced on water without a land base. (Experiments of this nature are being conducted.) Prior to such a development, land might seem quite scarce relative to the demand for food production, but after the development there might be far more land than would be needed. Intensive utilization of land diminishes the need for land itself and has a similar effect relative to scarcity. The point here is that scarcity is a relative concept.

2. **Modifications.** Land reflects what man has done *to* the land and *on* the land. A road system, utilities, schools and nearby shopping facilities reflect improvements that may be beneficial *to* the land. Buildings and fences are improvements *on* the land. All these modifications are economic characteristics important to value of real estate.

3. **Fixity.** Once capital and labor have been committed to improvement on land, that investment is relatively fixed—it normally cannot be economically relocated or dismantled. Such fixed investments may occur through drainage or irrigation investments related to the land itself, or through buildings and associated improvements.

4. **Situs.** Situs is an economic location factor rather than geographic. Quality of location is extremely important. Land adjacent to the expanding area of a city may be much more valuable than comparable land located only a short distance away but in an area that is developing less rapidly. Location is an extremely important characteristic in determining land value.

Summary

This chapter has attempted to build a foundation for the balance of the book. Buyers and sellers of farmland should start with a understanding of what real estate is and what rights and limitations go with ownership of real property. Valuation is a central focus for buyers and sellers of real estate. But there are two basic "values." The market value (called *objective value*), represents that price a willing buyer and willing seller could be expected to agree upon, assuming a typical operation for the farm. The other value is termed *subjective* and represents what the real estate is worth to a given individual. The discussion in this chapter traced through value characteristics and specifically considered the physical and economic characteristics of land that create value.

CHAPTER 2
CHARACTERISTICS OF FARMLAND AND FARMLAND TRANSFERS

To deal wisely in farmland, either as a buyer, seller, or manager, you should have a clear understanding of the characteristics of farmland and farmland transfers. These characteristics include, but are not limited to—
- how much farmland exists
- how the land is used
- who owns the land
- who buys farmland
- who sells farmland
- what are the sizes and prices of parcels sold
- what are the terms of financing
- what is the reason for, and how important are, real estate exchanges

In this chapter we describe in detail these important characteristics of the farmland market.

How Much Farm and Ranch Land Is There?

There are 2.26 billion acres of land in the United States (Figure 2.1). Of this amount, just over 1 billion acres or 46% consist of crop land, grass land, pasture and range. There are 465 million acres of crop land, and 598 million acres of grass land, pasture and range. While the total land area is fixed, the amount used in farming and ranching changes over time.

The increasing population of the United States adds to the pressure on agricultural land. Since 1954, the amount of land in farms has decreased at an annual rate of about one percent per year. The implication is clear. The supply of farm and ranch land in the United States is dwindling. The growing demand for land coupled with the dwindling supply has placed upward pressure on land values. However, each year some land is brought into farm production through irrigation, clearing and drainage, so that total acreage of *cropland* has not declined significantly in the last 20 years.

Despite the dwindling supply of farmland the average size of farms continues to grow. This is possible because of the reduction in the total number of farms. The remaining farms are not only larger, but also more efficient in food and fiber production.

How Is the Land Used?
Each year part of the nation's land is used for crops, part for pasture and part is idle. But the proportions may vary significantly, due in part to government programs which have occasionally taken land out of crop production. The major uses of crop land are identified in Table 2.1. Most crop land is harvested, although there are significant amounts of summer fallow, idle, and pasture land.

Changes over time in the amount of land used for crops are identified in Table 2.2. Significant regional differences are apparent. Since 1950, a significant drop in the land used for crops in the Northeast and Southeast sections of the United States has occurred. In contrast, the amount of land used for crops has grown in the Cornbelt and Delta states.

The important point is that the use of land is not static. Changes are occurring continuously and these changes have an impact on the price of farm and ranch land. Contrary to popular belief, the total amount of land used for crop production has not declined significantly over time.

Figure 2.1: Major Uses of Land in the United States[a]

Total acreage—2,264 mil.

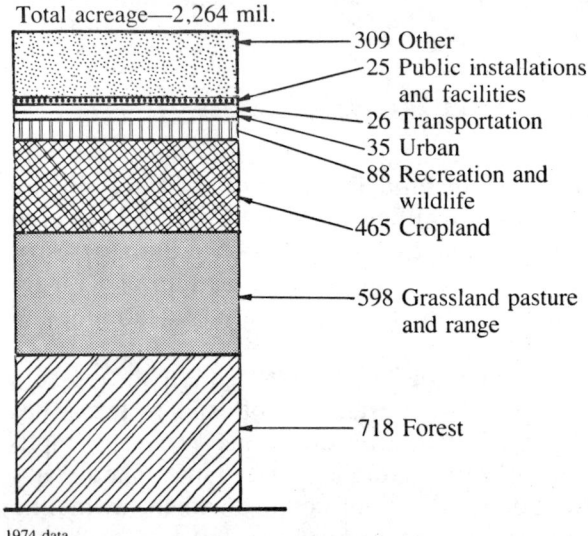

- 309 Other
- 25 Public installations and facilities
- 26 Transportation
- 35 Urban
- 88 Recreation and wildlife
- 465 Cropland
- 598 Grassland pasture and range
- 718 Forest

1974 data

[a]Source: 1979 Handbook of Agricultural Charts, U.S. Department of Agriculture, Handbook No. 561, October 1979.

Table 2.1 Major uses of cropland in the United States.[a]

Use	1959	1964	1969	1974
		—million acres—		
Total Cropland	458	444	472	465
Harvested	317	292	286	322
Failed	10	6	6	8
Summer fallow	31	37	41	31
Idle	34	52	51	21
Pasture	66	57	88	83

[a]Source: Census of Agriculture

Table 2.2 Index numbers of the amount of land used for crops by farm region (1967 = 100)[a]

Region	1950	1955	1960	1965	1967	1970	1975	1977
Northeast	122	117	106	102	100	94	97	99
Lake States	108	108	102	99	100	93	108	113
Corn Belt	102	105	103	95	100	94	109	112
Northern Plains	105	107	103	99	100	99	104	107
Appalachian	129	122	105	96	100	95	111	117
Southeast	163	151	116	96	100	98	115	123
Delta States	102	96	86	92	100	110	116	128
Southern Plains	128	127	114	108	100	100	112	113
Mountain	102	105	99	101	100	102	105	106
Pacific	101	102	101	100	100	99	108	105
United States	110	110	104	99	100	98	108	111

[a]The index numbers reflect the amount by which land used for crops was above or below the amount used in 1967. For example, in 1950 there was 22 percent more land used for crops in the Northeast region than in 1967. Source: Changes in Farm Production and Efficiency, U.S.D.A.

Who Owns the Land

Not all land is a part of the farmland market. In the U.S., about 40% of the land area is held by federal, state, and local governments. Hence about 40% of the land area is outside the market because federal, state, and local governments sell little of the land they own. The remaining 60% is held in private hands.

Families, rather than corporations or partnerships, own most of the farmland. Ninety percent of the owners of privately held farm and ranch land are classified as husband-wife or sole proprietor. (Table 2.3). However this group owns only 74% of the privately held farm and ranch land. Family partnerships and family corporations own 18% of the privately owned farm and ranch land, while nonfamily part-

FARMLAND

nerships and corporations and miscellaneous groups hold the remaining 8%.

Regional differences are important in farmland ownership patterns. For example, both family and nonfamily partnership and corporation ownership of land is far more common in the western parts of the U.S. than in other regions.

Considerable amounts of farm and ranch land are owned by people who do *not* have farming as their major occupation. (Table 2.4). Less than 60% of the farm and ranch land in the U.S. is owned by the people who claim farming as their major occupation. White and blue collar workers own over 20% of the farm and ranch land while retirees own 17%.

The occupations of farm and ranch land owners vary substantially from region to region. In the Appalachian and Southeastern regions of the U.S. less than 40% of farm and ranch land is owned by active farmers. In contrast, nearly 80% of the farm and ranch land in the Mountain region is owned by active farmers.

Who Buys Farm Land?

Farm and ranch land is bought by tenants, owner-operators, retired farmers, non-farmers, and absentee owners. Many tenant

Table 2.3 Type of landowner: Regional distribution of acres owned, *farm and ranch land*, by owner type.

Region	Owner type							
	Sole proprietor	Husband-Wife	Family partnership	Non-family: partnership	Family corporation	Nonfamily corporation	Misc.	Total
	Percent of acres							
Northeast	32.4	51.0	9.2	2.2	1.9	1.7	1.6	100.0
Lake	36.7	51.4	7.1	.6	1.8	1.2	1.2	100.0
Corn Belt	35.2	47.3	9.1	1.1	2.8	1.0	3.5	100.0
Northern Plains	39.5	38.8	11.6	.9	5.2	.9	3.1	100.0
Appalachian	40.9	40.0	11.7	1.9	1.8	.9	2.8	100.0
Southeast	52.1	22.7	12.0	2.0	3.1	4.1	4.0	100.0
Delta	44.5	30.1	13.4	1.5	4.7	2.2	3.6	100.0
Southern Plains	41.3	31.1	15.2	1.8	2.9	1.1	6.6	100.0
Mountain	26.8	31.8	13.9	2.9	16.7	5.6	2.3	100.0
Pacific	25.8	35.0	15.1	4.9	10.4	5.2	3.6	100.0
United States[1]	36.6	37.3	12.2	1.9	6.1	2.4	3.5	100.0

[1]Excluding Alaska. Source: 1978 ESCS Landownership Survey, U.S. Dept. of Agriculture, September, 1979.

Table 2.4 Occupation: Regional distribution of acres owned, *farm and ranch land*, by occupation.[1]

Region	Occupation					
	Farming[2]	White collar	Blue collar[3]	Retired	Other	Total
	Percent of acres					
Northeast	44.7	19.2	16.2	16.1	3.8	100.0
Lake	59.3	9.9	12.7	15.5	2.6	100.0
Corn Belt	48.0	13.8	10.4	22.6	5.2	100.0
Northern Plains	64.9	9.2	3.2	17.4	5.3	100.0
Appalachian	37.0	19.1	16.0	22.9	5.0	100.0
Southeast	36.5	20.6	13.5	23.6	5.8	100.0
Delta	41.1	17.5	12.9	23.8	4.7	100.0
Southern Plains	52.8	20.7	6.4	16.0	4.1	100.0
Mountain	77.5	8.2	3.6	7.5	3.2	100.0
Pacific	58.6	14.9	6.5	17.3	2.7	100.0
United States[4]	56.4	14.1	8.2	17.0	4.3	100.0

[1]Not including corporations and large partnerships. [2]Including farm managers and farm laborers. [3]Including private household and service workers. [4]Excluding Alaska.
Source: 1978 ESCS Landownership Survey. U.S Dept. of Agriculture. September. 1979.

operators actively seek to acquire their own farmland through purchases and they account for about 10% of the acres purchased each year. Existing operators desiring to expand are the major participants in the market. Owner-operators are responsible for about 60% of the total acreage purchased, but it would be a major mistake to consider farmers the only important participants in the market. (Table 2.5).

Local nonfarmers are also important purchasers of farm and ranch land, accounting for about 6% of total purchases. In addition, absentee buyers—defined as buyers living outside the county in which the land is purchased—are now purchasing from 15 to 20% of farm and ranch land sold. This of course has implications for sellers of farm and ranch land. It suggests that limiting the sales effort to local buyers ignores many likely buyers.

Data in Table 2.5 also show the great variation between region and between time periods in the relative importance of different buyers. For example, in the Pacific region in 1979, owner-operators bought 72% of the acreage sold. By 1980, however, this had dropped to 59%. Likewise in the Mountain States, absentee owners bought 41% of the acreage in 1979, but only 6% in 1980. The essential point is, to deal successfully in the farm real estate market, one should know the type of buyers in the market. Yet the market can be highly dynamic with major shifts in the type of buyers in short periods of time.

Within the category of absentee buyers, there is considerable concern and controversy about *foreign* ownership of farmland. In par-

Table 2.5 Farm and ranch land buyers: Percentage distribution of acres by type of buyer for years ending, March 1, 1979–80.[1]

Region	Tenant		Owner operator[2]		Retired farmer		Local non-farmer		Absentee[3]		Other	
	1979	1980	1979	1980	1979	1980	1979	1980	1979	1980	1979	1980
	Percentage distribution of acres											
Northeast	17	15	44	44	2	2	10	7	11	14	16	18
Lake States	17	17	54	56	2	2	8	8	11	9	8	8
Corn Belt	16	16	61	56	2	2	8	8	8	11	6	6
Northern Plains	19	16	59	64	4	1	5	6	8	9	7	4
Appalachian	10	12	48	48	3	2	11	15	16	15	12	8
Southeast	7	8	65	54	1	1	9	8	13	16	6	12
Delta	8	14	56	44	1	1	7	9	18	16	11	16
Southern Plains	8	13	59	48	1	1	7	5	19	29	6	6
Mountain	5	4	44	80	*	*	3	2	41	6	7	8
Pacific	14	5	72	59	1	*	4	8	6	24	4	3
48 States	11	10	55	62	1	1	6	6	20	13	7	8

[1]Percentages may not add to 100 because of rounding. Based on farmland sales reported in March survey. [2]Includes part-owner operators and full-owner operators. [3]Buyers living outside of county. *Less than 0.5 percent.
Source: Farm Real Estate Market Developments, U.S. Dept. of Agriculture, CD 85, August, 1980.

ticular, there is a fairly common fear that oil rich OPEC nations are purchasing U.S. farmland. The fear seems unfounded (Table 2.6). Foreigners hold less than one-half of one percent of all privately held U.S. agricultural land. While purchases have occurred nationwide recent acquisitions have been greater in California, Texas, Georgia, Nevada and Colorado.

By far the most common foreign purchases of U.S. agricultural land are individuals from countries which are close allies with the U.S. In order of the most acres purchased, these countries are the United Kingdom, Luxembourg, Canada, Netherlands and West Germany. Citizens or residents of OPEC countries own very litte U.S. agricultural land.

Who Sells the Land

Owner-operators are the major sellers of farm and ranch land, (Table 2.7). These sellers account for nearly 50% of the acres transferred. Retired farmers account for about another 11%, while about 15% of the acreage sold is the result of estate settlements.

There is also concern expressed at times about the takeover of farm and ranch land by nonfarm investors. But recall that these groups also sell land. By comparing information in Tables 2.5 and 2.7 we see that in 1979 absentee and local nonfarmers were *net* buyers of land; that is they bought more land than they sold. In 1980, the reverse was true, local nonfarmers and absentee owners sold more land than they bought. On balance, local nonfarmers and absentee owners have been net buyers of land over the last 10 years, but *net* purchases have been only 3–5% of the acreage purchased.

Table 2.6 Acreage of U.S. farmland held by foreign owners, February 1, 1979.[a]

STATE	ACRES	STATE	ACRES
Alabama	163,498	Montana	180,561
Alaska	337	Nebraska	65,559
Arizona	82,780	Nevada	155,577
Arkansas	40,046	New Hampshire	30,943
California	221,506	New Jersey	18,879
Colorado	180,265	New Mexico	193,606
Connecticut	303	New York	205,130
Delaware	837	North Carolina	152,096
Florida	202,101	North Dakota	15,053
Georgia	257,309	Ohio	7,638
Hawaii	2,940	Oklahoma	15,544
Idaho	9,905	Oregon	169,532
Illinois	37,891	Pennsylvania	153,655
Indiana	12,906	South Carolina	240,437
Iowa	24,077	South Dakota	15,242
Kansas	31,475	Tennessee	287,855
Kentucky	14,327	Texas	413,710
Louisiana	69,950	Utah	34,441
Maine	951,576	Vermont	39,932
Maryland	17,727	Virginia	55,209
Massachusetts	438	Washington	40,618
Michigan	42,032	West Virginia	3,580
Minnesota	18,241	Wisconsin	12,710
Mississippi	84,846	Wyoming	13,271
Missouri	44,178	Puerto Rico	780
TOTAL			5,033,429

[a]Source: USDA news release. 1637-50.

Buyers should recognize that the total quantity of farm and ranch land offered for sale varies with changing economic conditions. Regional totals for the number and amount of farm real estate sales for the years 1972 through 1980 are reported in Table 2.8. Notice that the number of sales dropped sharply from 1974 through 1980. Hence buyers of land are faced with a declining number of sales. Likewise, the total acreage offered for sale has dropped sharply since 1974. Substantial regional variations are apparent from the data presented in Table 2.8.

Economists have attempted to develop models to explain the ups and downs in the quantity of farm and ranchland sold. Some have suggested a close positive relationship between farm income and the quantity of farmland sold. While income is likely important, the changes in quantities of land sold do not closely correspond with the level of farm income. The lower quantity of land offered for sale

Table 2.7 Farm and ranch land sellers: Percentage of acres by type of seller for years ending March 1, 1977 and 1980.[1]

Region	Type of seller										
	Owner operator		Retired farmer		Estate		Local nonfarmer		Absentee[2]		Other	
	1979	1980	1979	1980	1979	1980	1979	1980	1979	1980	1979	1980
	...Percentage distribution of acres...											
Northeast	53	48	16	18	11	6	7	8	9	16	4	4
Lake States	59	50	13	19	10	10	7	8	7	11	4	3
Corn Belt	41	39	15	14	21	23	8	8	11	12	3	3
Northern Plains	45	39	14	16	22	18	4	4	11	19	5	5
Appalachian	37	34	11	12	21	21	12	10	11	18	7	5
Southeast	53	49	10	8	9	13	9	5	15	21	4	4
Delta	47	48	8	12	10	11	11	8	17	16	6	5
Southern Plains	43	39	9	15	23	10	8	7	13	26	4	2
Mountain	45	61	7	7	11	4	1	17	34	6	2	4
Pacific	61	49	15	7	11	6	2	4	10	25	1	9
48 States	47	48	11	11	16	11	5	10	18	15	4	4

[1] Percentages may not add to 100 because of rounding. Based on sales reported in the March survey. [2] Sellers living outside of county, including a small number of sales by lending agencies and by Federal, State and local Governments. Source: Farm Real Estate Market Developments.

Table 2.8 Number and acres of farm real estate sales by region, year ending March 1, 1972–80.[1]

Region	1972	1973	1974	1975	1976	1977	1978	1979	1980
	—Real estate sales (number)—								
Northeast	6,516	7,772	6,573	5,843	5,302	4,517	5,409	4,553	5,000
Lake States	12,162	12,821	12,766	11,882	9,462	8,710	7,754	7,680	7,549
Corn Belt	25,211	27,907	28,004	19,682	18,101	18,624	17,910	16,636	15,401
Northern Plains	8,252	9,495	9,438	6,627	5,718	5,572	5,504	6,094	6,674
Appalachian	14,253	17,154	15,312	11,375	10,974	10,716	10,756	9,870	9,888
Southeast	6,805	9,317	7,302	5,739	3,743	5,082	4,356	4,432	4,638
Delta	4,165	5,821	6,441	5,257	3,616	4,775	2,907	3,503	3,272
Southern Plains	13,323	14,323	15,338	10,093	9,266	6,195	6,585	5,895	6,037
Mountain	5,320	5,742	6,280	4,834	4,191	3,794	3,864	3,699	4,257
Pacific	6,845	8,282	8,012	7,558	5,629	6,562	5,713	5,811	5,632
48 States	102,852	118,634	115,466	88,890	76,002	74,547	70,758	68,173	68,348
	—Acres sold (in thousands)—								
Northeast	834	1,073	848	818	700	641	676	849	780
Lake States	1,861	1,910	1,991	1,747	1,344	1,167	1,140	1,225	1,111
Corn Belt	3,656	4,047	4,145	3,090	2,607	2,682	2,454	2,254	2,195
Northern Plains	2,575	3,542	4,209	2,101	2,047	1,967	1,926	2,172	2,750
Appalachian	1,824	2,144	2,036	1,456	1,361	1,629	1,635	1,287	1,216
Southeast	2,116	2,385	1,906	1,452	857	1,296	1,059	1,068	948
Delta	1,229	1,770	2,460	1,588	1,407	1,972	971	1,144	975
Southern Plains	4,623	6,130	8,497	3,734	3,882	2,720	2,647	1,582	2,875
Mountain	6,012	8,705	9,665	6,347	4,543	3,544	7,794	6,118	5,656
Pacific	1,848	2,841	3,285	2,320	2,274	3,025	2,171	5,107	2,631
48 States	26,578	34,547	39,042	24,653	21,022	20,643	22,473	22,806	21,137

[1] Voluntary and estate sales only. Source: Farm Real Estate Market Developments. CD-85, U.S. Dept. of Agriculture, August, 1980.

between 1973 and 1980 is likely related to the substantial capital gains occurring over this period. Owners may be reluctant to sell when the prospects of capital gains returns are high. (Comparisons between return on investment in farmland and other investments are presented in Chapter 3.)

Sizes and Prices of Parcels Sold

Unlike corporate stocks and some other forms of investment, farm and ranch land is not often available in small incremental units. That is, one cannot build a viable farming unit by purchasing one or two acres at a time. Rather, land is most often sold in tracts of substantial size and requires a substantial capital investment (Table 2.9). The average size of tracts sold is well over 100 acres in all regions of the country and has averaged over 1500 acres in some areas.

From 1973 through 1980 the U.S. average size of tract purchased has not greatly increased. In a number of areas, land sales are conducted by splitting existing farms into two or more tracts. The intent is to find more buyers who have sufficient financial reserves to buy the smaller tracts.

Since 1973, the price of farm and ranch land has gone up sharply in all regions of the country. The average sale price per tract has gone up accordingly. A major concern for most potential buyers of land is to acquire the necessary funds for purchase. For the U.S. as a whole, the average land sale is over $200,000—a sizable sum of money to save or borrow. Regional differences are important. The lowest average price per sale occurs in the Northeast, Appalachian and Lake States Regions.

The high average price per sale on farm and ranch land has discouraged many would-be buyers. Most lenders will not finance over 70 to 80% of the purchase price. Many buyers find it impossible to accumulate the remaining 20 to 30% of the purchase price. However, as shown in later chapters of this book, there are effective methods of overcoming this problem.

Perhaps the typical view of a farm and ranch land sale is that of a disposition of a complete farming unit resulting from retirement, estate settlements, or people leaving farming. However, there appears to be a growing tendency to sell only part of a farm (Table 2.10). More than 50% of the farm sales now involve less than complete units.

Financing Farmland Transfers

About 90% of farm and ranch land transfers now involve some sort of financing—a substantial increase from the 1950s and 1960s (Table 2.11). The other 10% involve gifts, inheritances, or payment in cash. While information in Table 2.11 is for the U.S., there are very few differences among regions. Hence a major concern of buyers nationwide is the terms of financing.

Over time, there has been a dramatic increase in the ratio of debt to purchase price (Table 2.11). In 1950, downpayments on the purchase

Table 2.9 Average size and price per acre for farm and ranchland sales, year ending March 1, 1973–80.[1]

Region	1973	1974	1975	1976	1977	1978	1979	1980
	\multicolumn{8}{c}{Average size (acres)}								
Northeast	138	129	140	132	142	125	187	156
Lake States	149	167	147	142	134	147	159	147
Corn Belt	145	148	157	144	144	137	135	143
Northern Plains	373	446	317	358	353	350	356	412
Appalachian	125	133	128	124	152	152	130	123
Southeast	256	261	253	229	255	243	241	204
Delta	304	382	302	389	413	334	327	298
Southern Plains	428	554	370	419	439	402	268	476
Mountain	1516	1539	1313	1084	934	2017	1654	1329
Pacific	343	410	307	404	461	380	879	467
48 States	291	334	272	271	274	308	335	309
	\multicolumn{8}{c}{Price per acre (dollars)}								
Northeast	546	694	812	790	823	1071	741	1145
Lake States	369	444	562	696	788	865	1050	1184
Corn Belt	556	687	790	1054	1345	1443	1582	1719
Northern Plains	169	211	286	350	412	412	410	459
Appalachian	442	542	560	722	689	789	1088	1199
Southeast	437	524	648	588	682	735	955	979
Delta	332	399	411	580	658	698	847	992
Southern Plains	222	241	276	303	374	400	484	586
Mountain	100	141	168	210	244	175	192	262
Pacific	397	448	622	691	783	797	469	1024
48 States	292	340	438	528	654	591	618	779
	\multicolumn{8}{c}{Average price per sale (1,000 $)}								
Northeast	75	90	114	104	117	134	139	179
Lake States	55	69	83	99	106	127	167	174
Corn Belt	81	102	124	152	194	198	214	246
Northern Plains	63	94	91	125	145	144	146	189
Appalachian	55	72	72	90	105	120	141	147
Southeast	112	137	164	135	174	179	230	200
Delta	101	152	124	226	272	233	277	296
Southern Plains	95	134	102	127	164	161	130	279
Mountain	152	217	221	228	228	353	318	348
Pacific	137	184	191	279	361	303	412	478
48 States	85	114	119	143	179	182	207	241

[1] Sales reported by farm realtors and others. Weighted by the number of voluntary and estate transfers of 10 or more acres reported by crop reporters by acre-size class. Source: Farm Real Estate Market Developments, CD-85, August, 1980.

of land average more than 40% of the purchase price. By 1980, however, this figure had declined to 22%.

Because of the crucial nature of the financing decision, several later chapters are devoted to this topic. Chapter 8 identifies how terms of financing influence farmland values while Chapter 9 outlines institutional and noninstitutional sources of funds for the purchase of farm and ranch land.

Table 2.10 Farm and ranch land transfers by method of operation before the sale.

Type of Operation	1974	1975	1976	1977	1978	1979	1980
			—percent—				
Complete farm before sale	53	45	49	43	45	42	40
Part of another farm before sale	37	45	42	47	45	48	50
Part-time farm before sale	10	10	9	10	10	10	10

Source: Farm Real Estate Market Developments, CD-85, U.S. Dept. of Agriculture, August, 1980.

Real Estate Exchanges

As an alternative to sale, there is a growing interest in "like-kind exchanges". These transactions are designed to *defer* but not avoid income tax on potential gain. The growing interest in non-taxable exchanges of farmland is a direct result of the rapid increases in land values and the substantial amounts of capital gains taxes that would be due if the property were sold.

Under current tax laws, no gain is recognized if property held for productive use or investment is exchanged solely for property of a like-kind to be held for productive use or investment. The term like-kind exchange merely means that real property must be exchanged for other real property. An exchange is defined as a reciprocal transfer of property rather than a transfer for money only. Paying cash in addition to exchanging property does not prevent the transaction from being an exchange:

Table 2.11 Financing on farmland transfers, U.S., selected years.

Year	Percent of transfers which involve financing	Ratio of debt to purchase price[1]
	—percent—	
1945	44	57
1950	58	57
1955	64	59
1960	67	65
1965	73	72
1970	78	73
1975	88	76
1979	90	79
1980	91	78

[1]For those sales which involve financing.
Source: Farm Real Estate Market Developments, DC-85, U.S. Department of Agriculture, August, 1980.

Example: Smith has a farm valued at $400,000 in Wisconsin with an adjusted basis of $100,000. Smith decides to leave the cold Wisconsin winters for retirement in Florida. Jones has an apartment complex in Florida valued at $350,000 with an adjusted basis of $200,000. Jones decides to leave Florida and acquire a farm in Wisconsin. Learning of each others' interests, Smith and Jones decide to exchange property with Jones paying Smith $50,000 in cash to compensate for the difference in value between the two properties. Neither property is mortgaged. In this example, the $50,000 would be treated as a taxable gain for Smith. All other gains are deferred. The new adjusted basis for Smith is $100,000 and for Jones is $250,000. By using the exchange mechanism, the parties have deferred the majority of the capital gains tax.

The example above points out several important aspects of real estate exchanges. For one, parties to the exchange are often separated geographically and do not know each other. This calls for an intermediary, thus a number of real estate exchange groups have been formed by realtors. In addition, the tax aspects of real estate exchanges can become extremely complex. Competent tax advisors should be used by parties entering into exchange agreements. Further discussion of the tax aspects of land purchases appears in Chapter 17.

CHAPTER 3
WHY BUY OR SELL FARMLAND?

"Buy land! They ain't making any more of the stuff"   Will Rogers

In every land sale transaction there is both a buyer and a seller. The reasons motivating the buyer differ from those motivating the seller. In this chapter, reasons are identified why individuals choose to buy and sell farmland.

Since there are both economic and personal reasons for buying or selling land, and both play an important role in most farm and ranch land transactions, it is well to identify them clearly. Too often people lose sight of the fact that personal reasons are a major determinant of land purchases and sales.

Why Buy Land?
The reasons behind the purchase of land are likely different for each individual. Some of the more important personal and economic reasons for buying land are identified below.

Personal reasons for buying land include:

Desired Occupation—Some people buy farmland primarily because they want to become operators of farm or ranch land. By owning farmland, they have control over one of the major assets needed to pursue their occupation. For owner-operators, it means a chance to work with their hands to produce products for which they have great pride. There appear to be a rather large number of people who would like to pursue farming or ranching as an occupation, but feel it is impossible because they do not have the financial capacity to purchase farmland. While working as a tenant farmer provides a similar occupation, it does not provide the same degree of managerial responsibility or the job security offered by land ownership.

Freedom of choice—Closely associated with a desired occupation incentive for land ownership is the *freedom of choice* incentive. Owner-operators express this as *being one's own boss*. By owning and operating the land, individuals make their own choices on how they work, when they work, and what crops to grow. In contrast, nonfarm employment may require one to "punch a time clock." Freedom of choice is also limited for tenant operators since they may face legal sanctions if they do not perform farming activities in a timely manner or if they use crop rotations not agreed to by the landlord.

Security of land to operate—Land operated under rental arrangements can be lost for a variety of reasons—disagreement with the landlord, death of the landlord and unwillingness of heirs to renew the lease, sale of the property to owners who wish to operate the property themselves, plus many others. Rather traumatic changes in lifestyle can and do occur as a result of the unexpected loss of rental property. Tenants forced off land they have operated for a number of years may be damaged both economically and emotionally. The strong competition for rental tracts has made the pressures more severe in some areas than others. Land ownership provides one mechanism for avoiding these risks. A number of former tenant operators have purchased land with the vow that now no one can "kick them off *their* land".

Pride of ownership—There is often substantial social status involved in landownership. Individuals who own their own land are often viewed as solid citizens with a strong stake in maintaining a viable economic community. Such individuals are often elected to serve on local boards, such as zoning boards and boards of directors of farm lending institutions. Land ownership is sometimes the key which unlocks the door to such positions. In addition to the social status of owning land, there is the personal pride of being able to say "it's mine". Such personal pride in land ownership has at times led to the accumulation of land beyond the amount needed to achieve economically efficient units. When this occurs, there can be a negative social connotation attached to land owners such as "land baron" or "landed gentry". Hence, the personal pride in land ownership at some level may be partially or totally negated by social castigation.

Quality of life—Rural America is often viewed as a place which provides a high quality of life. Rural residents are exposed to cleaner air and less noise pollution than their city counterparts. Many view the country as an ideal place to raise a family. To many, the economic rewards of a good paying job in the city are not as important as quality of life in the country. Farmland ownership provides the opportunity to take advantage of these perceived differences in quality of life.

Desire for self-reliance—"Things may get bad, but at least we'll never be hungry or cold". This statement reflects the self-reliance incentive for owning land. In an interdependent society in which 4% of the people grow the food necessary to feed the remaining 96%, there are a substantial number of people who are concerned about what would happen if the economic system fails. They recognize that one can be virtually self-sufficient—growing both food and wood for fuel—through land ownership. A number of people have in fact left well paying jobs because they take great pleasure in such self-sufficiency.

Recreational value—A number of land owners have purchased their land primarily for the recreational activities it can provide. Most often, purchasers of land for these reasons are urban residents who will not operate the land they purchase. Rather, they view the land as a place to escape the population density of city life for a weekend or evening. The farm may also serve as a convenient place to maintain horses or other pet animals. Often times the land serves as a hunting or fishing spot. For the individual who buys farmland for this reason, economic reasons for buying land may be of lesser importance.

Farming as a hobby activity—Many part-time operators view farming or ranching as a hobby activity. For these individuals, land is an essential ingredient to pursue their hobby. Most often, hobby farmers are employed in high paying jobs in nearby cities or have retired from such jobs. They farm, not because they expect to make a good living by doing it, but because they take personal pleasure in being able to raise livestock or crops.

Keep the land in the family—Occasionally, individuals purchase farmland out of a sense of duty to other family members. A fairly common example of this is the following: The father of a family dies or is incapacitated by health problems. The mother works to maintain the farm but has difficulty in doing so. Out of a sense of family duty, a son or daughter returns from a city job to take over and run the farm even though they may have preferred life in the city. Purchase of the home farm is seen as the best means of taking over the farming operation. In examples of this nature, the decision to purchase land is based heavily on personal reasons rather than on economic reasons. A less common version of this incentive occurs when one buys the "home farm" for sentimental reasons. Because of the death of parents, the farm is put up for sale. A son or daughter of the deceased may buy the land, not because they wish to operate it, but rather because they want to retain ownership of the "place where I grew up".

Show place farms—Some individuals or firms buy farmland because they want a "show place" for friends, acquaintances or business associates. Fences and buildings are usually maintained to the utmost perfection. Crops and livestock are expected to be top-notch at all cost. Machinery is likely to be the latest available and in top condition and repair. Most commonly, owners of such places do not operate the farm and are much more concerned with appearances than profitability.

While not carried to the extreme indicated above, show place farms are also used by agribusiness firms. For example seed, machinery, and agricultural chemical firms may acquire farmland to use as demonstration plots. Profitability of the demonstration farm may not be the major objective.

Institutional farms—Institutions such as prisons, mental health homes, and others often acquire farmland. The purchase of land for such purposes is seldom profit oriented. Rather, the land is purchased to provide gainful employment in a controlled setting for occupants of the institution.

From the preceding discussion it is clear that there are many personal reasons why individuals and institutions desire to own farm and ranch land. These personal reasons tend to vary with whether the purchaser is going to be an owner-operator or whether the purchaser is going to be a nonoperator landlord or hobby farmer. In most cases, there is more than one personal factor involved in the decision to purchase land.

Economic Incentives For Buying Farmland
The economic incentives for buying farm and ranch land can be broken into two major categories: (1) expected net income returns to land and (2) expected capital gains on land. The effect of buying farmland on the expected net income returns to land may depend upon whether the land is an *add-on* investment to an on-going operation. In fact, about 60% of the farmland transfers involve tracts which are to be added to an existing operation. In these cases, the purchase of additional land may affect the profitability of the existing operation, particularly if there are economies of size from the expansion. Add-on investments are likely to have little impact on expected capital gains on acres already owned.

When land purchases involve the addition of acreage to an existing operation, it is important to determine the impact on the entire operation. The additional land may well affect profits in the remainder of the business and, therefore, the purchase cannot be judged solely on the basis of returns on the added acreage. The effect on the remainder of the business may be either positive or negative. The following discussion focuses on the circumstances under which investment in land can make a positive addition to income for the remainder of the business.

Underutilized Labor—In calculating the income returns to land, one should deduct from gross income, all expenses associated with operating the land including a return to any unpaid operator or family labor. However, in many farming operations there is operator or family labor which is not fully utilized. In this case, the purchase of additional land can add to total profits in the business, not only because the added land generates a profit, but also because it provides a means for using available labor.

Underutilized Machinery—A number of farm operators have excess machinery capacity for the amount of land they operate. The purchase of additional land may allow for greater utilization of existing machinery. In these cases, the cost of machinery is spread over more acres which in turn generates a lower machinery cost per acre for those acres available prior to the additional purchase.

Economies of Size—Even if there are no underutilized resources in the existing operation, the addition of land may affect overall profitability of the business through economies of size. An example can clarify this concept. A farmer is operating a 500 acre cash grain farm in which a $50 net return per acre is generated. An additional 500 acres comes up for sale and if operated in the same manner as the existing farm could also generate $50 net return per acre. However, by combining the two farms, hiring additional labor, and buying larger, more efficient machinery, the net return can be increased to $55 per acre. This example reflects economies of size—one of the major motivating factors underlying the expansion of farm size in the United States. In this case the decision to buy should be based upon the impact on the entire business and not just the return on investment for the additional acreage purchased.

Farmland As a Hedge Against Inflation

Since 1940, investment in farmland has done very well as a hedge against inflation. As shown in Figure 3.1, the value of farmland has increased by more than the consumer price index in all but a few years since 1940. The forty year period from 1940 to 1980, in which inflation in land values exceeded the inflation rate in the general economy, has spawned a group of investors who frequently believe that this will always be the case. But from 1914 to 1920, farmland prices did not keep pace with inflation. And from 1920 to 1932 land values fell faster than did the consumer price index.

During the decade of the 1970's, the rate of increase in farmland values was almost double the inflation rate in the general economy. Yet farmland did not top the list of investments which outpaced the

Figure 3.1   **Annual Change in Farmland Prices and the Consumer Price Index**

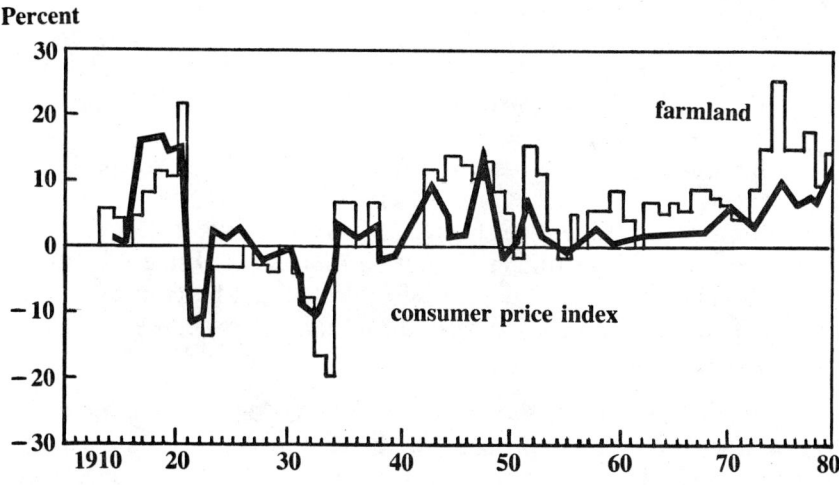

general rate of inflation. Gold, antique Chinese ceramics, old master paintings, and fine diamonds all increased in value more than farmland. However, none of the investments ranked ahead of farmland generate annual cash income. Investors who hold such assets may gain substantially through capital gains, but they must be able to *carry* these investments with income or wealth accumulated from prior investments. In addition the investments ranked ahead of farmland may experience extreme price fluctuations (gold in particular) or may have high transactions costs for sale (old master paintings, as one example).

Returns on Investment
Farmland purchasers often wonder if they are making the right decision when buying farmland. They are concerned that despite the large gain in real estate values in recent years, the cash returns have been relatively low. Should existing land owners invest further, or would it be advantageous to diversify into other investments? Lack of knowledge concerning returns on nonfarm investments makes the decision difficult. By comparing rates of return on investments, one can better decide on the investment plan to follow.

Simply stated, an annual *rate of return* measures the *monetary* rewards from an investment for a year divided by the dollar value of the investment. Monetary rewards can include a cash return and/or a capital gain return. Some investments are chosen because of the cash earnings they provide, with no expectation of capital gains. Passbook savings or certificates of deposit are common examples. Other investments yield no cash return but are chosen because of anticipated increases in value. Gold is a good example. Still other investments, like farmland, are chosen because there is an anticipation of both cash returns and capital gains on the asset.

In comparing returns on investment in farmland with other investments, care should be taken to measure the dollar amount of investment and the dollar amount of returns in a consistent fashion. For example, the dollar amount of an investment can be measured as the original *cost* or as the *current market value*. In addition, one can measure total investment (debt plus equity capital), or only equity investment. Unless a consistent standard of measurement is used, comparisons are meaningless.

In addition to the problem of consistency, there is a problem of accounting for differences in risk among different investments. The level of risk is not identical for all investments. In general, the higher the level of risk, the higher should be the expected rate of return on investment. Consequently, rates of return cannot be compared alone, but should be accompanied by indications of the degree of riskiness of the investment.

Historical Returns on Farmland Versus Other Investments
How does farmland compare with other long term investments which generate cash income and which can also appreciate or

depreciate in value over time? Let's compare the returns to farmland with stock market investments and investment in long term U.S. bonds. Detailed comparisons of these returns are illustrated in Table 3.1.

Annual income rates of return on investment in common stocks, long-term bonds, and farmland are shown in Section I of Table 3.1. For consistency, returns in each case are measured in current dollar values and investments at estimated current market values. On the average, the annual income rate of return to investment in farmland during the 1960's and 1970's was above the annual income rate of return to investment in the stock market, but below that of long-term bonds. During the 1950's, however, common stocks generated a higher annual income rate of return, than either bonds or farmland.

One should not compare investments without considering differences in risk among the investments. One measure of risk is the degree to which returns fluctuate over time. Statisticians measure the relative amount of fluctuation with a statistic called the *coefficient of variation*. For our purposes it is sufficient to know that the higher the coefficient of variation, the higher the investment risk. Notice in Table 3.1, the

Table 3.1 Returns on investment in common stocks, long-term government bonds, and farmland.

	I.			II.			III.		
	Annual income return			Capital gain or loss return[a]			Total Return		
Year	Common stock	Long-term bonds	Farmland	Common stock	Long-term bonds	Farmland	Common stock	Long-term bonds	Farmland
Annual Average	(percent)			(percent)			(percent)		
1950–54	5.85	2.61	4.94	14.58	.94	N.A.	20.43	3.55	N.A.
1955–59	3.94	3.38	3.16	14.99	−2.78	N.A.	18.93	.60	N.A.
1960–64	3.20	4.00	3.68	8.79	−.23	4.77	11.99	3.77	8.45
1965–69	3.18	5.01	4.46	5.54	−5.19	6.17	8.72	−.18	10.63
Annual									
1970	3.83	6.59	4.37	−14.94	−6.20	4.22	−11.11	.39	8.59
1971	3.14	5.74	4.26	18.11	11.95	8.11	21.25	17.69	12.37
1972	2.84	5.63	5.86	11.10	1.45	12.30	13.94	7.08	18.16
1973	3.06	6.30	10.29	−1.62	−8.60	23.50	1.44	−2.30	33.79
1974	4.47	6.99	6.46	−22.88	−8.52	12.91	−18.41	−1.53	19.37
1975	4.31	6.98	5.55	2.80	.14	13.70	7.11	7.12	19.25
1976	3.77	6.78	3.75	19.77	2.95	16.55	23.54	9.73	20.30
1977	4.56	7.06	3.42	−3.75	−3.97	9.02	0.81	3.09	12.44
1978	5.28	7.89	4.73	−2.11	−10.51	14.40	3.17	−2.62	19.13
1979[b]	5.46	8.74	5.10	2.32	−9.43	16.50	7.78	−0.69	21.60
1970–79	4.07	6.87	5.38	.88	−3.07	11.75	4.95	3.80	18.50
Coefficient of variation[c]	21.5	25.6	52.3	280.6	192.4	59.2	151.7	184.5	48.1

[a]Capital gains reflect primarily unrealized capital gains or losses. For long-term bonds one could achieve the annual income return by holding the bond to maturity. However, opportunity losses would occur if interest rates are rising over the time period the bond is held.
[b]Preliminary.
[c]Computed over the 1960–1979 period only because of missing data.

coefficient of variation in annual income rates of return has been much higher for farmland than for stocks or long-term bonds. This implies that investors in farmland face much more uncertainty about income returns than do investors in common stocks and long-term bonds.

The annual percentage increase or decrease in the price of common stocks, long-term bonds, and farmland are shown in Section II of Table 3.1. During the decade of the 1970's farmland increased in value at an annual average rate of 11.75%. In contrast stocks increased at an annual average rate of less than one percent. Long-term bonds actually declined in price because interest rates were rising.

It is clear that capital gains (losses) on common stocks and long-term bonds have been much more variable than for farmland. As measured by the coefficient of variation, annual capital gains on farmland have been more stable than the other two investments. For example, between 1970 and 1980 land values showed no decline from one year to the next. In contrast, the price of stocks and bonds declined from one year to the next in at least 5 of the 10 years. This means that investors in farmland have not had to face the risk of price declines nearly as much as investors in stocks or bonds.

Total annual rates of return can be measured by adding annual income rates of return and capital gains (losses) returns. Recognize, however, that capital gains returns are *unrealized* returns and hence are returns that cannot be spent without selling the investment. Total annual returns are shown in Section III of Table 3.1. During the decade of the 1970's, farmland, on the average, far outperformed stocks and bonds. Not only were total annual rates of return much higher (18.50 versus 4.95 and 3.80%), but the variability in these returns was also much lower. In summary, farmland has been a good investment.

Are Capital Gains on Farmland Justified?

A number of people have suggested the possibility of a major decline in farmland values. A major decline in asset values can occur under several conditions. First, asset values could fall because of a sharp and sustained drop in income from the asset. Second, income returns to assets could remain relatively stable, but past price increases resulting from speculative fever could be wiped out. Examples of this phenomenon are not uncommon in the stock market. Price increases not justified by growth in returns are the underlying cause.

To assess the potential for a major decline in farmland values, the rates of growth in land values are compared with the growth in returns to land (Table 3.2). Both capital gains and the growth in income returns to land are reported in nominal and real (adjusted for inflation) terms. On the average over the 1960–79 period, the growth in nominal and real income returns to land exceeded the capital gains rate on land. Hence, increases in land values which have occurred appear justified by the growth in income returns to land. However, the average increase in land values from 1975 through 1979 was substantially higher than the average growth rate in returns to land values and caused considerable concerns about "overpriced" land. Remember, however, that farmland should be viewed as a long-term investment.

The fact that investors view farmland as a long-term investment is evident from Table 3.2. For all years in which the growth rates in returns to land were negative (1964, 1967, 1968, 1973, 1975 and 1976), substantial capital gains were recorded. The explanation for

Table 3.2 Nominal and real capital gains on land and return to investment in land.

Year	Capital Gains on Farmland		Rate of Growth in Income Returns to Land[a]	
	Nominal	Real[b]	Nominal Returns[c]	Real Returns[b]
			percent[e]	
1960	1.7	.1	32.5	30.9
1961	5.2	4.5	18.9	18.2
1962	4.6	3.3	11.1	9.8
1963	6.0	4.4	0.0	−1.6
1964	6.3	5.2	−10.0	−11.1
60–64 Avg.	4.8	3.5	10.5	9.2
1965	7.9	6.0	50.8	48.9
1966	6.5	3.1	13.7	10.3
1967	6.7	3.4	−17.6	−21.0
1968	5.8	1.2	−1.1	−5.7
1969	3.9	−2.3	22.7	16.5
65–69 Avg.	6.2	2.3	13.7	9.8
1970	4.2	−1.0	2.8	−2.4
1971	8.1	4.7	2.7	−.7
1972	12.3	8.6	47.4	43.7
1973	23.5	14.1	98.2	88.8
1974	12.9	1.2	−23.1	−34.8
70–74 Avg.	12.2	5.5	25.6	18.9
1975	13.7	6.9	−7.8	−14.6
1976	16.5	11.3	−23.7	−28.9
1977	9.0	2.2	4.4	−2.4
1978	14.4	5.1	50.5	41.2
1979[e]	16.5	2.6	24.0	10.1
75–79 Avg.	14.0	5.6	9.5	1.1
60–79 Avg.	9.3	4.2	14.8	9.7

[a]Figures reported are based upon USDA estimates of return to "production assets". However, land is the major component of farm production assets.
[b]Nominal values adjusted for annual (January-to-January) changes in the Consumer Price Index.
[c]Returns to both debt and equity capital.
[d]Percentage growth figures are based upon year to year changes.
[e]Preliminary

this is that land is normally considered a long-term investment. Land values are determined in part by the expected growth rate in income. The wide fluctuations in the growth rate returns to land are likely to lead to expectations that are based on longer-term trends, and not simply one or two years' actual growth rates.

Some analysts draw a parallel between the stock market and farmland. In the 1950s common stocks increased rapidly in value, despite relatively low annual income returns for investors. In the 1960s, the capital gains rates were much lower. And during the decade of the 1970's, stocks appreciated very little. Similar to stocks in the 1950s, farmland in the late 1970s increased rapidly in value despite relatively low annual income returns to current market value.

Will farmland in the 1980s follow the same path as the stock market in the 1960s and 1970s? The answer to that question depends upon the growth rates in returns to farmland in the 1980s. A *sustained* drop in income returns to land would likely lower expectations and lower land prices. But, what is the likely future direction of growth rates in income returns to land? Several factors bear on the likely outcome. For one, there seems to be a growing awareness on the part of policy makers that government price support programs which bolster farm income are quickly translated into higher land values, thereby benefitting current land owners, but making it more difficult for beginning operators to get started in farming. This may lead to reduced government support programs for agriculture and thereby slow the growth rate of land values. Alternatively, a growing population, along with the desire for improved diets, continues to place pressure on the demand for agricultural land. In addition, a move toward synthetic fuels, alcohol in particular, could generate an expanded demand for agricultural products which in turn will impact on the value of land.

Why Sell Land?

As with buying land, there are both personal and economic reasons for selling land. The factors which finally result in the decision to sell differ with each individual. But some of the more common reasons why individuals sell land are described below.

Personal Reasons for Selling Land—

Quality of Life—Rural life appeals to different people in different ways. Some sell their land because they dislike the isolation of life in the country. They wish to live in urban areas where schools, shopping, cultural activities, and health facilities are likely to be more readily available than in rural areas. For these individuals, the positive aspects of city life more than offset the negative aspects which motivate some buyers to seek life in the country.

Too much work—Some farmers explain their choice of selling land with the phrase—"There must be more to life than work". This reason for selling is particularly true among some of the more labor intensive farming operations, such as dairy. For example, a number of dairy operators believe they have become slaves to an operation which requires their personal attention "seven days a week, 365 days a year". They may sell, not because economic reasons have forced them out, but rather because they prefer a lifestyle which provides more free time.

Retirement—One of the major reasons behind farmland sales is retirement of the operator. The retiring operator sells the land because he or she no longer has the desire or the physical capacity to continue operating the land. People who sell land because of retirement are often reluctant to take the final step because it likely means leaving behind a lifetime of work. Frequently the retiring operator sells the land to a son or other close relative, even if other buyers offer more for the land. This gives the retiring operator a better chance to revisit the farm and perhaps even to continue helping the new owner.

Poor health—Each year a significant number of owner-operators sell out as a result of poor health. The operator is reluctant to sell, but may have little choice in the matter. Health problems may dictate a change in occupations, and lacking the financial resources to hold the farm as a landlord, the owner may be forced to sell.

Death—The sale of land is sometimes essential to settle the estate of a deceased landowner. Heirs may not wish to continue operating the farm and therefore sell the land to divide the estate. In some cases, land must be sold to acquire the funds necessary to pay the estate taxes. A detailed description of transfers of land at death, including tax consequences, is provided in Chapter eighteen.

Economic Reasons for Selling Land

The economic reasons for selling land can be categorized as either voluntary or involuntary. Involuntary reasons for selling land include the sale of land to the state under the power of eminent domain, the forced sale of land to pay taxes, and foreclosure sales to pay delinquent loans. A description of each follows:

Eminent domain—As described in Chapter 1, one of the rights in land—the right of eminent domain—allows governmental units to acquire private land for the public good. The most common examples of this in agriculture are the taking of land for public roads and parks. The process normally begins with a representative of the governmental unit explaining the need for the land and offering the owner a specified amount. The owner and agent then bargain over the price. Recognizing the right of society to claim such land, the process normally ends in an agreement to sell. If not, the governmental unit begins legal procedures for condemning the land and compensating the owner for the land taken.

Tax sales—Governmental units have the right to levy taxes against farmland. If these taxes are not paid in a timely fashion, the taxing body can place a tax lien on the property. If the tax plus accumulated interest and penalty charges remain unpaid, the property is offered for sale. Buyers can acquire tax certificates for the property in question. The tax certificate generates full ownership rights if the delinquent owner fails to redeem the property in the time period specified by law. In some cases, the state or county government does not allow outside bidders, but rather bids off the property in the name of the state or county.

Foreclosure—As shown in Chapter 2, most farmland is acquired with the use of borrowed funds. To ensure payment, the lender (the mortgagee) receives a signed mortgage from the borrower (the mortgagor). If the borrower fails to pay, the lender can, after a specified grace period, foreclose on the property. Foreclosure is a judicial procedure which forces part or all of the land to be sold to satisfy the unpaid debt. Foreclosure often represents an unpleasant method of land sale, both for the land owner and for the lender who has to initiate the legal action to recover money loaned out.

Voluntary economic reasons for selling land focus on the returns from farming and the returns from other employment options.

Poor returns—Each year a number of farm operators sell their land because they have been unable to generate sufficient income to meet family living expenses and keep the operation financially sound. The poor returns may be a result of depressed commodity prices or poor yields caused by natural disasters such as wind, hail, and floods. Sometimes the poor returns are a direct result of poor management. No matter what the reasons for the poor returns, the most viable solution may be to sell the land with the seller discontinuing farming operations.

Better nonfarm returns—In many cases, farms have been sold because, while farm returns were adequate, the operator could achieve better returns on labor, management and capital elsewhere. Frequently, the motivating factor has been the wages or salaries that could be achieved in nonfarm occupations. Less frequently, landowners sell their land because they have other investment opportunities which are expected to yield a higher return.

Know What Motivates the Buyer/Seller

As a potential buyer of farm or ranch land you should have a clear understanding of what personal and economic incentives motivate your planned purchase. You should also know what factors motivate the seller. Is the sale motivated by personal or economic reasons? For example, you may be in a better position to negotiate a favorable deal if the seller is being forced out of farming by economic reasons or for other reasons has a need to complete the sale rather quickly. In these cases, the seller may be anxious to sell and may take a substantial reduction from the asking price. In contrast, an individual who is selling in contemplation of retirement may still have a rather strong attachment to the land and may be in no hurry to sell. As a buyer, you need to obtain a clear understanding of the factors motivating the seller and plan your purchasing strategy accordingly.

As a seller of farm or ranchland, you need a clear understanding of the personal and economic incentives which motivate the buyer. For example, suppose two potential buyers are interested in your property. One wants to expand farming operations while the other wants the farm for recreational or hobby farm activities. The seller may take advantage of this knowledge and split the property for sale into two parcels—one cropland, the other the buildings and noncrop land—and thereby obtain a higher total selling price. By knowing the factors motivating the buyer, the seller is in a more favorable position to obtain the best deal on the land sale.

CHAPTER 4
GATHERING DATA AND INFORMATION FOR EVALUATING VALUE OF FARM REAL ESTATE

Estimating value on farmland requires substantial information. Gathering and classifying details about a property sufficient to allow such analysis can be done by professionals employed to perform the task, or by individuals themselves. The goal of this chapter is to describe relevant information and details needed for the evaluation—factors that affect value of real estate. The major emphasis is to identify systematically relevant factors and describe how each impacts on value. Major sources of this data and information are enumerated in the latter portion of the chapter.

General Considerations

Numerous considerations of land value relate to the geographic area and economic conditions and expectations. Such considerations are especially crucial in unfamiliar geographic areas.

Area Itself

The task of evaluating a farm and surrounding area is made much easier if you begin with some maps and plats. A county highway map should be obtained. This is available through the county highway department (or may be printed in the front part of a county plat book). The county highway map indicates towns, types of roads, and section numbers within each township. The county plat book, a second item needed, is typically much enlarged, relative to the county highway map, to show boundaries and ownership of each farm.

City and village boundaries and such landmarks as railroads and major drainage ways are shown, facilitated by a full page typically devoted to a single township. Beyond these two tools, a property plat is needed that shows the property boundaries drawn to scale according to the legal description. If possible this should be done on an aerial photograph (usually available through the Agricultural Stabilization and Soil Conservation Service (ASCS) office). Features of the property such as topography, fences, land usage, etc. can all be recorded on this property plat as the subject property is reviewed.

Location is extremely important for real estate. Areas have a character of their own and generally show little change over time. Experienced land appraisers and farm loan representatives can classify various geographic areas within a larger territory, even though an untrained person may not be able to readily see the difference. Some farm real estate lenders follow a practice of requiring their appraisers to classify areas within which an applicant is located, e.g., a 1, 2, 3, etc. area. Then each property is also classified. For example, you might have a 1 area (best possible) but a *C* farm within that area. Assume for this example than an *A* farm is the highest possible classification. Or conversely, you may have an *A* farm in a *3* area. A given property that is better than those around it will not normally sell as well as when the surrounding properties are equal to or better than the subject property. A prospective buyer should evaluate the general condition and desirability of a given area as well as the specific farm. With this background, we turn to a number of specific considerations to aid in evaluating an area.

Type of farming in an area should be reviewed to determine what production predominates. If 9 out of 10 farms are dairy operations, that suggests dairy farms will likely sell better than poultry farms or hog operations in this area. A professional appraiser will evaluate what a "typical" operation on the subject farm would be, based on the types of operations in the community. Such analysis assumes that a "typical" buyer would most likely conform to community practices. For a professional appraiser, the ultimate goal is to estimate a net income from the operation based on the "typical operator" approach. This idea is developed in Chapter 5.

Trends in the area and community impact on future value of properties within the area. If an area is changing, for better or worse, one needs to assess the potential impact. A generally deteriorating area would not be a good place to own property. Those decisions are easy. But going back to the dairy example above, what impact would there be if dairying is being discontinued in the area? Maybe it is a crop area where the cropping programs are changing. Or perhaps changes are brought about by an expanding metropolitan area or an industrial plant locating in a previously all-agricultural area. Trends in an area may be dramatic—or they may be subtle. Whichever, they should be evaluated.

Markets are often taken for granted, when in reality they differ substantially from area to area. Produce is turned into cash through some marketing system. Prices vary significantly for a given commodity—e.g., corn, depending on location. East central Illinois farmers now have access to one of the major grain market outlets in the midwest. With its huge volume and transportation system to move corn into national and international markets, considerably higher prices are paid for corn than farmers can receive in western Iowa and Nebraska. So access to a market for whatever might be produced is only one facet of the concern—the other is what kind of price that market pays.

Similar concerns exist for livestock and livestock products. On occasion a farmer relocates geographically and does price budgeting based on experience in another geographic area. Such a mistake could be serious. Check out market outlets and prices received for production in the area you are considering. Be sure to include transportation and storage costs in your analysis.

Roads and transportation refer to the type of roads serving the property. Farms on highways and hard surface roads are more desirable than if located on dirt or gravel roads. Relationship to major highways and interstate expressways giving easy access for travel to major metropolitan centers might be especially important to some individuals, but might have minor overall impact on price. It may be more important to learn of any planned or proposed highway construction that may be forthcoming and that would affect the farm directly. Road widening projects can be especially burdensome if fences and buildings are affected.

Schools and churches are a basic concern to anyone considering a move. Quality and location of the schools plus transportation arrangements are important. Choice of denomination is the main interest with churches.

Shopping and service facilities, like schools and churches, are a central concern when evaluating the desirability of a living location. Lengthy travel may be required for major shopping. This might be true for clothing, other major household and personal items and medical care. And it may also hold for machinery dealerships and other service businesses related to the farm operation. Both cost and inconvenience increase with distance to shopping locations.

Ethnic and religious backgrounds of people in a given area are well known to local residents. But such is not the case for an outsider coming in. Questioning local residents is the best way to evaluate this item.

Off-farm employment opportunities have become increasingly important. With about half of the income of the farm population now earned through off-farm employment, it follows that access to such employment may impact on value of farms. This is especially true for small, part-time farm operations.

Climatic Factors

Precipitation—Both the amount and monthly distribution of moisture are crucial. Casual observation will not reveal the answer to this question. There are United States maps and often state maps that show amount of annual precipitation by area of the state, which may vary substantially. For example, in the state of Nebraska annual rain fall varies from approximately 15″ in the extreme western portions to over 35″ in the extreme southeast corner.

Growing season is indicated by average frost free days. It is useful to also have the average dates of the first and last killing frosts. Elevation is important in certain areas, depending on the type of farming operation. Crops requiring long growing seasons may not be suitable in higher areas and in more northern locations. Large bodies of water, like the Great Lakes, often impact on temperatures and length of growing periods. Also, be aware that low lying fields are sometimes subject to an earlier frost than higher land immediately adjacent.

Temperatures, along with precipitation and growing season, determine adaptability of crops. Typical length, intensity, and timing of especially high temperatures should be determined. Hot dry winds, for example during the short pollination period for corn, can substantially reduce yields.

Climatic hazards of greatest concern include hail, drought, and flood. The task is to assess the frequency and intensity of such hazards for a given area. Hail insurance is widely available to protect against hail damage, with the cost being a function of payments for losses that insurance companies have experienced over time for a given area. Drought hazards are perhaps the more difficult to evaluate. An area may look lush and green at a moment in time, giving no indication that with some regularity lack of moisture will create serious problems. The sandier a soil is the more subject to drought it becomes. So while drought for a given area may not be a problem, it could be for a given farm. Flood hazards are associated with rivers and streams but even then may not be apparent on a visual inspection. Flood plains have often been determined by public authorities. There are usually historic records to show frequency of flood problems.

Economic

The third set of general considerations is economic in nature. The two financial or economic reasons for owning real estate are: 1) an annual net operating income return to land, and 2) an increase in value of the asset throughout the period of ownership.

Inflation is a central issue—both as to its impact on increasing the net farm income stream over the years and on increasing the value of the property. Expectations with respect to inflation become a key input into analysis of real estate ownership. This will become quite clear in the chapters that follow. So as one begins gathering data regarding real estate, it is necessary to decide what inflationary expectations you have.

Expected Net Income Returns to Land—Undoubtedly the most important element in the price of land is the expected net income returns to land. Methods of calculating net income returns to land are presented in Chapter 6. But here it should be made clear that net income returns to land are *not* the same as net farm income. Net farm income reflects a return to land as well as to all other factors of production such as labor and management.

This distinction is important because during the 1950's and 1960's land values increased while net farm income remained relatively stable. Many writers referred to this as a "paradox" which they attempted to explain. In fact, there was no paradox since (as was shown in Chapter 3) the growth in net income returns to land exceeded the growth in land values. Total farm income over this period remained relatively stable because while income returns to land were growing, returns to other factors of production (especially labor) were declining. This left total farm income relatively stable.

The higher the expected income returns to land the greater is the price that buyers are willing to pay. There are a multitude of factors which influence the expected net income returns to land. The more important of these are evaluated in the balance of this section.

Price of outputs—The price received by farmers for the commodities they produce influences the net returns per acre and hence the value of land. Prices received by farmers increased substantially over the 1978–79 period. Likewise, there was a surge in the value of farm real estate during this period.

It is important to recognize that in many areas, there are a variety of farm products produced. Deterioration in the price of one commodity may have little impact on land values if there are other commodities for which prices are improving. For example, in 1979 the price of hogs dropped substantially while the price of dairy products increased substantially. In areas such as Southern Wisconsin and Minnesota where both hogs and dairy products are important, land prices increased substantially through this period. Dairy farmers were in a good position to bid more for land while hog farmers were less inclined to do so. However, since only a small portion of the land transfers each year, the improved dairy prices affected a sufficient number of producers to keep the demand for land strong.

Cost of inputs—The cost of inputs used in the farm production process influences net returns per acre of land and hence the value of land. In recent years, the most rapid increases in input costs have been for energy and borrowed capital. Unless these increased costs are offset by higher commodity prices, they tend to reduce net returns to land and hence the price of land.

The importance of input costs in the net returns of farm operators has resulted in government price support programs which are based on cost of production estimates. For example, the 1977 Food and Agriculture Act provided for target prices to cover the direct costs of production plus a return to land. The return to land is a crucial variable in cost of production estimates. If the return to land is high enough to increase expected profits, producers will have an incentive to bid up the price of land, which will raise the cost of production, which will raise the target price, which will raise the price of land—in a never ending cycle. To the extent that these phenomena occur, we have a rather strange outcome in that higher costs of inputs lead to higher land values.

Government farm programs—Federal programs oriented toward the agricultural sector are referred to collectively as "farm programs." The most important of these have been *farm commodity programs* designed to stabilize and enhance commodity prices and farm income. Other components of Federal farm programs include *crop insurance, soil conservation,* and *credit programs.* In the aggregate, these programs have improved farm income and reduced risks.

Benefits from farm programs—the actual dollar payments, the reduced risks, and the heightened expectations—have motivated farmers and ranchers to acquire more land. Program benefits have thus been capitalized into land values. Expectations of continued farm programs heighten the expected net income returns to land and hence the price of land.

There is, however, a growing awareness on the part of policy makers that farm program benefits are quickly capitalized into land values. This benefits existing landholders but may increase the difficulty of entering into farming. Some policy analysts are urging that future policies be designed to prevent inequities. If this objective can be successfully implemented, future farm programs may have less impact on land values than past programs.

Expected capital gains on land—Purchasers of farmland expect to achieve both an annual income return and capital gains return from land investments. As shown in Chapter 3, capital gains have been an important component in the total return to land. Buyers recognize that future capital gains have value, and they are willing to bid this value into the price of land.

Expectations of continued inflation exist for several reasons. For one, some people believe that land prices will increase simply because they have increased in almost all of the post-depression years. Others believe that land prices will continue to increase because of inflation in the overall economy. Both reasons basically reflect the "greater fool theory." This theory suggests that land prices will continue to rise as long as there is a "greater fool" who will pay more for the land than the previous owner. The greater fool theory ignores the relationship between the income returns on land and the value of that land.

Economic theory suggests that the price of an income earning asset is related to the amount of income earned by that asset. Hence another explanation of expected capital gains is the expectation that net returns to land will increase over time. And as shown in Chapter 3, expectations of increased net returns per acre are more likely to be based upon long-term trends rather than month-to-month changes in the net returns to land.

Soils and Land Useability

Soils are basic to production from farm land. Understanding the inherent production capacity, the highest and best use of land and limitations applicable to the highest and best use is essential to analyze a subject farm and for comparing that farm with others that have sold.

Key soil characteristics include the following:

Texture relates to the size of soil particles, ranging from a very coarse sandy type to a very small soil particle that characterizes a clay type soil. Water movement through the soil and water holding capacity are affected largely by texture. Sandy soils may be so porous that water runs through them like through a funnel. In contrast, clay soils may be so tight that heavy rains run off instead of soaking into the ground. With an adequate water supply for irrigation, sandy soils can often achieve high productivity. The ideal texture is a loam type soil—containing some clay and some sand.

Depth and subsoil—A soil profile includes topsoil and subsoil stratas. The depth and quantity of topsoil are important to plant growth since this is where most of the organic matter and fertility is stored and made available to plants. Topsoils are typically darker and more permeable than deeper soil layers. A deep topsoil allows ample room for root growth. Oftentimes, however, the topsoil may be very shallow or possibly eroded away. Productivity will thus be severely curtailed. Subsoil permeability is important to allow easy root penetration. Some crops are deep rooted and especially depend on growth in the subsoil area.

Productivity and ease of handling—Productivity is described in terms of physical output from the soil—in bushels, tons, pounds, etc. Many factors combine to generate productivity—not just soil factors but also climatic conditions and soil management. Some soils are more responsive to fertility treatment and good management practices than others. Past production history gives a good indication of productivity but alone is not sufficient. One needs a basic understanding of the soils on the farm—including approximate acreages of the basic soil types.

Ease of handling refers to how easy the soil is to till and work during the season. A soil with high clay content in the surface soil may stay quite wet in the spring and then dry out hard as a brick. Skillful management is required to get the most out of such soils. Stones and rocks in soils also present difficult problems of handling and can be especially hard on machinery. Size and shape of fields affects the efficiency of an operation—especially when larger machinery would typically be used and the fields are quite small in size and irregular in shape. Different soil types with different handling characteristics within a given field can also create problems. For example, a clay soil may remain wet while the rest of the field is ready for tillage or harvest operations.

Topography and durability—Topography denotes slope and

relates directly to how the land can be farmed. It affects runoff, soil erosion, drainage and use of machinery. Durable soil is that which can withstand intensive use without eroding or otherwise deteriorating, and thus is closely related to topography. Percentage of slope is often used to accurately describe topography—that is the number of feet rise in 100 feet of horizontal distance.

Land Classification
The study of soils is a complete field unto itself. Therefore, buyers and sellers of real estate have access to considerable soil information. The task is often one of being able to profitably use what is available. The following classifications are useful.

Land capability classes are used by the Soil Conservation Service (SCS). This rating system is not based on value or yield but rather on limiting factors. It is useful and is a classification already completed for many farms. The classes are as follows:

Classes I–IV: Land suited to cultivation
Class I: Soils with few limitations that restrict their use.
Class II: Soils with some limitations that reduce the choice of plants or require moderate conservation practices.
Class III: Soils with severe limitations that reduce the choice of plants, require special conservation practices, or both.
Class IV: Soils with very severe limitations that restrict the choice of plants, require very careful management, or both.

Classes V–VIII: Land not suited to cultivation
Class V: Little or no erosion hazard but have other limitations that are impractical to remove and make soils unsuitable for cultivation.
Class VI: Soils with severe limitations that make them unsuited for cultivation.
Class VII: Soils with very severe limitations that make them unsuited for cultivation.
Class VIII: Soils with limitations that prevent their use for commercial plant production. Such uses as wildlife preserves, watershed protection, and recreation are possible.

To better understand their classification, SCS personnel should be consulted.

Soil surveys and land classification—Soil surveys in the U.S. were begun just prior to the turn of the century in 1899, in work by USDA and the state agricultural experiment stations. Soils have been inventoried and mapped using soil terms that first designate a *soil series* and secondly a soil type which describes the texture of the surface soil. For example, Muscatine Silt Loam designates the Muscatine series and a surface texture of silt loam. Soil maps and bulletins for each state are available through the experiment station of the Land

Grant University. Help on interpretation of soil maps for a given farm is available through the SCS or the Cooperative Extension Service in local areas. From soil maps that delineate the soil types on a farm, it is possible to calculate the approximate acreage of each soil type.

Productivity ratings—A more recent development in many states is a system of soil productivity rating. In Illinois, for example, productivity indexes have been assigned to the many soil types ranging downward from an index of 100 which is the top productivity rating. Crop yields for the various crops under differing levels of management are associated with the productivity indexes. By capturing in a single index number a wide variety of characteristics affecting soil productivity, it is possible to accurately evaluate information on soils without having a strong soils background.

Land usage—Regardless of what soil data are available, every farm evaluation should include an inventory of the total acres, reflecting current land usage. An appropriate classification would be: crop land, pasture, timber, roads, building sites and other waste. A road of 66 feet width takes out one acre for each ¼ mile on each side of the road. For example, a square 160 acre tract with roads on all four sides would have approximately four acres out for roads. The most accurate way to evaluate land usage is to get an aerial photograph of the subject property from the local Agricultural Stabilization and Conservation Service (ASCS) and outline the boundaries and various land use categories on the map. Tracing paper can be used if the actual aerials cannot be purchased. Estimates of acreages can be made with nothing more than a see-through ruler which has 100 squares per square inch. However, more precise measurements are possible with ASCS equipment.

Land Use, Production and Treatment History

Cropping history—The way in which the land has been operated in the past is one indication of the current condition of the land. It may reflect what is typical for the area and provides an indication of how the cropping program might look under new ownership. Quality of past management is partially revealed as the past cropping program is compared to how the land should be operated.

Yields, carrying capacity on pastures, quotas and allotments—The production results should be known and evaluated. The local ASCS office may have considerable detail on some of these items, depending on what government programs have been in effect and the extent of participation by the subject farm. County average yields and possibly even yields from the property may be available through the ASCS office. The owner and/or occupant and surrounding neighbors can provide much insight into past production levels on the farm. Quotas are extremely important if the farm and area are suitable for crops or livestock currently subject to a government quota system.

Tobacco is a classic example. Tobacco allotments are required to allow marketing the product. And since allotments may be bought and sold independently of land transactions, the value of a subject property is affected substantially by the crop base or quota in existence.

Fertilizer, lime and herbicide treatment history—The productivity level is obviously affected by the fertilizer and weed control programs. These are also an indication of the level of past management. The owner and/or occupant should be a good source here. But observation and discussion with others may reveal additional insight. Inquire about availability of any recent soil tests.

Government farm programs utilized on the subject property at present or in recent years should be thoroughly explored. The local ASCS and SCS offices should have this information. While checking out the past involvement, you should check out programs that might be applicable for the future. See further discussion on government farm programs in Chapter 14 as well as the earlier section in this chapter.

Buildings and Improvements

The value that buildings and improvements add to farm real estate is difficult to determine. A systematic gathering of details and information about them is the first step. Buildings and facilities are considered first, followed by the non-building improvements.

Buildings and facilities—The home and all buildings should be inventoried. This can be done in a table format. List the buildings down the left hand side, and then in columns across the table record such data as building dimensions, square footage, type of construction or foundation, floor, walls, and roof. Note the general condition and write down additional details for more complex structures. Special attention should be given to evaluation of physical and functional obsolescence.

A major advantage of going over each structure in some detail is that a better appreciation of the improvements is achieved. A home can add a great deal of value and may have an important impact on satisfaction to be gained from living on a property. Therefore, it should be thoroughly inspected—the kitchen, bathroom(s), and heating plant being key items. Ask the question: Would it be pleasant to live in this home?

A professional appraiser would likely estimate the cost to reconstruct each building new and would then estimate the depreciation of the existing building. By subtracting depreciation from new reconstruction cost, a contributory value is determined. A building might have considerable physical life left but functionally be totally obsolete. For example, consider a machine shed so low that today's modern machinery cannot be stored in it. If there are no other uses for the building it may be judged functionally obsolete and would thus have a very low contributory value.

The efficiency and volume of operation that the buildings and facilities permit should be evaluated. Is there provision for crop drying and storage and is it efficiently organized? If there are livestock and feed processing facilities, the desirability of such improvements should be considered in light of current technology. Just because someone invested $250,000 five years ago, for example, is no guarantee that those facilities have much value today. They may have less value than the remaining cost value of the improvements. Always compare what actually exists to what you would construct at the current time period.

One last note on buildings. Be alert for any recent construction, either new or remodeling, completed within the past few weeks or months. The reason is that if payment is not made for such buildings or remodeling, the builder may file a mechanics lien against the property—a lien that takes priority even over those who might have purchased the property in the meantime. Suppliers or contractors typically have 60–90 days to file the lien. As a purchaser you could become subject to such mechanics liens arising out of improvements made shortly before sale of the property with such a lien not showing up on the title examination.

Non-building improvements—Here we want to gather data on such items as fences, tile and drainage systems, irrigation structures and equipment that are attached and included as part of the real estate, and wells and water supply equipment. These items are important in terms of what value they add to the property and how they relate to potential production. You need rather extensive details on these improvements to avoid problems after assuming ownership and for tax records, since improvements can be separately depreciated for tax purposes. Our suggestion is that you get full details on the improvements on the initial inquiry.

Fences—Record on the property plat the kind and condition of all fencing. Be sensitive to how fencing on this unit compares to that on surrounding farms. Explore with the occupant what arrangements exist, informally or as a matter of formal fenceline agreement, relative to division of responsibility for construction and maintenance of partition fences. If fenceline responsibility has been established by formal fenceline agreement, a summary of that agreement should appear in the abstract. Informal agreements, existing by virtue of convention or tradition in the community, should be communicated to the buyer even though such arrangements may not be legally enforceable. Any special arrangements for the maintenance of flood gates or any other difficult to fence areas should also be discussed.

Tile and drainage ditches—Adequate drainage for farms that have a natural wetness problem is essential for top production. Your task becomes one of discovering the extent of any drainage problems, what has been done with tiling and drainage ditches to solve the problems, and how well the artificial drainage is working. A starting point on tile would be tile maps or information about the location of tile lines including dates of installation, size and types of tile, and outlet locations. SCS personnel can help evaluate the adequacy of the system for the soils being drained. Later on if you own the property, repairs on tile lines may be expedited if it is known with certainty where the mains and laterals are located. In addition, information about the tile system may be helpful in claiming investment tax credit and in establishing the initial depreciation schedule in terms of useful life (period to recover investment) and depreciation value.

Drainage ditches need to be evaluated in terms of current condition, how well they are doing the job, and who is responsible for keeping them up. Their existence would strongly suggest the possibility of a drainage district. If so, explore the financing conditions of the district, tax obligations it creates, and provisions for upkeep and maintenance of the ditches in the district.

Wells and water supply—Evaluate the adequacy of water for domestic use, livestock use, and irrigation if applicable. Get some description of well depths, pumping equipment and pumping capacity of each. If water comes from non-well sources, evaluate fully the water rights available to the property and any cost associated with them.

Irrigation structures and equipment—A good supply of water for irrigation is of utmost importance in areas of low rainfall. Start by evaluating the water source and its future potential. Some areas of the country are using wells that are gradually exhausting huge underground water reservoirs. Others pump out of underground supplies that are replenishable. Still others use river water or water that is caught and stored in above ground reservoirs and then distributed. Next evaluate the entire distribution system associated with irrigation. Evaluate its condition, adequacy and efficiency. Specialized assistance may be needed if you are not acquainted with the type of irrigation used on the property being considered.

Additional Considerations

Boundaries should be confirmed at the time you first visit the farm. To do this the legal description should be obtained and a plat drawn according to the description, using the plat book and aerial photographs as aids (see Chapter 10 for a complete discussion on legal descriptions). By using the plat book to locate bench marks (roads, towns, etc.) and the car odometer, it is possible to identify boundaries of the subject farm. There should be no *guessing* on this part of the information gathering phase. In addition give careful scrutiny for potential problems of a legal nature. With the boundaries confirmed it is possible to observe if the property has access to a public road.

You can determine if individuals (other than the land owner) may be making use of the property in some fashion. If such use continues for a specific period, typically ten years, the users may have acquired a right to continue the use even over the landowner's objections. Thus, the use of land by neighbors as a regular passageway for more than ten years could result in an easement giving those individuals a right to continue using the property for that purpose. Inquiry of knowledgeable persons may be advisable if any indications point to the possibility of such past use.

The location of boundaries is critically important. In many states, an erroneously located boundary that is respected for a period of years, again typically ten years, may become the true boundary if formal objection is not made during the ten-year period. A visual check should be made for obvious indications of how adjacent property owners have been respecting the boundary and also whether buildings or other improvements may be encroaching on the subject farm. Such items do not show up in the local records and hence do not arise during the course of the regular title search of the property conducted later by an attorney.

Easements and right-of-ways—This can be a limitation on land use created by prior owners or the current owner of the land. It can also be limited by the right of eminent domain on the part of governmental agencies or, in the case of utilities, by virtue of authority obtained from governmental agencies. An easement is a non-possessory interest one party holds in the land of another and in which the first party is accorded partial use of such land for a specific purpose. What one needs to know is to whom the easement has been granted, what area is included in the easement, whether it goes with land, and what rights or uses the easement permits. Easements for utilities are common, such as for gas lines and electric lines. Easements for ingress and egress are not uncommon, where the easement will grant permanent passage over certain property. Land near airports is often subject to an easement preventing the construction of improvements above a certain elevation. While such a limitation may not be of great significance so long as an agricultural use is contemplated, a limitation of that type could have a substantial effect upon land value if it rendered unacceptable a land use with high potential development value.

Mineral rights—People often forget that purchase of land includes rights to what is below the surface as well as the surface and certain air rights above it. Mineral rights, which refer to rights to oil, coal, and other minerals, can be sold separately from the land. It is not uncommon for sellers to retain partial or complete mineral rights, which means that if minerals are ever discovered and mined on the property, they or their heirs would get the revenue—not the current land owner. The status of ownership to mineral rights is best evaluated by an attorney who would check deeds and other documents filed at the county courthouse.

Zoning and land use restrictions—An important overall consideration is the zoning status or any general land use restrictions. Such restrictions may arise in various ways. The governmental subdivision, usually the county in the case of farmland but possibly a regional or city authority, may have enacted a formalized set of zoning restrictions.

In some areas agricultural land is exempt from most zoning limitations. But in others, constraints are imposed upon certain types of agricultural development, notably those involving buildings. Intensive livestock operations may be limited or curtailed by zoning regulations. Another example, that impacts even more directly on land values, is in urban areas where certain lands are zoned for agricultural use only, prohibiting housing or industrial development that might generate a much higher price for the land if it were allowed.

Information about the zoning status of the tract in question can be obtained from the local office charged with administering zoning laws. In some instances a specific office has that responsibility. In others the information can be obtained from the city or county recorder's office or the office of the administrator for the local governmental subdivision.

Property taxes and assessed valuation—Concerns focus on the relative tax burden in the general area affected by the school districts and other public bodies supported primarily by property taxes. Both current and anticipated tax dollar needs should be assessed. Another major factor affecting individual taxes is the amount of taxable property in the district. Industrial companies also help bear the taxes. On the other hand, publicly owned lands diminish the base and put more burden on individual property owners. Beyond the tax burden to be shared by all property owners is the share of taxes on the subject property relative to comparable properties. A history of past taxes plus an assessment of future tax dollar needs provides useful information.

If not previously indicated, prospective owners should explicitly inquire as to the existence of drainage, fire protection, irrigation or other special improvement districts. These may be separate taxing bodies and can create considerable tax burden. Full details of any such districts should be fully evaluated.

Deed restrictions by present or past owners—Limitations of various types are occasionally imposed by property owners on subsequent uses—and users—of the land. Such restrictions may preclude types of development thought to be undersirable or may impose specific limitations on improvements placed on the land. Thus, land use restrictions might specify the minimum square footage for units in a housing development, a minimum setback distance for dwelling units from streets or highways, restrictions on types of construction material that can be utilized, specification on types of sewage disposal systems, just to mention a few of the major types of restrictions in general use.

Some land use restrictions may be viewed as positive in enhancing the value of the property in the future by preventing uses on adjoining tracts that might reduce value. On the other hand, restrictions may be viewed as a factor to depress the value of land by ruling out uses of the land that would add additional value. Thus if the highest and best use for a tract of land is for high rise apartment units, limiting the tract to single family residential housing would likely limit the value of the land for development purposes.

Availability of utilities—If development of the land is contemplated in the near term, it is also important to check the availability of utilities to the site. The cost for obtaining access to sewer systems (or the cost for constructing a sewage disposal plant), the availability of electricity and the presence of a dependable municipal water supply are all important development factors. Likewise, if development is contemplated, it is important to consider the philosophy and attitude of the governmental subdivision most likely to be involved in the development process, including the matter of costs expected to be borne by the land developer.

Even for land acquired for agricultural purposes, these points may be significant inasmuch as a higher use than agriculture could emerge in a period of a few years especially if the tract in question is near a metropolitan area, a recreational or resort area of some significance or a major transportation or traffic artery which is likely to spawn development.

Lease arrangements—If the property is not being operated by the current owner, check the existing lease arrangements. If someone has a valid lease (oral or written) that extends to one or more crop years in the future, the property would have to be sold subject to the lease.

Disease problem on the premises—The buyer may also wish to inquire as to any disease problems on the premises. Some disease organisms may be viable for a period of several months or even years after the last outbreak of a disease. Those diseases should be discussed with the buyer.

In most states, the legal liability of landowners for the spread of disease depends heavily upon knowledge that the disease existed. Therefore, if a farm tenant brings clean livestock on the premises that have previously been infected with a contagious disease, the landowner is generally not liable unless the farm tenant was misinformed by the landowner as to the status of disease on the premises. Thus, if the tenant failed to raise the question, the landowner would generally not be liable for spread of the disease to the tenant's livestock.

Immediate surroundings—Property values are influenced by undesirable businesses or industrial operations. There are many examples of eyesores or nuisances that might impact unfavorably on properties in the immediate area. Inexpensive housing that presents a "junky appearance" creates a much different impact than an attractive, neat, well-planned housing development. Activities or industries that create noise or odor pollution can be a special nuisance. Some of these items may not be immediately apparent—such as if it takes the wind blowing from a certain direction to create odor problems. Be thorough in your evaluation of those immediate surroundings—they are not likely to change much in the near future.

Sources of Information and Assistance

Gathering and analyzing all the data and information needed to analyze a farm land tract is a major task. For land purchasers with the least experience—and even some with considerable expertise—professional help will be well worth the cost. If a professional is used the main concern will be how and where to locate a competent individual. After that, the task will be to understand and fully appreciate all aspects of the report presented. Beyond that it is important to know of basic information sources available to either the professional or layperson. No such list can be exhaustive but it can be suggestive. First, who are the professionals?

Professional appraisers—We use the term "professional appraisers" to denote those individuals who have met the rigorous qualifications of one of the professional appraisal societies to become accredited by the various societies. What assurance is available on integrity, training, experience, skill, independent judgment and reliability of someone qualified to render a professional judgment on the value of a given property? The best evidence is accreditation that professional appraisal societies bestow on those individuals who have met the education and experience criteria required. Passage of several hours of examination is a prerequisite along with approval of judgment and procedure demonstrated through several appraisals submitted to the examining committee. All societies require strict adherence to a prescribed code of ethics as a condition for retaining the accreditation title.

Most prevalent in farm appraising are appraisers accredited by the American Society of Farm Managers and Rural Appraisers—called, Accredited Rural Appraisers (ARA). A list of accredited appraisers is available through the American Society of Farm Managers and Rural Appraisers, Inc., P.O. Box 6857, Denver, Colorado 80206. Another prestigious accreditation is awarded by The American Institute of Real Estate Appraisers of the National Association of Realtors. The MAI title (Member of the Appraisal Institute) requires the applicant be age 30 or over, have a minimum of five years of appraisal experience, submit three acceptable narrative appraisal reports covering different classes of real estate property, and pass two written examinations—each eight hours in duration. The Society of Real Estate Appraisers awards three designations to qualified candidates: 1) SREA (Senior Real Estate Analyst), 2) SRPA (Senior Real Property Appraiser) and 3) SRA (Senior Residential Appraiser). Again, this society maintains standards for qualification as an accredited member.

Agricultural Stabilization and Conservation Service (ASCS) and Soil Conservation Service (SCS)—These two county offices are very important sources of data about a given farm and about local agriculture in general. You will likely find them located in the county seat town. Both are government offices. The SCS administers all conservation programs and is an integral part of the soil mapping program that spans the entire U.S. If the previous farm owners or operators have worked with SCS there will be even more specific detail about the subject farm. Questions related to drainage, soils, irrigation, ponds, terracing and contouring and forestry—just to mention the most likely areas of help—should start at the SCS office.

The ASCS office is charged with administering most other farm related government programs, except those credit programs handled through the Farmers Home Administration. Their role and involvement vary over the years as government programs change. But all allotments, grants, and various types of payment programs are centered here. Commodity credit corporation (CCC) loans are also handled by ASCS. It may be possible to obtain such specific information as county yield averages, yields and acreages of various crops for the subject farm, past levels of crop payments on the subject farm and all existing programs and requirements that affect the subject farm. The government is likely to be very much involved in agriculture over the foreseeable future. Contact at the ASCS office is needed to monitor how the programs will impact on a given farm. As mentioned earlier, aerial photographs of farms are extremely useful and can be ordered through ASCS.

Advisers—Through a combination of federal, state, and local funding and the land grant university within each state, an educational link is provided between the college of agriculture at the land grant university and the people of the state. It is a rich source of educational materials and even individual counseling on many topics. Seek the county adviser and learn of services and help available.

County Court house and other public offices—The relevant offices vary from state-to-state, but by knowing what to look for you should have little difficulty. Deeds are filed at a Recorder of Deeds office, and indexed in grantor and grantee books. The existing owner would have been the grantee at the time ownership began and thus it is reasonably easy to locate the deed. Confirm the legal description and read the deed to learn of any restrictions or limitations imposed on the current owner. Check especially for evidence on easements and mineral rights. Discover the state's system for attaching revenue stamps on deeds. It may be possible to get an idea of how much the existing owner paid for the property. That is useful background information as you negotiate a current transaction price. In this same office, real estate mortgages should be on file. The existing owner would be the mortgagor if a mortgage was placed on the property. Amounts, timing, and lending institutions involved may all be useful.

A visit to the county treasurer's office and/or the county assessor's office allows you to explore past assessed valuations on the property in question and the tax history. Separate building and land assessed valuations are available. These county officials are also good sources of information regarding special factors that may affect future value of the property and ideas as to what future tax rates may do. Check here also to explore about special taxing districts such as for irrigation, fire protection, or drainage. If these government officials are unclear as to whether the subject property is located in a special taxing district or if they are unfamiliar with the taxing base and tax rate, they can at least suggest other officials to see.

County officials responsible for county zoning ordinances can clarify present and proposed zoning regulations. They can help determine if one or more specific operational plans would be in conformity with existing zoning requirements.

Visit the county highway officials to learn of road development plans. In some counties there is a county surveyor and/or a county engineering department. Be especially sensitive to long range "rumors" about developments that might affect the subject property.

Bankers and other agricultural lenders stay very close to the agriculture of a community and the success farmers are having. One or more lenders—especially a real estate lender—should be on your list. Other input suppliers or market firms may well be added to your list of visits, but the perspective of a lender adds a dimension of knowledge different from others.

Neighbors and others living in the immediate vicinity are especially helpful if you are not acquainted with a local area. Learn about the reaction of others toward the farm you are considering—especially its strengths and weaknesses. Get to know the community and its people.

Visual inspection of the area and subject farm—A systematic and thorough evaluation is needed. Appraisers often prefer to start at the trade center which serves the subject property. Many of the considerations are detailed in the first part of this chapter. Drive to the farm via a route most likely for people living on the farm. Before entering the farm, drive around it and locate all boundary lines, and become familiar with the immediate surroundings. Use of the plat book described earlier as well as the property plat (hopefully on an aerial photograph) will get you well oriented. Next enter the farm and sytematically walk over the entire farm. Soil types should have been evaluated earlier. That background information facilitates visually examining the various soil types, degrees of erosion, drainage and other problems and opportunities associated with the soils and the land. Make notes on the property plat regarding the fencing, topography of various fields, tillable and non-tillable land including wooded areas, permanent and renovated pasture.

Evaluate the buildings and improvements next. You may want to locate the improvements on a plat of the farmstead to facilitate a better recollection later on.

Interview of occupant—The data gathering phase is enhanced substantially with a good interview. Generate a list of the many questions and concerns that need to be discussed. Throughout this chapter reference has been made to items that should be explored with the existing occupant. Learn how this individual reacts to this farm and this community.

Ownership of farm real estate is serious business. You can not have too much information about a property. To make a sound evaluation and analysis followed by judgment on value of a specific property, gathering of all relevant information is a necessary prerequisite.

In Chapter 5, the next phase of activity that professional appraisers would go through is described—namely the analysis of the data and estimation of value, using three approaches to that value.

CHAPTER 5
DETERMINING THE FAIR MARKET VALUE OF LAND

In Chapter 1, two basic values of real estate were identified. One, the so called "Objective Value" is the fair market value or value in exchange. The other is the "Subjective Value"—the value of a property to a specific person. This chapter describes the process that professional appraisers use to estimate fair market value of land. The next chapter identifies the subjective value. Until a property is exposed to the market by a willing seller and is ultimately purchased by a willing buyer, with neither being under abnormal pressure, a market price is not determined. Prior to this all that can be done is to estimate the fair market value. Whether you hire a professional appraiser or attempt to estimate that market price yourself, you need to understand the process that is used.

Three approaches to estimating fair market value are: 1) market data approach; 2) income approach and 3) cost approach. An appraiser typically uses two or three of these approaches, arriving at an indicated value of the property with each approach. The last step in the appraisal process is to correlate the two or three values and make one final estimate of value for the property. Differences in values estimated by each of the three approaches have to be reconciled. This is done largely by evaluating the relevance of each approach as to appropriateness for indicating value of the farm. Through judgment and a weighting process, a final estimated value is determined.

Why three approaches? The answer relates back to the complexity of what value is and how it is determined. The *market data approach* is perhaps the most useful—and really the most logical. It involves determining the value of a property by comparing that property to several similar properties that have actually sold—called "comparable sales." The logic is that a prudent person would not pay more for a given property than it would cost to buy another equally desirable similar property; nor will a well informed seller sell a property for less than other similar properties are selling for. This approach is most valuable when recent comparable sales in the immediate geographic area can be documented. The problem comes in finding comparable sales and then in adjusting for differences between the sales and the subject property being evaluated. Dates of sales are especially important so that time can be taken into account as the market is continually changing.

The *income approach* assumes a property is being purchased as an investment on which a return will be generated. The task, therefore, is to estimate the amount of average net income that can typically be produced by the property, and then capitalize that income stream into a value. For example, if one has a target of a six percent return, and the per acre net returns are $120 per acre, the capitalized value would be $120 ÷ .06 = $2,000. Several major problems exist with this approach. One is that income may be only one of several reasons for purchase of the property, e.g., capital appreciation is ignored. Another is the difficulty associated with choosing a proper capitalization rate. Lastly, using an average income is misleading, since net incomes per acre have continued to increase, and thus the flow will not be uniform.

The *cost approach* is the third approach to value. It stems from appraisal of real estate comprised predominantly of buildings and is a value predicated on the assumption that value of the whole cannot exceed the value of each part, described as the current cost of replacing an improvement less depreciation from deterioration and functional and economic obsolescence. It is easy to see how this approach to value would apply to a home in a new subdivision. Sale value would be closely approximated by cost of building the home plus the price of the lot. But with buildings normally a minor portion of the total value of farms, the cost approach has been broadly interpreted to mean a breakdown of the land into various categories of tillable land, pasture, and waste, plus the depreciated value of buildings. Thus the cost approach on farms is often termed inventory and cost value.

In the balance of this chapter, detail is provided on the process used to estimate value by each of the methods. The intent is not to provide a basic appraisal handbook. Rather, the goal is to allow an understanding and appreciation of what is involved with each approach.

Market Data Approach

The objective is to indicate the value of a property, sometimes referred to as the "subject" property, by comparing it to several other comparable farms that have sold recently in the area. The task is to find other sales that are comparable and evaluate and analyze each sale to determine how it differs from the property to be valued. Based on these differences, adjustments are made to provide an "indicated value of subject property" for each sale that is considered comparable.

Development of sales—Professional appraisers continually gather sales data. The courthouse records normally provide the best source of sales information. Someone systematically gathering these data might obtain the following items: 1) the book and page of recorded instruments and date of recording; 2) date of sale; 3) names of grantor and grantee (vendor and vendee); 4) legal description; 5) monetary consideration and/or amount of transfer fee stamps; 6) terms of sale, including down payment and interest rates, if available; 7) sale restrictions or reservations, and 8) personal property included in the transaction.

Several other sources of sales data should be cultivated. Real estate brokers, bankers, farm mortgage lenders and lawyers are often very helpful. Other key contacts for information on land transactions include assessors, farm managers, farm owners and tenants. Many of these contacts may not have all the details about a sale but can alert one to a sale that should be investigated.

After developing a list of sales in a given area, the next step is to reduce the list to those farms which are comparable to the subject farm. Criteria should be established for what makes a farm comparable. Key factors that determine comparability include the following: 1) type of farm; 2) date of sale; 3) location; 4) farm size; 5) the land and its productive capacity; and 6) buildings and improvements. A preliminary review and knowledge of the farm sales may eliminate some as not comparable, but for the most part that decision should be postponed until after further investigation.

Verify and investigate sales—Appraisers require confirmation of the sale transaction details from someone having personal knowledge about the sale, such as the buyer or seller, the real estate broker, or the lender involved. As the appraiser verifies what has already been learned about the sale, additional details may be acquired from those intimately involved. To be used as a comparable sale in the establishment of value for the subject farm, it is essential that each sale be an arm's length, bona fide sale. What is meant by these terms?

Bona fide sales are genuine, market value transactions where the property has been adequately exposed in the market to all potentially willing buyers. It is a price arrived at in free and open negotiations between a well informed seller under no compulsion to dispose of the property and a well informed buyer who is under no special pressure to buy the property in question. Examples of sales that would not qualify include family transfers or sales between relatives, foreclosure sales or those short of foreclosure but forced by a lender, sales prompted by non-payment of taxes due, and sales made under the stress of condemnation or zoning controversies. "Arm's length transaction" is a term commonly used to imply that the transaction is a non-family, bona fide sale where the market place with buyers and sellers has been able to function freely, and no family or other close relationships are present that might influence the selling price.

In verifying sales, the following check list is useful: 1) date of sale; 2) sale price; 3) whether the farm sold for cash or on contract; 4) if on contract, the terms; 5) value of any personal property included in the purchase; 6) whether an existing mortgage was assumed or taken subject to by the buyer; 7) whether any other financial obligations were included in or affected the sale price; and 8) possession date and expiration of any existing leases. In addition to gathering facts it is also important to learn why individuals buy or sell a given property. It is the behavior of people that makes the market. Thus, one's effectiveness in estimating market value of property is heavily dependent on the ability to predict behavior of people toward that particular farm.

Verification of a sale with the buyer is especially useful since each comparable sale should be evaluated rather thoroughly. Oftentimes, the buyer is excited about the purchase and welcomes a chance to show the property. The soils and productive capacity plus the buildings and improvements are best evaluated by actual inspection. Use of aerial photographs and soil maps often allows fairly close estimates on tillable acres and general productive capacity of the farm. Once the potential comparable sales are verified and inspected, it is reasonably easy to select those which are most comparable to the subject farm and thus usable in establishing a value on the subject property.

Comparison of Comparable Sale to Subject Property—The goal, with each farm judged as comparable, is to start with its sale price and make dollar adjustments for differences between it and the subject farm. That generates an indicated value of the farm to be valued. Assuming four to six comparables are used, this gives a range of indicated values for the subject farm. Naturally, some comparables are more relevant and applicable than others and the values from those farms will ultimately be relied on more heavily. Success of the comparison relies on accurate evaluation of the differences and similarities between each comparable sale and the property being valued. A table of dollar adjustments is typically presented in a summary of the appraisal report narrative evaluation.

Let's briefly review how this is done. Keep in mind that the process is applied only to those farms already judged to be similar to the subject farm and thus useful as a comparison to estimate market value of the farm. Each of the adjustment criteria obviously captures many considerations in one adjustment figure. Note that all adjustments are on the comparable sale to make it like the subject property. *Time* is perhaps the most straight forward. An adjustment for inflation or deflation since the date of purchase to the current time should be made. Indexes of land values on a state basis are readily available in *Farm Real Estate Market Developments,* published annually by U.S. Department of Agriculture, and can be used for the adjustment, in the absence of any more specific or localized indications of how land values have changed over time. For example, consider appraisal of a farm in 1980 where there is a comparable farm that sold for $2,200 in 1977. The formula is:

$$\frac{\text{Index of current year}}{\text{Index of earlier year}} \times \text{earlier sale value} = \text{estimated current value}$$

For example it is $\frac{404}{283} \times \$2,200 = \$3,141$, which is the estimated current value for the sale. The time adjustment is the difference between $2,200 and the indicated current price, or a $941 adjustment.

Size adjustments attempt to modify the dollar impact of the difference in size between the two tracts. For example, a 40 acre comparable sale may be included in trying to estimate value of a 160 acre farm. Past observations may indicate that in this particular geographic area a 40 acre tract will typically sell for $50 more per acre than a 160 acre tract. The adjustment would thus be a minus $50 on size, to bring the comparable in line with the subject.

Location is a major factor in real estate sale prices and adjustment should be made on whether the location of the comparable is better or worse than the farm being valued, and by how much per acre. Again the adjustment should be made to make the comparable like the subject. So if location is poorer for the comparable by $35 per acre, a plus $35 adjustment should be made.

Land encompasses the quality, productivity and usefulness of the soils. Such factors as growing season, water availability and quality of drainage are also considered. Percentage of tillable land, pasture, and waste should first be compared. The percentage of tillable land is a more significant factor than is often recognized. The quality and desirability can next be considered in light of how buyers would react to the different soils. By observing buyer preferences it is possible to estimate the dollar adjustment needed on the comparable sale to make it like the subject.

Improvements are the last major adjustment. Buildings and improvements are generally valued on replacement cost less depreciation. Depreciation includes physical as well as functional inadequacy and economic obsolescence. But building values are rarely fully reflected in the market value of a farm.

The adjustment of the comparable sale to the subject farm involves, therefore, estimating the difference in value of the improvements on the two properties and then deciding what percent of the difference in improvements contributes to the market value. Even on new buildings the contributory value may be only 50 to 60% of the cost value. At any rate a judgment should be made on how much and in what direction the comparable sale needs to be adjusted to make it like the subject property.

Other adjustments may be needed. Any other major differences between the comparable sale and the farm being valued that have not been accounted for, should be incorporated. For example, a sale may include part or all of a growing crop in addition to the farm land. Even terms of the sale may require adjustments to make them similar, especially in the case of a contract if the contract terms depart from arm's length arrangements that may generate a higher or lower selling price.

Table 5.1 Illustration of a Typical Analysis and Adjustments to Sale Prices of Four Comparable Farm Sales to Give Indicated Values of a Subject Farm.

Sale No.	1	2	3	4
Sale Price/Acre	$1,700	$1,075	$1,336	$1,624
Time	-0-	+300	-0-	+200
Size	−60	+60	-0-	-0-
Location	-0-	+140	+160	+60
Land	−80	+200	+40	−100
Improvements	+80	−200	+40	−60
Other	-0-	-0-	-0-	−100
Net Adjustments	−60	+500	+240	-0-
Indicated Value of the Subject Farm	1,640	1,575	1,575	1,624

Final estimate of value by market data approach—Following the steps suggested above, a value of the farm will have been indicated for each of the comparable sales analyzed. The next task is to consider all of the sales and the resulting indicated values, and from that to make a final judgment on the estimated value of the farm. Certain sales may be judged to be more comparable than others and thus greater reliance may be placed on some of the sales than on others. But the final result is an estimate of the fair market value based on the market data approach.

Income Approach

The underlying theory of the income approach, as it is typically used in farm appraising, is that the reason for purchase and ownership of land is to obtain an income stream that will continue into perpetuity. Therefore, the value of the property can be determined by converting the anticipated future income stream into a present worth—the value of the asset. For a goal of earning a ten percent annual return on an investment in land with an expected net return of $100 per acre, it is easy to grasp that the value would be $1,000. Value determined by capitalizing a net return is obtained by the formula:

$$\text{Value} = \frac{\text{Net Return to Real Estate}}{\text{Capitalization Rate}},$$

which in this case is $100 divided by .10 = $1,000. The income approach is most applicable when the income stream is the main reason buyers want to buy land and few non-income factors are present to motivate the purchase. A problem with this approach is that it does not consider inflation and the possible increase in value that might result. However, since the capitalization rate requires judgment and is flexible, it may be adjusted to capture some of these considerations. Furthermore, it is usually done on a before-tax basis. These shortcomings are overcome with the analysis in Chapter 6. The balance of this section develops the factors to be evaluated in arriving at an estimate of value based on income. The process involves estimating net returns and deciding on a proper capitalization rate, so the formula given above can be applied to estimate a value.

Establish typical rental terms and share arrangements—The first task is to determine whether land in the community would typically be operated by a tenant or an owner operator. If tenant operation is typical, normal rental arrangements for the community should be determined. In some areas, landlords pay 50% of the fertilizer, seed, and other production expenses and receive 50% of the crop. In other areas a 40%—or possibly a one-third—share of the expenses and income for the landlord would be more common. Sharing arrangements on revenue and expenses should be explored in detail. The goal is to determine the return to land for the owner. If owner-operator control predominates in the area, the revenue and expenses should be budgeted on that basis. Another possibility is rental under a cash rent arrangement. Normally, if tenant operations exist, the share rent approach is prevalent enough that it would be used rather than cash rent due to substantial variation in cash rents from year to year and even within a year.

Estimate gross returns for tenant or ownership pattern chosen— The process of estimating net returns to land depends on calculations for "typical" management and operations, based on what exists in the community. Thus a "typical" rotation should be assumed without inserting a higher level of management expectation than is "average" for the community. A careful breakdown of tillable acres, pasture, and waste should be used in determining a typical cropping program. Likewise, on yields the estimate should reflect a "typical" operation. An appropriate target would be to consider an average of yields over the past three to five years. Prices are the next major consideration. No one knows for sure what prices will average in the future. A typical approach is to utilize an average of prices over the past three to five years combined with some judgment of what lies ahead.

Returns from livestock would normally be considered only for owner-operated farms equipped for intensive livestock operations under a livestock share lease. In such cases, a typical livestock operation should be assumed. Dairy areas are often owner-operated and might require budgeting of a livestock operation. Roughage and pasture would be marketed through the dairy operation with much of the other feed grown also marketed through livestock.

In summary, one of three types of gross income would be budgeted: 1) landlord gross income (share rental); 2) owner-operator gross income; or 3) landlord gross income (cash rental). The next task is to estimate expenses that should be deducted from the gross income to generate a net return to the land.

Estimate expenses—Keep in mind that the goal is to estimate a net return to the land. That means a charge for management and an interest charge on the investment in non-real estate assets should be made as well as an adjustment for other expenses to the land owner. The estimate of expenses should reflect the "typical" operation that was assumed for calculating gross income. This expense estimate should reflect what a typical owner-operator or landlord would spend to produce the estimated gross income and to maintain the improvements.

Landlord expenses (share rental) would typically include the following, depending on the type of rental arrangement: 1) *Preharvest and harvest expenses*—many different rental arrangements exist but normally the landlord shares in such costs as seed, fertilizer, herbicide and irrigation. Harvest and storage costs are also usually shared but often on a different basis than pre-harvest costs. 2) *Taxes*—at least real estate and possibly special assessment and some personal property taxes would be paid by the owner. 3) *Repairs and maintenance* of all real property and possibly some sharing of expense on personal property are generally considered as a land owner expense. 4) *Insurance*—at least on buildings and improvements and possibly other types of insurance such as liability insurance, a portion of the coverage of jointly owned personal property and hazard insurance such as hail coverage are generally paid by the land owner.

5) *Depreciation* on all depreciating assets for which the owner-landlord has responsibility is borne by the land owner. For example, on a livestock operation with confinement facilities, depreciation might be a major consideration. 6) *Management*—since the goal is a net return to the land, a deduction for management input should be made. A common management charge is one-half the prevailing professional farm management fee. Naturally, a cash rental arrangement would require less management input and less of a fee than share rental arrangements.

Landlord expenses (cash rental) would be similar to the share rental, except that the landlord would bear no preharvest or harvest expenses. A lesser management fee would be charged. Insurance, taxes, repairs on buildings and depreciation would typically remain about the same.

Owner-operator expenses are the most difficult to estimate. All expenses that an owner-operator would incur should be budgeted and subtracted from the corresponding gross income. Livestock operations are more difficult to budget than grain operations, but the process is similar. Typical cost items include:

1) *Preharvest expense*—this includes all the operating costs of preparing the land and planting the crops, such as fuel, seed, fertilizer and livestock expenses. 2) *Harvest costs*—cover all costs associated with getting the crops harvested, stored, and marketed. 3) *Taxes*—there will be real estate and special assessment taxes plus any personal property taxes. 4) *Maintenance and repair expenses*—apply to both the fixed improvements and the personal property. 5) *Depreciation* is closely related to item (4) as it represents the annual loss in value of depreciable fixed improvements and personal property. Buildings and equipment are normally the major items.

6) *Insurance*—most owners carry fire and extended coverage on improvements. In addition, insurance costs involve liability, hail and workers' compensation insurance. 7) *Interest on investment*—on the non-real estate investment (equipment, livestock, furnishings and other personal property). A part of the gross return should be allocated as a return on capital invested in non-real estate items, regardless of whether there is money borrowed on these items. 8) *Labor*—a part of the gross return is for the labor required to earn it. Hired labor cost would be a direct deduction, but equally important is an allocation to the unpaid family and operator labor. The months of labor provided can be figured and multiplied times the given labor rate. 9) *Management*—like labor, management is a resource that does not earn a direct return. But the gross revenue should be allocated in part to the management resource. A good estimate for management expense is an amount equal to the prevailing professional farm management fee for the area.

In summary, the appropriate expenses are deducted from the gross revenue to provide an estimate of the annual *Net Return to the Land*. It is this series of expected net returns—called a stream of income, that creates the value being estimated. This is the figure that is used in the numerator of the equation $\text{Value} = \dfrac{\text{Net Return to Real Estate}}{\text{Capitalization Rate}}$.

Capitalization Rate—The last step in the income approach is choosing the proper capitalization rate that, when divided into the net return figure, generates the earnings value. Thus it is the capitalization rate that converts the future income stream into value. Minor changes in the capitalization rate have a major impact on the value determined. Therefore, the selection of the capitalization rate is most important. Unfortunately, there is no precise method for selecting the capitalization rate, but rather judgment is required. Conceptually, the capitalization rate is the opportunity cost for potential buyers—that is the highest return the buyer could earn in a long term investment other than the land, assuming similar risk characteristics. The rate is often termed the cost of capital, and it is assumed this opportunity cost is higher than the borrowing cost.

If a buyer is acquiring the land purely to receive the net income stream, theoretically that buyer would not buy the land unless the yield on it was at least as good as the yield that exists in the best alternative investment. Thus, the estimated alternative yield would conceptually be used to discount the income stream from the farm land. The value determined by the approach described above generates an income value that may have limited connection with market value, if in fact people are buying for reasons other than the income stream. In recent years the potential inflation value has been a major motivation and would not be captured in the procedure just described.

That has led to widespread use of a second approach to selection of a capitalization rate, called the market capitalization rate. The theory is that actual rates of capitalization being exhibited in the market on the comparable farm sales provide the best evidence available on the relationship between net farm earnings and value. Since the value is known for these sales transactions, the net income can be estimated and the capitalization can be determined (refer back to the basic valuation formula). For example, if a farm sold at $1,500 per acre and you estimate the net farm return at $45 per acre, the estimate of the capitalization rate used by the buyer is:

$$\text{Value} = \frac{\text{Net Return}}{\text{Capitalization Rate}}$$

$$\$1,500 = \frac{\$45}{X}$$

or solving for X we have $\$1,500X = \$45$ or $X = \frac{\$45}{\$1,500} = .03$ which is a three percent capitalization rate. By estimating this rate for each of the comparable sales used in the market data approach, a series of rates is determined from buyer behavior. One can then estimate with some precision what capitalization rate would be applied by potential buyers. The value thus estimated by using the market capitalization rate reflects an estimate of market value based on the income approach, and is the value that can then be used along with the value estimated by the cost approach (discussed below), to arrive at a final estimate of value on the property.

Cost or Inventory Approach

The cost approach is based upon 1) the assumption that a buyer will not pay more for an item than the cost of an acceptable replacement or substitute, and 2) the assumption that value of the whole is no greater than the summation value of the parts. There is, however, some confusion over the difference between cost and value. Cost is the outlay required to construct the improvement and value is what a buyer is willing to pay for the improvement. Buyers are often unwilling to pay 100% of the cost of an improvement that is already in existence—even if it is new. Depending upon the type of improvement, and its overall desirability and utility value, market value may be increased only 40 to 75% of the cost of an earlier improvement added. The bottom line is what value potential buyers believe a given improvement has added. The problem in determining contributory value becomes greater with increased age of the improvements.

The cost approach involves determining the value of the improvements and adding the value to the land. This method works better on non-farm commercial properties where improvements are the major portion of the real estate value. For a new commercial building, the logic is that total cost of the structure plus the market value of the land provides an upper limit on what rational buyers would be expected to pay for the property. If the contributory value were not 100% of cost, the cost value would over-estimate the market value. On some properties, the cost approach may be the only practical way of estimating value, e.g., non-profit entities where no comparable sales exist, such as a church or a fire station.

Some appraisers have concluded that the cost approach has limited usefulness in agriculture. Others argue it is a useful check on market value and the earnings value if the cost concept is broadened to concentrate on the inventory of land in addition to the value of buildings. This requires that land be classified and valued by category and that all buildings be inventoried and valued.

Valuing buildings—Two approaches are possible in valuing buildings—a reproduction value and a replacement value. Reproduction value assumes a replica of the existing structure would be reproduced, and thus an estimate is made of that cost under current conditions. This is somewhat unrealistic since owners today would likely not reproduce an older building with an identical structure if it were destroyed, but rather would replace it with a building more adapted for today's use. Therefore, replacement value has come to be the dominant building value.

Once the replacement cost is determined, the next step is to estimate the depreciation for the existing building. Depreciation consists of physical deterioration, functional obsolescence and economic obsolescence. Thus, depreciation represents loss in value of the asset due to all causes. Depreciation represents the difference between the cost of replacing a building and the amount that the building contributes to the market value of the farm.

Estimating depreciation involves an element of judgment. Physical deterioration relates to wear and tear and general disintegration resulting from age, use and structural defects that develop, and represents a loss in physical ability of the building to perform those services and functions for which the building was designed. Functional obsolescence relates to a loss in value because the building can no longer serve its intended function. A machinery storage shed with doors too narrow and the ceiling too low to accommodate modern equipment is not a functional building without substantial modification, however good the physical condition. Developments in technology may leave a hog house, for example, almost useless. Economic obsolescence relates to loss in value because of a changing economic environment. Large, stanchion type dairy barns may remain in an area that no longer has any dairy. Such buildings may add little to the present value of the farm. In areas that formerly involved livestock production but that have now turned exclusively to grain, livestock buildings add little to property value.

Once the new cost of a replacement building has been estimated, and the depreciation calculated, what remains is the contributory value of the buildings.

Inventory of land—The next step is to break down the total acreage of the farm into various categories, such as tillable land, pasture, timber, farmstead, roads and waste. At least on the tillable acres, a further breakdown is required to group soils with similar value into categories. Based on knowledge of how the various categories of land sell, an estimated market price is established for each grouping. The ideal circumstance is to locate sales of unimproved land comparable to each category. Usually, sufficient sales of unimproved land exist to permit an inventory and valuation of the land.

By adding together the land values and a summation of the building values, the "cost" value is estimated. Often, this value is higher than the values estimated by the market data and income approach.

Determining Final Estimate of Fair Market Value

With three separate value estimates, the final task is to make a final judgment on the fair market value. A typical pattern would be for the earnings value to be the lowest and the cost the highest. The values should, however, fall within a reasonably close range. If not, a critical analysis is needed to determine the factors causing the wide variation. The next task is to reason through which of the approaches would be most relevant in determining a market value on the farm being valued. From that point, a judgment can be made giving the final estimate of market value.

Summary

The objective of this chapter has been to describe the three approaches to estimating the market value that a professional appraiser would likely use. The value being estimated is the proxy for the "objective" value described in Chapter 1—an attempt to estimate the most likely value a willing buyer and willing seller would arrive at in an arm's length sale transaction. The next chapter discusses the question of what value a particular farm has to a specific individual. This is the "subjective" value referred to in Chapter 1.

CHAPTER 6
WHAT IS THE LAND WORTH TO YOU?

What price to pay—what price to sell for? No other single element of a land transaction involves more thought and discussion than price. Buyers usually want to pay the lowest price possible. Sellers want to achieve the highest price possible. How can buyers determine the price to offer for farmland? The previous chapter outlined three different approaches (income approach, market data approach and cost approach) for establishing the fair market value of a given property. These approaches are used by appraisers in establishing what they believe to be a fair market value for land. Unfortunately, appraisal reports are not always available. And even if they are, such reports seldom provide estimates of value under a set of conditions which may differ from those expected by the appraiser. Buyers also need to be able to determine a subjective value; the value to them.

This chapter is divided into two major sections. The first section is devoted to a discussion of the time value of money. The time value of money is important in determining the value of land. Readers already familiar with time value of money concepts can skip this section. The second section identifies the procedures necessary to calculate the value of land using the net present value approach. Information is provided to show how alternative expectations lead to different values for land.

Time Value of Money
Cash flows associated with the purchase and ownership of land occur over an extended period of time. Returns from production are spread over many years. Capital gains, if any, are realized only when the land is sold. And if borrowed funds are used to finance the purchase, the payments of principal and interest are spread over a number of years. However, the price of land is typically fixed at the time of purchase. Consequently, the current price should reflect the value of cash flows which occur over an extended period of time.

It should be recognized that a promise or expectation of receiving a dollar at some future date is not equivalent to receiving a dollar today. There are several reasons for this. For one, a dollar invested today could earn income over time so that investors would prefer a dollar today to a dollar at some future date. Second, people may have a preference for current consumption rather than future consumption. They would prefer to spend now rather than spend later. Finally, a promise of payment at a future date runs the risk that the payment will not be received. For these reasons there is a time value attached to money.

Investors in land are faced with comparing dollar amounts that occur at different points in time. The relevant comparison is between the outlays required and the returns to be generated. The absolute amount of the returns must exceed the outlays, or a person would be better off to put the money in a savings account that draws interest. That raises the question of how much the money would grow to at compound interest, and how to compare that amount with the returns over time from the investment.

The question of what size loan payment is required to amortize a long term loan is another question that can be solved by applying time value of money concepts. The goal may be to pay an equal amount each time that will be divided between interest and principal in such a way that the lender will earn a given amount on the principal balance remaining from time to time, and yet will allow the loan to be paid off completely with the last payment. A question somewhat in reverse of the above arises when a person wants to know how much to save in regular, equal savings deposits in order to end up with a predetermined amount at a specified time in the future, given a rate of interest. All of these questions relate to the time value of money.

There are two basic functions associated with assessing the time value of money: (1) compounding and (2) discounting. Compounding is used to determine future values while discounting is used to determine present values.

Compounding—Compounding is a mathematical procedure for determining the *future* value of a sum of money to be received now. The amount to be received may be either a single amount or a series of amounts to be received at specified intervals.

Compounding a single amount—The *future* value of a *single amount* of funds invested now can be calculated as follows:

(6.1) $V_n = V_o (1+i)^n$

$$\begin{aligned}
\text{where: } V_n &= \text{the future value of the investment} \\
V_o &= \text{the present value of the investment} \\
i &= \text{the rate of interest per period} \\
n &= \text{the number of periods over which the investment occurs.}
\end{aligned}$$

A common application of this formula is to determine the expected future value of land. For example, suppose you purchase land for $1,500.00 per acre. What is the expected value of this land at the end of 20 years if it inflates in value at the rate of 10% per year? Using the previous formula it can be seen that:

$$V_n = \$1,500 \, (1+.10)^{20}$$
$$V_n = \$1,500 \, (6.727)$$
$$V_n = \$10,090.50$$

Because the value of $(1+i)^n$ can become very tedious to calculate, tabular values are presented in Appendix A, Table 1. Reference to this table reveals that the value for $i = 10\%$ and $n = 20$, is 6.727. One can easily compute the expected future value of land using alternative expected inflation rates or alternative ending dates for the investment.

Compounding a series of equal payments—In some cases it is desirable to calculate the expected future value of a *series of equal payments*. For example, suppose you can save $200 per month for the next five years to apply toward a down-payment on land. If the money is accumulated in a savings account which pays 6% annual interest, how much will be accumulated by the end of 5 years? The future value of an equal-payment series could be treated as the sum of 60 single lump sum payments, (i.e., $V_n = 200(1+i)^1 + 200(1+i)^2 + \ldots 200(1+i)^{60}$). A short cut formula to calculate the future value of an equal-payment series is:

(6.2) $\quad V_n = A(EFIF_{i,n})$

where: A = the amount of each equal payment
$(EFIF_{i,n})$ = the future value equal payment interest factor where i is the interest rate per period and n is the number of time periods. Tabular values for different interest rates and lengths of loan are provided in Appendix A, Table 2.

The future value of an equal payment of $200 per month compounded monthly at 6% annual interest (½ percent per month) is:

$$V_n = 200(EFIF_{i=.005, \, n=60})$$

Finding the value for $(EFIF_{i=.005, \, n=60})$ in Appendix A, Table 2 we see that:

$$V_n = 200(69.757)$$
$$V_n = \$13,951.40$$

Notice that in this example i is an interest rate per month, and n is the number of months. This illustrates that i and n are periodic (annual, semiannual, quarterly, monthly) values; they need not be on an annual basis.

FARMLAND

Sinking funds—A variation of the future value of an equal payment series is the sinking fund problem. A sinking fund problem requires one to determine the amount of equal periodic payments required to accumulate a specified amount by a specified future date. The problem is to solve equation (6.2) for the value of A, or

(6.3) $A = V_n/(EFIF_{i,n})$

For example, suppose you want to accumulate $10,000 at the end of 5 years. How much would you need to deposit in an account at the start of each of the next five years if the account pays 6% compounded annually? Using equation (6.3) we see that:

$A = \$10,000/(EFIF_{i=6\%, n=5})$
$A = \$10,000/(5.637)$
$A = \$1,773.99$

Discounting—Discounting is a mathematical procedure for determining the *present* value of a sum of money to be received at some time in the future. The amount to be received may be either a single amount or a series of amounts to be received at specified intervals. Hence, discounting is the inverse of compounding.

Discounting a single amount—The value *now* of a single amount of funds to be received at a future date can be calculated as follows:

(6.4) $V_o = V_n \left(\dfrac{1}{1+i}\right)^n$

where V_o, V_n, i, and n are as previously defined.

A common application of this formula is to determine the present value of a single future payment. For example suppose you get an offer from someone to buy your land for $2,000 per acre to be paid two years from now. What is the present value per acre of that offer if you use an 8% discount rate? Using formula (6.4) we see that

$V_o = \$2,000 \times \left(\dfrac{1}{1+.08}\right)^2$
$V_o = \$2,000 \times .857$
$V_o = \$1,714.00$

Because the value of $\left(\dfrac{1}{1+i}\right)^n$ can be tedious to calculate if n is large, tabular values are presented in Appendix A, Table 3. For the values i = 8% and n = 2, confirm that the interest factor is .857.

Discounting a series of equal payments—In many cases it is necessary to calculate the present value of a series of equal payments to be received over time. For example, suppose someone agrees to buy your land for $200 per acre down, with the balance of principal and interest to be in equal annual installments of $150 per acre over the next 20 years. What is the present value per acre of this series of payments if you use a 10% discount rate? Assuming the downpayment is to be received immediately, the present value of that amount is $200 per acre. It need not be discounted since it is received at the present time. However, the $150 to be received over the course of the next 20 years must be discounted to determine its present value. The formula for determining the present value of an equal payment series is

(6.5) $\quad V_o = A\,(EPIF_{i,n})$
where: $(EPIF_{i,n})$ = the equal payment interest factor where i is the interest rate per period and n is the number of time periods. Tabular values for different interest rates and number of time periods are provided in Appendix A, Table 4.

Therefore, the present value of the equal payment series is:

$V_o = A(EPIF_{i=.08,\ n=20})$
$V_o = 150(9.818)$
$V_o = \$1,472.70$

Capital recovery—A variant of the present value of an equal payment series is the capital recovery (loan amortization) problem. A capital recovery problem is to determine the amount of *equal* periodic payments required to completely repay the principal and interest by a specified date. The problem is to solve equation (6.5) for A, or

(6.6) $\quad A = V_o/(EPIF_{i,n})$
where now A = the equal periodic amortization payment required to completely repay the principal and interest on a loan
V_o = the initial loan amount
i = the interest rate per period on the loan
n = the number of periods in the loan

For example, suppose you borrow $50,000 at 9% interest to be repaid in equal annual installments over 15 years. What is the amount of each payment? Using equation (6.6) we see that:

$A = 50,000/(EPIF_{i=.09,\ n=15})$
$A = 50,000/8.559$
$A = \$5,841.80$

Additional examples of calculating loan amortization payments are given in chapter 7.

Net Present Value of Land

The price *you* can be justified in paying for land depends upon a variety of factors including:
- your expectations of the net returns to land
- your expectations of the expected change in the net returns to land
- your expectations of the rate of change in land values
- the terms of financing you can arrange
- your marginal tax rate
- your length of planning horizon

These factors vary from one individual to the next, so the price one can be justified in paying for land also varies with each individual.

Many of the diverse determinants of the price of farmland can be examined in a rather comprehensive formulation—the net present value of land. The net present value of land represents that price which could be paid for land to achieve a desired rate of return, given expectations about income, capital gains, tax rates, and terms of financing.

The net present value of an acre of land is determined in the following manner:

Net Present Value = Discounted Net After-Tax Return + Discounted Terminal Value of Land After Capital Gains Taxes

Discounting must be used to arrive at the net present value because the income flows occur over a number of years and the capital gains, if any, occur when the land is sold. Net present value expresses the value of those returns in today's dollar.

Components of the Net Present Value Calculations

Information needed to calculate the net present value of an acre of land includes:

Expected net after-tax return per acre—The expected net after tax return per acre represents the returns per acre of land after paying income taxes and other costs of production. To illustrate, assume that you are considering the purchase of land which you intend to rent out on a cash rent basis. You estimate that you can rent the land for $120 per acre. Property taxes and insurance are $20 per acre leaving $100 before income taxes. Assuming you are in the 32% tax bracket, your expected net after-tax return per acre is $68.00.

For an investor who plans to operate the farm, calculations of the expected net after tax return per acre require estimates of gross receipts minus gross expenses per acre. Expenses should include any additional expenses of machinery or labor associated with making the investment.

This procedure for estimating the net after-tax returns per acre is illustrated in Table 6.1. Notice that many of the items may vary from one potential buyer to the next.

Because of the critical importance of properly estimating the net after tax return per acre, a line by line description of the procedures for completing this table are presented in the appendix to this chapter. In our example, the net after tax return per acre of land is $66.75. In the examples which follow, a wide range of values for the expected net after tax returns per acre is used to reflect the wide variety of returns to land that exist throughout the country.

Expected growth rate in the net after-tax return—Values calculated for the net after-tax return per acre are at a particular point in time. It is unlikely that these values will remain constant over time. In an inflationary economy, it is quite reasonable to expect the dollar amount of the net after-tax return per acre to increase over time. The higher the expected growth rate, the greater the price one should be willing to pay. As shown in Chapter 3, the growth in income returns to land was substantial from 1960 through 1980.

Expected rate of growth in land values—Previous discussion established that expected capital gains on farmland are an important component of land values. The higher the expected rate of growth in

Table 6.1   Procedures for calculating the expected net after-tax returns per acre of land[1]

Item	A Before Purchase	B After Purchase	C Net Change
1. Number of acres owned	320	400	80
2. Expected gross receipts	$105,600	$132,000	$26,400
3. Expected cash operating expenses	52,800	66,000[2]	13,200
4. Machinery depreciation	8,640	10,800	2,160
5. Building and other depreciation	2,880	2,800	0
6. Nonland imputed returns	8,960	11,200	2,240
7. Imputed labor and management return	8,320	8,320	0
8. Personal exemptions	4,000	4,000	0
9. Tax deductible items (3+4+5+8)	68,320	83,680	15,360
10. Taxable income (2−9)	37,280	48,320	11,040
11. Income tax payment	9,056	13,955	4,899
12. Taxable income attributable to land purchase (10−6−7)	20,000	28,800	8,800
13. Tax payment attributable to land purchase (12÷10) × 11	4,858	8,318	3,460
14. Expected net after-tax return (2−3−4−5−6−7−13)	19,142	24,482	5,340
15. Expected net after-tax return per acre (14÷1)			$66.75

[1] A line by line description of the procedures for completing this table is contained in the appendix to this chapter.
[2] Does *not* include interest expenses associated with financing the new purchase.

land values the higher the price one is willing to pay for the land. However, as explained in Chapter 3, the rate of growth in land values is not independent of the rate of growth in income returns to land. If the price of the asset increases more rapidly than the income returns to that asset over a long period of time, the asset is becoming overpriced. While this does not appear to have happened with agricultural land, it is important that you not set your expectations of capital gains at a level that is inconsistent with your expectations regarding the growth in net after-tax cash flows.

Terms of financing—Most land purchases now involve the use of borrowed funds. The terms of financing—the downpayment required, the interest rate, the length of repayment—influence the cash inflows and outflows associated with the purchase. Therefore, to calculate the net present value of land, one must first determine the terms of financing on the tract to be purchased.

All the examples of net present value presented later in this chapter are based on the assumption that all loans are: 30% down, equal annual amortization payments, 10% interest and a 30-year loan. Chapter 8 identifies how modifications in the terms of financing influence the price that can be paid for land.

Discount rate—Calculation of the net present value of an acre of land requires a discount rate for discounting future cash inflows and outflows back to the present. Selecting an appropriate discount rate is crucial in determining the price you can pay for land. The discount rate selected should reflect three components: (a) after-tax rate of return you would require if there were no inflation in the economy and no risk associated with investing in assets like land, (b) the inflation rate in the general economy to maintain the purchasing power of the equity capital invested, and (c) the risk premium to compensate you for the risks associated with land purchase.

In practice, the three elements of the discount rate are often combined into one number to give an overall discount rate adjusted for risk. Because the discount rate will likely differ from one individual to the next, our examples will show the net present value of land under differing sets of values for the discount rate. Used in this manner, the net present value of land is the price that you can pay to achieve a given rate of return.

Tax rates—Net after-tax cash flows per acre are determined in part by the marginal tax bracket of the individual investor. For investors in high tax brackets, the after-tax cost of borrowed funds is less than that for investors in a lower tax bracket.

Capital gains taxes also influence the net present value of land. Under current tax laws, in general only 40% of the long term gains on real estate assets are subject to taxation upon sale of property. If land which has appreciated in value is not sold but is passed on by inheritance there may be no capital gains taxes paid. For example, real estate valued at $400,000 with an income tax basis of $100,000 could be transferred at death with the new owner having a "stepped-up" basis of $400,000. (See Chapters 17 and 19 for an indepth discussion of this point) The net present value calculation requires some estimate of the cash outflows associated with the transfer of land.

Planning horizon of the investor—Farm real estate is considered by most as a long term investment. But the length of time one expects to hold the land (i.e., 15 years, 30 years, or until death) has some bearing on the net present value of that investment. Unless stated otherwise, the following examples assume a planning horizon of 30 years.

Examples of Net Present Value Calculations

A variety of values can be used in net present value calculations. The following examples highlight the impact on land values for different levels of net after-tax cash flow per acre, inflation rates, and tax rates.

No inflation—Potential investors in farmland are often concerned with the degree to which inflationary expectations are bid into land values. To address that issue we can begin by calculating the net present value of land under the assumption that investors expect no increase in land values and no increase over time in the net after-tax cash flow per acre. A later section removes this assumption.

Table 6.2 shows the net present value of an acre of land for alternative levels of expected annual net after-tax returns ranging from $20 per acre to $100 per acre. These values were chosen to reflect a wide range in the productivity of land. To set some perspective, however, land in Central Illinois has had an average of $80 to $100 per acre net after-tax returns over the 1975–80 period. In contrast, wheat land in the High Plains may yield only a $20 to $40 per acre net after-tax return.

The net present value per acre of land is also highly dependent upon the discount rates used in the calculations. For example, if the expected annual net after-tax return is $80 per acre and a 3% discount rate is used, the net present value of the acre of land is $1,506. The value of $1,506 per acre represents the price you can pay to get a 3% annual return on equity capital invested assuming your expectations of inflation and net after-tax return per acre are correct. To obtain a higher return on your investment you would need to pay a lower price for the land.

Land in Central Illinois which generated an $80 net after-tax return per acre during the 1975–80 period was selling in the range of $3,000–$4,000 per acre. But as shown in Table 6.2, one could only pay $1,506 per acre to get a return of 3%. It seems investors would be unlikely to accept a return below 3% when the yields on money market certificates or certificates of deposit are much higher. What accounts for the wide discrepancy between current land values and the net present values reported in Table 6.2? The reason for the discrepancy

FARMLAND

is that net present values reported in Table 6.2 fail to account for expectations of increases in net after-tax returns or expectations of capital gains. Yet most buyers of farmland expect the net after-tax returns to grow over time and they expect the land to appreciate in value. The rate at which buyers expect returns to increase and land to inflate in value influences the amount they are willing to pay for land.

Inflation in income and land values—The net present value calculations for land under alternative sets of inflationary expectations are shown in Table 6.3. In all cases, the growth in the after-tax returns per acre is assumed to be slightly above the inflation rate in land values. While other relationships between the growth in net after-tax returns and capital gains may be explored, the one depicted is about comparable to what occurred between 1960 and 1980 (see Chapter 3).

Net present values per acre of land calculated under an assumption of a 5% expected annual growth in the after-tax returns and 4% expected annual increase in land values are shown in Panel A of Table 6.3. For example, if you expect $80 net after-tax returns per acre, you expect this to grow 5% annually, and you expect the inflation rate in land values to be 4% per year, then you would be justified in paying $3,408 per acre to get a 6% return on your investment. In contrast, the net present value with no inflationary expectations was shown in Table 6.2 to be only $1,194 per acre. Clearly, inflationary expectations are bid into current land values.

The impact of higher inflationary expectations on the value of land are shown in Panels B and C of Table 6.3. In each case, the net present values which most closely match currently observed land values are those where the after-tax return (discount rate) is one to two percentage points higher than the expected growth rate in the net after-tax return and the expected capital gains on land. This shows that there are many different sets of inflationary expectations that are consistent with current land values. But the important point is that the return on investment tends to be only slightly above the level of inflationary expectations, whatever they are.

Table 6.2 Net present value per acre of land assuming no inflation.[1]

Expected Annual Net After Tax Returns Per Acre	Price you can pay per acre to get after-tax return of:				
	3%	6%	9%	12%	15%
$ 20	$ 376	$ 298	$ 253	$ 223	$ 201
40	753	596	506	446	402
60	1,129	895	760	669	603
80	1,506	1,194	1,013	892	804
100	1,882	1,492	1,266	1,115	1,005

[1]Calculations are based upon the assumption that the buyer is in the 28% tax bracket, has a 30 year planning horizon, and finances the purchase with a 30% downpayment and a loan for 30 years at 10% interest.

Table 6.3 Net present values per acre of land for alternative levels of annual net after-tax returns, inflation rates of land, and discount rates.[1]

If you expect:			Price you can pay per acre to get a return of:			
Annual net after tax returns per acre of:	Annual rate of growth in net after tax returns of:	Annual rate of inflation in land value of:	6% after income taxes	9% after income taxes	12% after income taxes	15% after income taxes
		Panel A				
$ 20	5%	4%	$ 852	$ 524	$ 388	$ 313
40	5%	4%	1,704	1,047	775	627
60	5%	4%	2,556	1,571	1,163	940
80	5%	4%	3,408	2,094	1,551	1,253
100	5%	4%	4,260	2,618	1,938	1,567
		Panel B				
$ 20	8%	7%		1,264	657	456
40	8%	7%		2,528	1,315	912
60	8%	7%		3,792	1,972	1,369
80	8%	7%		5,057	2,629	1,825
100	8%	7%		6,321	3,287	2,281
		Panel C				
$ 20	10%	9%			1,164	641
40	10%	9%			2,328	1,283
60	10%	9%			3,492	1,924
80	10%	9%			4,656	2,565
100	10%	9%			5,820	3,460

[1]Calculations are based upon the assumption that the buyer is in the 28% tax bracket, has a 30-year planning horizon, and finances the purchase with a 30% down payment and a loan for 30 years at 10% interest. Calculations also assume the land is sold at the end of 30 years with a tax rate of 50% in that year as a result of the increase in income due to the inclusion of 40% of the capital gains as taxable income.

Capital gains taxes—Net present values reported in Table 6.3 are calculated under the assumption that the land is sold at the end of the planning horizon, and that the taxpayer is in the highest possible tax bracket as a result of substantial capital gains. In other words, the capital gains taxes are the maximum amount possible under current tax laws the top income rate in 1982 drops to 50%.

There are a variety of circumstances under which the capital gains taxes could be less. For example, the total taxable capital gain may not be sufficient to push the taxpayer into the highest possible tax bracket. In addition, the use of an installment sale of the land could spread the gains over a number of years thereby reducing the overall level of the gains as discussed in Chapter 17. Alternatively, the seller may realize the entire capital gains in one period, but use income averaging to reduce tax liabilities. Finally, the investor may not plan to sell the land at all, but rather leave it to heirs at death. Under current tax rules, the heirs would generally receive a "stepped up" basis equal to the value used for federal estate tax purposes and no capital gains taxes would be due on the property, unless the property was sold by the heirs for more than the new basis.

To explore the impact on the net present value of land, we used alternative assumptions about the extent of the capital gains taxes. Results are reported in Table 6.4. Panel A of Table 6.4 is identical to Panel C of Table 6.3 and is calculated under the assumption that capital gains taxes are the maximum amount possible: 20% of the full gain in land values (40% × 50% maximum income tax bracket).

Net present values reported in Panel B of Table 6.4 are calculated under the assumption that the investor pays no capital gains taxes. This implies that the land is not sold but is transferred to the heirs with a "stepped up" basis. Using a 12% discount rate, the values increase by 26% over comparable estimates in Panel A. However, the calculated values are well above currently reported land values. This seems to imply that either the after-tax rate of return is more than 2% above the expected inflation rate or that investors do not expect to be able to avoid the capital gains taxes. Using a 15% discount rate, the net present values of land are about 8% higher in Panel B (no capital gains taxes) than in Panel A (maximum capital gains taxes). The important point is that investor's expectations about capital gains taxes on land do have an important bearing on the price they are willing to pay for land.

Table 6.4 Net present values per acre of land for alternative levels of annual net after-tax returns, capital gains tax rates, and discount rates.[1]

	If you expect:		Price you can pay per acre to get a return of:	
Annual net after tax return per acre of:	Annual rate of growth in net after tax return of:	Annual rate of inflation in land values of:	12% after income taxes	15% after income taxes
PANEL A (50% Tax Rate × 40% of gain taxable = 20% of full gain)				
$ 20	10%	9%	$1,164	$ 641
40	10%	9%	2,328	1,283
60	10%	9%	3,492	1,924
80	10%	9%	4,656	2,565
100	10%	9%	5,820	3,207
PANEL B (0% Tax Rate × 40% of gain taxable = 0% of full gain)				
$ 20	10%	9%	$1,469	$ 692
40	10%	9%	2,938	1,384
60	10%	9%	4,407	2,076
80	10%	9%	5,875	2,768
100	10%	9%	7,345	3,460

[1]Calculations are based upon the assumption that the buyer is in the 28% tax bracket, has a 30-year planning horizon, and finances the purchase with a 30% down payment and a loan for 30 years at 10% interest.

Income tax rates—A given tract of land might be expected to generate a *before-tax return* of say $100 per acre independent of the way it is financed or the tax bracket of the investor. However, the amount of income tax paid on that income can vary substantially from one investor to the next. Consequently, the net *after-tax return* for a person in a low income tax bracket may be $80 per acre while that for a person in a higher tax bracket may be only $60 per acre. Note that this occurs even if no debt financing is used to finance the purchase.

Likewise, since interest is a tax deductible expense, the impact of the interest deduction varies with the income tax bracket of the investor. It is important to note that the after-tax returns per acre used in Tables 6.2 through 6.4 do *not* account for the tax savings of interest payments, although these tax savings are taken into account in computing net present values. Hence, for an investor in a high income tax bracket the net after-tax returns per acre would be lower, but the tax reduction associated with interest expenses would be higher than for an investor in a lower income tax bracket. The nature of this tradeoff is shown below.

Assume that a given tract of land generates $111.11 per acre before taxes. Net after-tax returns per acre (excluding the effects of interest deductions) are shown below for different income tax brackets.

$$\begin{aligned}
\$111.11 \times .72 \ (28\% \text{ tax rate}) &= \$80.00 \\
111.11 \times .61 \ (39\% \text{ tax rate}) &= 67.77 \\
111.11 \times .52 \ (48\% \text{ tax rate}) &= 57.78
\end{aligned}$$

Net present values reported in Table 6.5 illustrate the impact of higher income tax rates on the net present value of land. While the annual net after-tax return per acre drops as the income tax rate increases, this is offset to a large extent by the fact that interest payments on the loan are tax deductible. For example, when a 12% discount rate is used the net present value per acre of land declines by only about one percent in moving from a 28% tax bracket to a 39% tax bracket. In moving from a 28% tax bracket to a 48% tax bracket, the decline in net present value is just over 2.6 %. The differences between tax brackets become more pronounced when one uses a higher discount rate, such as 15%.

One might conclude from the information in Table 6.5 that low income tax bracket investors might be able to outbid high income tax bracket investors in farmland. Such conclusions however, may be unwarranted. High income tax bracket investors, because of substantial income from other sources, might be able to negotiate a loan with better terms than low income tax bracket investors. The values reported in Table 6.5, however, were calculated under the assumption that all investors obtain the same loan terms.

Likewise, the high tax bracket investors might find it much more feasible from a cash flow standpoint to make the loan payments. While

Table 6.5 Net present values per acre of land for alternative levels of annual net after tax returns, income tax rates and discount rates.[1]

	If you expect:		Price you can pay per acre to get a return of:	
Annual net after tax returns per acre of:	Annual rate of growth in net after tax returns of:	Annual rate of inflation in land values of:	12% after income taxes	15% after income taxes
		(28% tax bracket)		
$80.00	10%	9%	$4,656	$2,565
		(39% tax bracket)		
$67.77	10%	9%	$4,597	$2,395
		(48% tax bracket)		
$57.78	10%	9%	$4,533	$2,227

[1]Calculations are based upon the assumption that the buyer has a 30-year planning horizon, finances the purchase with a 30% downpayment and a loan for 30 years at 10% interest. The capital gains taxes are assumed to be 20% of the full gain.

net present value calculations indicate the price that can be paid to achieve a given rate of return, they do not reveal the cash flow feasibility of the purchase. Because of the importance of cash flow feasibility, the next chapter in this book is devoted to that topic.

APPENDIX

Line by line procedures for completing Table 6.1 are described below.

Line 1—Number of acres. This line should reflect the number of acres you own. Rented acreage should not be included on this line. If this is your first purchase of land, the number of acres owned before the purchase, Column A, would be zero.

Line 2—Expected gross receipts. This line should reflect the *expected* gross receipts from your farming operations. The value for this line is crucial to calculating the net present value per acre of land. Several strategies might be followed in estimating the future value of gross receipts.

First, one could estimate gross receipts under a "pessimistic plan" to see what would happen under adverse conditions. Minimum expected prices for commodities (government support prices might be used for covered commodities) could be used in making this projection. This approach would help you assess the value of land under difficult conditions. However, pessimistic plans should have a low chance of occurring and are therefore of limited value.

A second approach is to use the "current evidence" plan for projecting future gross receipts. Under this approach you should use all available evidence from outlook reports that are now a feature of popular farm publications. Trade magazines and information from unversity extension personnel might also be used. The goal is to estimate the level of gross receipts for the next year.

A third approach is the "status-quo-plan" with estimates of future gross revenues based upon your own records on gross receipts on existing operations. Information from tax returns might be used. The advantage of this approach is that it is relatively easy to apply and it should be more realistic than the pessimistic plan.

A fourth approach is the "optimistic plan". Here you estimate gross receipts under the assumption that prices and yields will be highly favorable to you. Like the pessimistic plan, the optimistic plan should have a low probability of occurring. Unfortunately, there appears to be a tendency for purchasers to be overly optimistic, particularly if returns have been highly favorable in recent years.

As a potential investor in farmland, several of the above plans should be tested. This allows you to see how changes in gross receipts will affect the price of land.

Line 3—Expected cash operating expenses. This line should reflect the expected amount of *cash* operating expenses associated with your farming operations before and after the purchase. A good source of information on this item is previous year's schedule F or other business schedule of your tax returns. Interest expenses associated with the purchase should *not* be included in Column B. These interest expenses are accounted for directly in the net present value calculations.

Line 4—Machinery depreciation. This line reflects the amount of depreciation associated with the machinery used in producing agricultural products. Since the bottom line of this table is the returns to land, it is important to account for the costs associated with machinery used in the production of crops. The inclusion of machinery depreciation allocates the cost of machinery over the useful life of the machines. Past tax returns are a good source for this information.

At times, the purchaser of land is able to farm the additional acreage being purchased with the existing stock of machinery. In this case, one might argue that line 4 should be the same for Columns A and B. However, one should recognize that existing machinery will likely wear out long before the land is sold or passed to heirs. If there is no increase in the depreciation charge on machinery under the new purchase, then one must assume that existing acreage will continue to subsidize the machinery input throughout the life of the new purchase. In our example in Table 5.1, we assumed the machinery costs per acre would increase with the increase in acreage owned.

Line 5—Building and other depreciation. This item is treated in a manner similar to machinery depreciation. In our example we assumed that there would be no new buildings needed to handle the additional land. Hence the net change in building depreciation was set at zero. If livestock depreciation is relevant, it should be included on this line.

Line 6—Nonland imputed interest. Agricultural operations typically combine land inputs with other inputs such as machinery, crops and livestock inventories, and other nonland items. Since these are an integral part of the farming or ranching operation, one must allocate part of the total return to these assets. This line provides an estimate of the imputed returns to investment in these assets. The value is found by multiplying the value of nonland assets by a return that you believe is appropriate.

Line 7—Imputed labor and management returns. Labor and management are important inputs to the production process. Failure to account for returns to unpaid labor and management would result in capitalizing these returns into the value of land. This would lead to over valuation of the land you intend to purchase.

Line 8—Personal exemptions. This line identifies the number of exemptions you can claim for income tax purposes, multiplied by the allowable deduction per exemption. In our example we assume 4 exemptions.

Line 9—Tax deductible items. This line is merely an accounting of tax deductible items. The value is found by adding the amounts in lines 3, 4, 5, and 8.

Line 10—Taxable Income. This line is an accounting line to determine taxable income. The value is found by subtracting values in line 9 from values in line 2.

Line 11—Income tax payment. This line reflects the amount of taxes due on the taxable income reported in line 10. Consult an appropriate tax table to determine the amount of taxes due.

Line 12—Taxable income attributable to land. Only a fraction of your taxable income is attributable to returns generated from land. The balance is attributable to taxable income associated with labor, management, and nonland inputs. The value for this line is found by subtracting the amounts in lines 6 and 7 from the amount in line 10.

Line 13—Tax payments attributable to land. Line 13 is used to calculate the amount of taxes paid which are attributable to land. The value is found by dividing line 12 by line 10 and then multiplying this fraction by line 11.

Line 14—Expected net after-tax returns to land. This line reflects the expected total after-tax returns to land. The value is found by subtracting lines 3 through 7 and line 13 from line 2.

Line 15—Expected net after-tax returns per acre. This line places the expected net after-tax returns on a per acre basis. It is calculated by dividing the amount in line 14 by the amount in line 1.

If desired, you could compute the expected net after-tax returns per acre for each year you plan to hold the land being purchased. However, the computational burden becomes greater. A short-cut approach is to compute the value for one year and then form some expectation as to the growth rate of this value over time.

CHAPTER 7
CASH FLOW FEASIBILITY OF FARMLAND PURCHASES

The purchase of farmland is a major capital expenditure that requires careful financial planning. It is important to recognize that an investment can be highly profitable, but yet infeasible from a cash flow standpoint. For example, you may have a fantastic opportunity to acquire land at a very reasonable price. However, the seller wants cash. If you do not have the cash and you cannot borrow the money on reasonable terms, you must forego a highly profitable investment opportunity.

Usually, the cash flow feasibility of investment in farmland is not as clear cut as the above example indicates. Rather, there may be a rather fine line between an investment which is feasible and one which is infeasible in a cash flow sense. Therefore, the investor should use all the financial planning tools available to estimate cash flows and thereby avoid cash flow problems.

Estimating Cash Flows

The cash flows associated with the purchase of farmland involve two major components: the cash inflows and outflows associated with the production of agricultural commodities, and the cash flows associated with financing the purchase. Procedures for estimating both components are reviewed below.

Cash flows associated with financing—Cash flows associated with financing the purchase of land include the downpayment, principal and interest payments and any service charges, stock purchase requirements or loan origination fees. Potential investors need to know how to estimate the cash flows associated with each of these items.

Estimating the downpayment required on the purchase of land is straightforward. For example, if a tract of land sells for $200,000 and there is a 20% downpayment required, the purchaser must be able to come up with $40,000 for the downpayment. The large cash flow requirement for the downpayment is a major stumbling block for many would-be investors in farmland. Alternatives for overcoming these problems are discussed in a subsequent section in this chapter.

If a land purchase involves financing, loan amortization payments represent a major cash flow item. Potential buyers of land (as well as sellers under a land contract) should be aware of the various repayment plans available and know how to calculate the payments required under each plan. Modification in loan repayment plans can dramatically change the cash flow feasibility of an investment opportunity. Therefore, the potential investor should calculate how different repayment plans might affect the cash flow feasibility of an investment.

Too often it seems that potential investors rely on their lender to tell them what the amount of loan payments will be for a specific set of loan terms. The lender's office, however, is not the most appropriate place to study the cash flow feasibility of an investment. The cash flow feasibility of an investment should be studied carefully by the buyer. Understanding the mechanics of calculating loan payments is essential.

There are a variety of loan repayment plans in existence. Procedures for calculating the amount of loan payments under the more common repayment plans are given below.

Full amortization, equal payment—A full amortization, equal payment loan requires the borrower to pay equal periodic installments that reduce the loan balance to zero by the end of the loan period. While the total amount of each payment is equal, the portion of each payment attributable to interest declines over time while the portion attributable to principal increases. The formula to calculate the amount of each equal payment was given in Chapter 6, but is repeated here for convenience.

(6.5) $A = V_o(EPIF_{i,n})$

where A = the amount of each equal payment

V_o = the initial amount borrowed

$(EPIF_{i,n})$ = the equal payment interest factor where i is the interest rate per period, and n is the number of time periods over which the loan is to be repaid. Tabled values for different interest rates and lengths of loan are provided in Appendix A, Table 4.

Several example problems may help clarify the calculations.

Problem 1 Assume you are offered 120 acres of land for the total purchase price of $180,000. The seller wants 20 percent down, with the balance due in 10 equal annual installments with interest at 12 percent. What is the amount of each equal annual payment?

Solution   First calculate the total amount borrowed:
$180,000 \times .8 = $144,000$
Apply formula (6.5)
$A = $144,000/(EPIF_{i=12\%, n=10})$
Looking in Appendix A under i = 12%, n = 10 we find the value of 5.650, so
$A = $144,000/5.650$
$A = $25,486.72$
By paying $25,486.72 at the end of each of the next 10 years, the loan balance will be reduced to zero by the end of the 10th year.

Problem 2   Same circumstances as problem 1, except the loan is to be repaid in quarterly installments.

Solution $A = $144,000/(EPIF_{i=3\%, n=40})$
Notice that now the rate is 3 percent and the number of time periods (quarters) is 40.
Looking in Appendix A, Table 4 under i = 3% and n = 40 you find the value 23.115
$A = $144,000/23.115$
$A = $6,229.72$
Therefore, quarterly payments would total $6,229.72.

The preceding examples are based upon the assumption that the interest rate is fixed throughout the length of the loan. However, Federal Land Banks, and to a limited degree, life insurance companies, offer farm mortgage loans with a variable interest rate. How are the amortization payments calculated in these cases?
   Assume that you obtain a $200,000 loan for 30 years at 9% interest from the Federal Land Bank. The loan requires annual amortization payments, but the interest rate is variable. Assume that at the end of 2 years, the interest rate is increased to 10%. To calculate the amount of each equal amortization payment, calculate the payment according to equation 6.5. Assuming the loan in this example requires equal annual amortization payments, the calculations would be:

$A = $200,000/(EPIF_{i=9\%, n=30})$
$A = $200,000/10.274$
$A = $19,466.61$

So each of the first two payments would equal $19,466.61.

Suppose that starting in Year 3 the interest rate increases to 10%. To find the new amount of each equal amortization payment, first calculate

FARMLAND

the remaining balance of the loan outstanding. Calculations are shown below:

End of Year	Payment	Interest	Principal	Balance
0	—	—	—	$200,000.00
1	$19,466.61	$18,000.00	$1,466.61	198,533.39
2	19,466.61	17,868.01	1,598.60	196,934.79
3	21,159.86	19,693.48	1,466.38	195,468.41

At the end of year 2, the loan balance is $196,934.79. Amortization of that amount over 28 years at 10% interest gives $21,159.86. This would be the amount of each equal payment, at least until the interest rate is changed again.

$$A = \$196{,}934.79/(EPIF_{i=10\%,\ n=28})$$
$$A = \$196{,}934.79/9.307$$
$$A = \$21{,}159.86$$

Notice that the future cash flows on this loan contract cannot be known with certainty because of the variable interest rate.

Partial amortization, equal payments—A partial amortization, equal payment loan requires equal payments each period, but the loan balance is not reduced to zero by the end of the loan period. Rather the loan requires a "balloon" payment for the remaining principal at the end of the loan. An example may help clarify the procedure.

Assume you borrow $300,000 at 10% interest, equal annual amortization payments, a 30 year payout, and a balloon payment for the balance at the end of 5 years. The amount of payments each year is identified below:

End of Year	Payment	Interest	Principal	Balance
0	—	—	—	$300,000.00
1	$ 31,823.49	$30,000.00	$ 1,823.49	298,176.51
2	31,823.49	29,817.65	2,005.84	296,170.67
3	31,823.49	29,617.07	2,206.42	293,964.25
4	31,823.49	29,396.42	2,427.07	291,537.18
5	$320,690.90	29,153.72	$291,537.18	0

Notice that while equal annual payments are based upon a 30 year payout, the entire balance of the principal becomes due at the end of year 5 as a lump sum or "balloon" payment.

In financing the purchase of farm and ranch land, balloon payment loans are perhaps used most frequently for land contracts and life insurance company loans. While terms of loans vary widely, a balloon payment at the end of 5 years appears to be a common timing for the "balloon" payment on life insurance company loans. The balloon payment loan is also used on land contracts, particularly where the seller wants to get the money out of the sale in a relatively short time period.

In practice, the borrower seldom has the funds necessary to pay off the loan balance when the balloon payment becomes due. Rather the borrower refinances the loan with the existing lender or seeks new financing. A word of caution: the lender may have no legal obligation to refinance the loan. This could be devastating to the borrower if the balloon payment becomes due in a period of tight money. Borrowers who enter into these contracts should recognize the risks involved. It is helpful for the buyer to negotiate for the right to refinance over a two or even three year period to avoid the tight money problem.

Equal principal—Some loans require equal payments of principal, plus interest on the remaining balance. This makes the total payment (principal + interest) decline over time. This type of loan amortization plan is most commonly used in land contracts and, to some extent, by life insurance companies. The typical procedure is to require that a given percentage of the principal be repaid each year.

Assume you borrow $100,000 at 10% interest for 20 years with principal reduction to be 5% each year. The amount of each payment for this loan is shown below. The disadvantage of this loan payment plan is that the highest payments occur early in the loan period, or at a point in time when the borrower is less likely to have the cash flow necessary to make the loan payments.

End of Year	Principal	Interest	Total Payment	Balance
•				$100,000
1	$5,000	$10,000	$15,000	95,000
2	5,000	9,500	14,500	90,000
3	5,000	9,000	14,000	85,000
•	•	•	•	•
•	•	•	•	•
•	•	•	•	•
20	5,000	500	5,500	0

Variable amortization payments—Variable amortization payment plans allow the amount of loan payments to vary with changing economic conditions. For example, a variable amortization plan might call for a standard payment of $10,000 per year. The $10,000 would be varied up or down depending upon an index of economic conditions. For farm loans, the index might be based upon an index of farm income in the region. To insure prompt repayment to the lender, the borrower might also be asked to establish a debt reserve to provide payments to the lender if the income index is low. In practice, variable amortization payment plans have not been widely used in the United States, probably because of the difficulty in setting up a reliable index of economic conditions that accurately reflects economic conditions for the borrower.

FARMLAND

Loan service charges can also be an important component of the cash flow associated with financing the purchase of farm and ranch land. Service charges vary widely among lenders and in different areas of the country. For example, in April of 1980, service charges on Federal Land Bank loans ranged from zero in some areas to 8% in other areas. Potential investors need to be informed of these charges in evaluating the cash flow feasibility of purchase. Up-to-date information on service charges can be obtained from lenders in your area.

Will the Land Pay For Itself?

In Chapter 6 we introduced the concept of the expected net after-tax return per acre. Recall that the net after-tax return per acre of land was defined as the expected net return per acre after paying income taxes and other costs of production. Procedures for calculating the net after-tax return per acre were identified in Table 6.1

If borrowed funds are used to finance the purchase of land, an important issue is whether the purchased land will generate sufficient returns to make the loan payments. That is, will the land purchased generate a surplus or deficiency in making the loan payments? If there is a deficiency, what is the amount of the deficiency, how can it be made up, and how long will the deficiency last? To answer these questions, the surplus or deficiency on the purchase of land must be calculated.

To determine the surplus or deficiency in making loan payments on the purchase of land, you need to determine (a) the per acre amount of the annual payments (principal plus interest) on the loan, (b) the expected after-tax returns per acre, and (c) the tax savings associated with interest payments on the loan. The surplus or deficiency per acre is calculated by subtracting the sum of items b and c from item a.

The net present values of land for different levels of expected net after-tax returns per acre and different rates of inflation in land are presented in Table 7.1. Also presented are the first year surpluses or deficiencies in returns necessary to make the loan payments.

Net present values of land for different levels of expected net after-tax returns and different discount rates are identified in Panel A of Table 7.1. Procedures for making these calculations were described in Chapter 6. The information in the remainder of the table is based upon the assumption that the buyer pays the net present value to acquire the land.

Panel B of Table 7.1 identifies the loan amount per acre under the assumption that the buyer finances 70% of the purchase price. It is assumed that the loan will be repaid over a 30 year period with equal annual amortization payments.

Panel C reflects the first year deficiency in making the loan payment associated with the loan amount in Panel B. To illustrate how to read Table 7.1, consider the following. If you paid $852 per acre for land generating $20 net after-tax returns per acre (Panel A) and financed the purchase with a loan of $596 at 10% interest for 30 years (Panel B), you would be short $26.57 per acre on the land purchased in making the loan payment in the first year. This deficiency would need to be made up from other sources of income or from savings.

Table 7.1 Net present values per acre of land, loan amounts, and first year deficiencies—moderate inflation.[1]

	If you expect:		Price you can pay per acre to get a return of:		
Annual net after tax returns per acre of:	Annual rate of growth in net after tax returns of:	Annual rate of inflation in land value of:	6% after income taxes	9% after income taxes	12% after income taxes
PANEL A (Net Present Value Per Acre)					
$ 20	5%	4%	$ 852	$ 524	$ 388
40	5%	4%	1,704	1,047	775
60	5%	4%	2,556	1,571	1,163
80	5%	4%	3,408	2,094	1,551
100	5%	4%	4,260	2,618	1,938
PANEL B (Loan Amount Per Acre)					
$ 20	5%	4%	$ 596	367	272
40	5%	4%	1,193	733	543
60	5%	4%	1,789	1,100	814
80	5%	4%	2,386	1,466	1,086
100	5%	4%	2,982	1,833	1,357
PANEL C (First Year Deficiency Per Acre)					
$ 20	5%	4%	$ −26.57[20]	$ −8.61[8]	$ −1.19[2]
40	5%	4%	−53.14[20]	−17.22[8]	−2.37[2]
60	5%	4%	−79.70[20]	−25.84[8]	−3.56[2]
80	5%	4%	−106.27[20]	−34.45[8]	−4.75[2]
100	5%	4%	−132.84[20]	−43.06[8]	−5.93[2]

[1]Calculations are based upon the assumption that the buyer is in the 28% tax bracket, has a 30-year planning horizon, and finances the purchase with a 30% down payment and a loan for 30 years at 10% interest. Capital gains taxes are assumed to be 20% of the full gain.

To illustrate how to obtain the values in Panel C, let's trace through the calculations for the value −$26.57. First calculate the amount of the annual amortization payments (procedures for calculating amortization payments were outlined earlier in this chapter). A loan of $596 per acre for 30 years at 10% interest would require annual amortization payments of $63.26. Interest payments the first year would be $59.60. Since these interest payments are tax deductible, tax savings of $59.60 × .28, or $16.69 per acre would be gained. (The buyer was assumed to be in the 28% income tax bracket). Subtracting the total loan payment of $63.26 from the net-after tax return of $20 per acre plus the tax savings of $16.69 per acre gives:

FARMLAND

$20.00 + $16.69 - $63.26 = -$26.57$

The other numbers in Panel C are calculated in a similar fashion.

While the total loan payment remains the same over time, the tax savings from the interest portion change. Likewise the buyer expects the net after-tax returns to grow at an average rate of 5% per year. Consequently, the deficiency is eventually removed. The superscripts in Panel C reflect the number of years needed to remove the deficiency. In our example, the initial deficiency was $26.57, but the deficiency would be removed in 20 years.

The right hand side of Table 7.1 shows that as the price you pay per acre declines, the first year deficiency also declines. But even at prices well below current market prices, there is likely to be a deficiency in first year loan payments. However, the deficiency is relatively small and can be removed in several years.

By recent standards, the expected inflation rates used in calculating the net present values in Table 7.1 are rather moderate. What happens to the deficiencies if higher inflationary expectations are used? Table 7.2 provides an answer to that question.

As shown in Table 7.2, if you expect a 10% growth in the net after-tax returns per acre, and you use a 13% discount rate, the net present values per acre are slightly above comparable values reported in Table 7.1. Assuming you borrow 70% of the purchase price, the loan amount per acre is slightly higher as is the first year deficiency. The important point is that the length of time necessary to remove the deficiency is substantially *shortened*.

In Table 7.1 it was shown that a first year deficiency of $26.57 per acre on land generating $20 net after-tax return was expected to grow 5% per year. In Table 7.2, you can see that a first year deficiency of $29.75 on land generating $20 net after-tax returns per acre would be removed in 10 years if the returns grow at the rate of 10% per year. Hence, a greater first year deficiency is removed in less time. In general, the higher the expected growth rate in net after-tax returns and land values, the greater is the deficiency per acre, but the shorter will be the period necessary to remove the deficiency.

The implications of the deficiencies are clear. Young farm operators may find it very difficult to purchase land, and that problem increases if rates of inflation are expected to remain high. But if young operators can survive the first few years, they should accumulate substantial wealth due to inflation in land values.

In contrast, well established operators or non-farm investors with substantial cash reserves can better afford to "carry" the deficiency through the early years of the loan.

The deficiencies identified in Tables 7.1 and 7.2 also have implications for the demand for land. Deficiencies per acre are lower on land which has lower expected net after-tax returns. Suppose two tracts of land of 160 acres each come up for sale. Suppose further that tract

Table 7.2 Net present values per acre of land, loan amounts, and first year deficiencies—high inflation.[1]

	If you expect:		Price you can pay per acre to get a return of:	
Annual net after income tax cash flow per acre of:	Annual rate of growth in net after income tax cash flow of:	Annual rate of inflation in land value of:	13% after income taxes	15% after income taxes
PANEL A (Net Present Value Per Acre)				
$ 20	10%	9%	$ 910	$ 641
40	10%	9%	1,820	1,283
60	10%	9%	2,730	1,924
80	10%	9%	3,641	2,565
100	10%	9%	4,551	3,207
PANEL B (Loan Amount Per Acre)				
$ 20	10%	9%	$ 637	$ 449
40	10%	9%	1,274	898
60	10%	9%	1,911	1,347
80	10%	9%	2,549	1,796
100	10%	9%	3,186	2,245
PANEL C (First Year Deficiency Per Acre)				
$ 20	10%	9%	$ −29.75[10]	$ −15.05[7]
40	10%	9%	−59.49[10]	−30.10[7]
60	10%	9%	−89.24[10]	−45.16[7]
80	10%	9%	−118.99[10]	−60.21[7]
100	10%	9%	−148.73[10]	−75.26[7]

[1]Calculations are based upon the assumption that the buyer is in the 28% tax bracket, has a 30-year planning horizon, and finances the purchase with a 30% downpayment and a loan for 30 years at 10% interest. Capital gains taxes are assumed to be 20% of the full gain.

A is land which generates $20 net after-tax returns per acre and a first year deficiency of $29.75 per acre. The total first year deficiency to be covered from other sources is thus $4,760. In contrast, tract B is land which generates $80 net after-tax returns and a first year deficiency of $118.99. The total first year deficiency is thus $19,038. While both investments might generate the same *rate* of return on investment, there are likely to be many potential buyers who could make up the deficiency on tract A, but not on tract B.

As a result of these cash flow problems, the seller who can split the sale of a farm into smaller tracts might have more potential buyers than if the land is offered as one large tract. This could enhance the price to the seller and make the purchase more feasible for the buyer who supports the deficiency from other sources of income. As shown in Chapter 2, the average size of tract sold has declined slightly over time.

Is There Adequate Cash Flow Available For Debt Servicing?

In the preceding section the issue addressed was whether the land being purchased would generate sufficient returns to make the loan payments. In most cases it will not. Therefore, the next issue to address is whether there is a sufficient cash flow from the entire operation to make the loan payments. That is, will there be an adequate cash flow from existing operations, non-farm income, savings or other sources to make the loan payments on the purchase?

To answer these questions, you need to develop a cash flow projection. The basic concept in a cash flow projection is simple—all cash inflows *must* equal all cash outflows. While the concept is simple, there are a variety of formats that can be used for the cash flow projection. Lenders, however, like to see the cash flow projection structured so that the bottom line is the "cash flow available for debt servicing." They can then compare this figure with the amount of principal plus interest payments required. Lenders also rely heavily on past tax records in preparing such cash flow statements, primarily because tax records may be the best data available.

A cash flow projection which can be used to estimate the cash flow available for debt servicing is shown in Table 7.3. The table is patterned after forms used by the Federal Land Banks and life insurance companies. The table is designed to record historical values as well as projections of future cash flows.

Lines 1 through 12 measure the total cash inflows during the year. Most of the historical information is readily available from past tax records. Projected values need to be developed carefully and should not be overly optimistic or pessimistic. Alternative procedures for making projection estimates were identified in the Appendix to Chapter 6.

Lines 13 through 24 measure the total cash outflows during the year. Historical farm cash expenses come directly from the Federal tax forms. Other expenditure items should be readily available, with the possible exception of family living expenses. A family living expense budget should be used, if possible, but in most cases the value is estimated. Total cash outflows (line 24) must equal total cash inflows (line 12).

Lines 25 through 30 provide information on debt service requirements and the amount of funds available for debt servicing. Procedures for measuring the components of cash available for debt servicing vary from lender to lender. In our example, it is assumed that any additions to savings or money used to purchase stocks and bonds could have been used to make scheduled payments of principal and interest. Likewise, any prepayments of debt are funds available for scheduled debt servicing.

In our example, there was $56,656 available for debt servicing

Table 7.3 Cash flow projection of cash available for debt servicing.

	Historical			Projected	
	1978	1979	1980	1981	1982
INFLOWS					
1. Beginning cash balance			$ 3,400	$ 3,130	$ 3,000
2. Gross farm profits (line 32, schedule F)			91,480	114,350	124,000
3. Wages, salaries, tips, etc. (line 8, 1040)			3,450	3,800	4,300
4. Interest income (line 9, 1040)			2,426	850	900
5. Dividend income (line 8, schedule B)			324	375	400
6. Sales of capital assets (land, machinery, breeding stock, etc.)			1,300	4,000	2,000
7. Sales of stocks & bonds			2,726	3,000	3,500
8. Withdrawals from savings			0	20,000	0
9. Other cash income: (specify)			1,200	1,200	1,200
10. Short and intermediate term loans			35,000	46,000	53,000
11. Long term loans			0	200,000	0
12. CASH INFLOWS (add lines 1 thru 11)			141,306	396,705	192,300
OUTFLOWS					
13. Farm cash expenses (line 57–line 56, 1040)			58,245	87,500	92,250
14. Purchases of capital assets (land, machinery, breeding stock, etc.)			8,500	225,000	5,000
15. Income and social security taxes			7,245	6,500	7,200
16. Family living withdrawals			13,280	14,600	16,000
17. Additions to savings			4,000	0	0
18. Purchases of stocks and bonds			4,275	4,000	4,000
19. Other (specify) _____			—	—	—
20. Ending cash balance			3,130	3,000	3,000
21. CASH OUTFLOW BEFORE PRINCIPAL PAYMENT			98,675	340,600	131,450
22. Principal paid on short and intermediate term loans			35,000	46,000	49,000
23. Principal paid on long term loans			7,631	10,105	11,850
24. CASH OUTFLOWS (21 + 22 + 23) (Must Equal Line 12)			141,306	396,705	192,300
DEBT SERVICING					
25. Total interest paid			5,750	21,005	23,600
26. Principal paid (22 + 23)			42,631	56,105	60,850
27. TOTAL CASH AVAILABLE FOR DEBT SERVICING (17 + 18 + 25 + 26)			56,656	81,110	92,450
28. Scheduled debt payments (principal + interest)			43,750	73,200	84,450
29. Balance remaining			12,906	7,910	8,000
30. CASH AVAILABLE/DEBT SERVICE (27 ÷ 28)			1.29	1.11	1.09

in 1980 (line 27). Scheduled debt payments totalled $43,750 leaving a balance of $12,906. The ratio of cash available to debt service (line 30) was 1.29 indicating a rather strong cushion in the ability to meet debt servicing requirements.

Projected cash flows under the expanded operation are shown in the right hand columns of Table 7.3. The projected cash flows suggest tht debt servicing requirements can be met. However, the ratio of cash available to debt servicing drops from 1.29 to an average of 1.10 in the two years following the purchase. Hence, there is much less cushion in case of poor income conditions.

The example was chosen to reflect a borderline case. For example, the Federal Land Banks typically like to see a balance of at least 10% of total annual income after subtracting debt service and total living expenses. They believe this degree of cushion is needed because of the risks associated with agricultural production and income.

Negotiating To Improve Cash Flow Feasibility

Many land deals have gone by the wayside because of perceived cash flow problems on the part of the buyer. The two biggest hurdles are the downpayment and the annual debt servicing requirements. There are, however, ways to reduce these problems. Innovative financing techniques combined with the ability to negotiate successfully can help you overcome these cash flow problems.

Down payment—Cash downpayments on the purchase of farmland are undoubtedly the major obstacle for most would-be land buyers. However, in sales involving seller financing, the seller may be willing to accept a lower downpayment in exchange for a higher price. For example, the seller may want 25% down, but would accept 10% down for a higher price. But how much more can you afford to pay for land to obtain a lower downpayment? That question is addressed in detail in Chapter 8. For the moment, however, you should recognize that if you do not have the necessary cash to make the downpayment, your ability to negotiate a lower downpayment may be one of your few chances for completing the deal.

In many cases, however, land is not sold on contract and the financing is provided through an established lending institution. The two major farm real estate lending institutions, Federal Land Banks and life insurance companies, both have legal limits that prevent them from providing financing beyond a certain level. In the case of Federal Land Banks, the maximum loan is 85% of appraised value.

For life insurance companies, some state laws restrict the size of farm real estate loans made by life insurance companies to 75% of the appraised value of the property. You can attempt to negotiate up to these legal limits, but in practice the average loan-to-appraised value at these institutions is below the legal limit.

The Farmers Home Administration (FmHA) also provides financing for the purchase of farm real estate. Working in cooperation with other lenders, FmHA can provide up to 100% financing for the purchase of farm real estate. A more detailed description of the lending institutions providing farm mortgage loan funds is presented in Chapter 9.

One method of overcoming the problem of a large downpayment is by stretching it out or otherwise reducing the "up-front" money needed. By stretching out the downpayment, you buy time to come up with the cash needed. A variety of techniques can be used.

You might be able to stretch out your downpayment to match crop marketing patterns. For example, you have an opportunity to buy land for $200,000 on January 1. The seller wants 20% down, or $40,000. You can only come up with $20,000.00 now. Through negotiation with the seller, you might be able to delay payment of half of the downpayment until December when income from the crops is available. Caution! If you go this route, be sure that you are not just postponing a cash crunch until the end of the year. Be sure you have sufficient funds to make the first year's payment.

A similar approach is to ask the seller to accept installments on the downpayment in exchange for a security interest in other assets (machinery, livestock or crops) that you already own. For example, suppose a seller wants a downpayment of $30,000, but you can come up with only $15,000 now and $5,000 over each of the next 3 years. You can offer the seller a security interest in your machinery to ensure that you will meet your obligations.

Another variant to stretch out the downpayment is to rent with the option to buy. For example, suppose you are considering land which is currently selling for $2000 per acre and would cash rent for $80 per acre. You might negotiate with the seller to rent the land at $125 per acre ($45 per acre over current rates) over the next 3 years with the option to buy at $2,400 per acre. If the option to buy is exercised, $135 per acre of the rents ($45 × 3 years) is applied toward the downpayment. Under this procedure you would lock in a fixed price on land, and would accumulate over 5% of the purchase price to be applied toward the downpayment. If the land increased beyond the option price you would gain additional equity through inflation. If the land failed to increase in value you would likely not exercise the option to buy and would, therefore, forfeit $135 per acre. Substantial risks are involved, but the procedure could help you come up with the money needed for the downpayment. Note: As discussed in Chapter 16, under a lease with option-to-purchase, you claim investment tax credit on otherwise eligible items when the option is exercised. However, prior use of assets by the taxpayer may preclude claiming investment tax credit.

In the housing sector, it is not uncommon to find cases where the buyer of a new home acquires part of the downpayment through "work equity." That is, the buyer provides labor for painting, landscaping, insulating, or other parts of construction, thereby increasing the value. The same approach could be used to improve the value of real estate to be purchased. Caution! Be sure that you have a legally binding contract and permission to make such improvements if they occur before you take possession of the property.

Another method of reducing the downpayment is to negotiate a second mortgage with the seller. For example, suppose you are considering the purchase of land for $3,000 per acre and the Federal Land

Bank will provide you with a loan for 70% or $2,100 per acre. You can only come up with $500 per acre, but you convince the seller to take back a second mortgage for $400 per acre for the balance. This lowers your downpayment, but increases your debt servicing requirements. Before attempting to arrange a second mortgage, check with your first mortgage lender. If you attempt to arrange a second mortgage without the knowledge or consent of the first mortgage lender, the first lender may nix the entire deal.

Selling off a portion of the property purchased may also help ease the cash downpayment problem. For example, suppose you are arranging for the purchase of a $200,000 piece of property. The seller is willing to provide a deed to the property for 25% down and will carry back the remaining 75% of the purchase price on a 20 year loan. Your primary interest is the crop land and you find a buyer who will buy the homestead for $30,000 payable in cash on the day of closing. Since the total downpayment is $50,000, your prearranged sale of the farmstead would reduce your downpayment to $20,000. The key here is to be sure that your purchase contract provides you with a deed to the entire property or at least the homestead portion so that you can sell the homestead as of the day of the closing.

In the above example, what would happen if the financing was to be provided by a Federal Land Bank or life insurance company? In this case you should get an estimate of the dollar amount of financing the lender would provide on the entire property and the amount the lender would provide with the farmstead excluded. If the homestead is to be sold, you would not be able to borrow the full $150,000 because this would place the loan-to-appraised value above the legal limits. But the reduction in loan amount might not be proportional to the reduction in the sale price when the homestead is excluded. Check carefully with your lender to determine the amount of financing that would be provided with and without the homestead.

Assume for the moment that you purchased the above property for $200,000, borrowed $150,000 from a Federal Land Bank, kept the homestead, and now two years later you can buy additional land. To obtain equity for the downpayment on the new land, you want to sell the homestead on the original purchase. Since your mortgage with the Federal Land Bank was for the entire property, you cannot sell the farmstead unless you get a "partial release" from the Federal Land Bank. The terms of such a release should be discussed with an FLB loan officer. If the land has appreciated in value, you might be able to obtain a release with no payment on principal. Alternatively, the FLB may require some principal reduction to grant the release. Before committing yourself to the new purchase, check with the FLB to determine the terms of the release.

Debt servicing requirements—Many of the techniques for overcoming the downpayment problem could increase the cash flow problems associated with debt servicing. Trading-off downpayment problems for debt servicing problems may not be a good idea. The astute purchaser attempts to solve both problems. As with downpayments, there are a variety of techniques that can be used to alleviate cash flow problems resulting from debt servicing requirements.

Lengthening the term of loan—The most obvious procedure for reducing a burdensome debt servicing requirement is to lengthen the terms of the loan. For example, suppose you borrow $100,000 at 10% interest. Equal annual amortization payments for different lengths of loans are shown in Table 7.4. Moving from a 10-year to a 20-year loan substantially reduces the amount of each annual payment. This can be particularly important in land contract sales if the seller wants the money out of the property in a short period of time. In these cases, the buyer can often negotiate a balloon payment loan with annual payments based upon, say, a 20-year payout, but the balloon portion becomes due in 5 or 10 years. This allows the buyer to gain equity in the property and also adequate time to seek financing from institutional lenders to pay off the balloon payment when it becomes due. *CAUTION:* The buyer may want to arrange the contract such that, for adequate compensation, the seller will continue to provide financing for the balloon portion of the loan. The provision protects the buyer against a scarcity of funds at the time the balloon payment becomes due.

Table 7.4. Equal annual amortization payments for a $100,000 loan at 10% interest.

Length of Loan	Annual Payment
10 years	$16,273
15 years	13,148
20 years	11,745
25 years	11,017
30 years	10,608
35 years	10,369
40 years	10,225
50 years	10,086

As shown in Table 7.4, there is very little advantage to extending the loan payment period beyond 30 years. Payments on a 40-year loan of $100,000 at 10% interest are only $522 per year lower than a 30-year loan for the same amount and interest rate. If debt servicing requirements are a problem for a 30-year loan, lengthening the loan offers little improvement.

Increasing loan payment plans—Current lending practices typically require the borrower to repay in installments which are equal or decline over time. Yet the income generated from the land being purchased is likely to increase over time. Hardly anyone expects the dollar amount of income per acre to be the same in 1990 as it is today.

Hence, while the loan payments remain the same or decline through time, the income from which payments are to be made is likely to rise. The main problem for the borrower is in the early years of the loan. An increasing amortization loan payment plan could help relieve the cash flow strain.

For purpose of illustration, assume you are purchasing land for $1,571 per acre with 30% down on a 30-year loan at 10% interest (see Table 7.1, $60 net after-tax cash flow, 9% return). The loan for $1,100 (70% × $1,571) would require equal annual amortization payments of $116.62 per acre. However, there would be a deficiency of $25.84 per acre in the first year; that is the land being purchased will not generate sufficient income to make the loan payments.

Amortization payments and related information for an equal annual amortization loan are illustrated in Table 7.5. Notice that the deficiencies in returns to meet loan payments are high in the early years. However, since income is expected to increase by 5% per year, there is a substantial surplus by the end of the loan period.

An amortization plan which matches growth in income with an increasing amortization payment could help alleviate some of the cash flow problems associated with land purchases. An example of such an increasing amortization plan is depicted in Table 7.6. This amortization schedule requires annual payments starting at $93.29 and increasing to $190.97 by the end of the loan. The payments start at 80% of the equal annual amortization amount and increase at the rate of 2.55% per year, or about one-half as fast as the expected growth in returns. As before, the loan is completely retired in 30 years.

Under this amortization plan, there is only a small deficiency in year 1. Consequently, the borrower should find it much easier to meet scheduled amortization payments under this plan than the equal payments plan shown in Table 7.5.

While the borrower may favor the amortization plan depicted in Table 7.6, what about the lender? Because of the low initial payments, the loan balance would actually increase in the early years of the loan. The increased risk for the lender, however, may be minimal.

The ratio of loan balance to purchase price starts at .70 and increases to .774 at the end of year 11. However, the ratio of the loan balance to current market value declines steadily over time assuming the inflation rate of 4% per year is achieved. The risk of loan delinquency could be reduced with this type of amortization schedule because the borrower would not face substantial deficiencies in the early years of the loan.

One should not conclude from the example in Table 7.6 that an increasing amortization plan can always remove a deficiency in meeting loan amortization payments. The increasing amortization plan can, however, modify the amount and timing of the deficiencies. The example in Table 7.7 illustrates this point.

Assume you buy land generating a $60 per acre net after-tax return

Table 7.5 Annual payments and related items for a 30-year loan of $1,100 at 10% interest, amortized with equal annual payments.

End of Year	Annual amortization payment	Surplus or deficiency in income per acre[1]	Loan Balance	Ratio of loan balance to purchase price	Ratio of loan balance to current market value[2]
1	$116.62	$ −25.84	$1,092.67	.696	.669
2	116.62	−19.87	1,085.32	.691	.639
3	116.62	−16.77	1,077.23	.686	.610
4	116.62	−13.53	1,068.33	.680	.582
5	116.62	−10.13	1,058.55	.674	.554
6	116.62	−6.57	1,047.78	.667	.527
7	116.62	−2.86	1,035.94	.660	.501
8	116.62	1.03	1,022.92	.651	.476
9	116.62	5.10	1,008.59	.642	.451
10	116.62	9.35	992.83	.632	.427
11	116.62	13.80	975.50	.621	.404
12	116.62	18.45	956.43	.609	.380
13	116.62	23.30	935.45	.596	.358
14	116.62	28.37	912.38	.581	.336
15	116.62	33.66	887.00	.565	.314
16	116.62	39.19	859.08	.547	.292
17	116.62	44.95	828.37	.528	.271
18	116.62	50.97	794.59	.506	.250
19	116.62	57.24	757.43	.482	.229
20	116.62	63.78	716.55	.456	.208
21	116.62	70.60	671.59	.428	.188
22	116.62	77.70	622.13	.396	.167
23	116.62	85.09	567.72	.362	.147
24	116.62	92.78	507.88	.323	.126
25	116.62	100.78	442.04	.282	.106
26	116.62	109.09	369.63	.235	.085
27	116.62	117.73	289.97	.185	.064
28	116.62	126.70	202.35	.129	.043
29	116.62	136.01	105.97	.068	.022
30	116.62	145.66	0.00	.000	.000

[1]Calculated by adding the net after-tax cash flow ($60 per acre increasing 5% per year) and the tax savings on interest payments (28% of the interest paid). From this total subtract the annual amortization payment.
[2]Land values are assumed to increase 4% per year.

for $2,556 per acre and obtain a loan for $1,789 per acre (See Table 7.1). The first year deficiency on the loan would be $79.70 per acre, but with income growing 5% per year the deficiency would be removed in 20 years. Using an increasing amortization plan starting at 80% of the equal amortization payment would lower the first year deficiency from $79.70 to $41.74 per acre. In fact the deficiencies in the first 12

Table 7.6 Annual payments and related items for a 30-year loan of $1,100 at 10% interest, amortized with increasing annual payments.

End of year	Annual amortization payment	Surplus or deficiency in income per acre[1]	Loan balance	Ratio of loan balance to purchase price	Ratio of loan balance to current market value[2]
1	$ 93.29	$−2.51	$1,115.99	.711	.683
2	95.67	1.72	1,131.92	.721	.666
3	98.11	3.04	1,146.99	.730	.649
4	100.61	4.43	1,161.08	.739	.632
5	103.18	5.91	1,174.00	.745	.614
6	105.81	7.47	1,185.59	.755	.597
7	108.51	9.11	1,195.64	.761	.579
8	111.28	10.85	1,203.93	.767	.560
9	114.11	12.68	1,210.21	.771	.541
10	117.02	14.60	1,214.21	.773	.522
11	120.01	16.61	1,215.62	.774	.503
12	123.07	18.72	1,214.11	.773	.483
13	126.21	20.93	1,209.32	.770	.463
14	129.42	23.23	1,200.83	.765	.442
15	132.72	25.63	1,188.18	.757	.420
16	136.11	28.13	1,170.89	.746	.398
17	139.58	30.72	1,148.41	.731	.375
18	143.14	33.41	1,120.11	.713	.352
19	146.79	36.19	1,085.33	.691	.328
20	150.53	39.05	1,043.33	.664	.303
21	154.37	42.00	993.30	.633	.278
22	158.31	45.02	934.32	.595	.251
23	162.34	48.11	865.41	.551	.224
24	166.48	51.25	785.47	.500	.195
25	170.73	54.44	693.29	.441	.166
26	175.08	57.67	587.54	.374	.135
27	179.54	60.91	466.75	.297	.103
28	184.12	64.15	329.30	.210	.070
29	188.82	67.36	173.41	.110	.035
30	190.97	73.19	0.00	.000	.000

[1]Calculated by adding the net after-tax cash flow ($60 per acre increasing 5% per year) and the tax savings on interest payments (28% of the interest paid). From this total subtract the annual amortization payment.
[2]Land values are assumed to increase 4% per year.

years of the loan would be lower under the increasing amortization plan. This could help the borrower avoid delinquency in the early years of the loan. In our example, the deficiencies are not removed, but they are more equally spread throughout the length of the loan. Procedures

Table 7.7 Annual payments and related items for a 30-year loan of $1,789 at 10% interest.

End of year	Equal Payment Plan			Increasing Payment Plan		
	Annual payment	Surplus or deficiency in income per acre[1]	Loan balance	Annual payment	Surplus or deficiency in income per acre[1]	Loan balance
1	$189.80	$ −79.70	$1,778.37	$151.84	$ −41.74	$1,816.33
2	189.80	−73.86	1,766.41	155.74	−38.71	1,842.25
3	189.80	−70.89	1,753.25	159.68	−38.64	1,866.79
4	189.80	−67.78	1,738.77	163.76	−38.56	1,889.72
5	189.80	−64.54	1,722.84	167.93	−38.44	1,910.76
6	189.80	−61.16	1,705.32	172.21	−38.31	1,929.62
7	189.80	−57.63	1,686.05	176.61	−38.15	1,945.97
8	189.80	−53.95	1,664.85	181.11	−37.97	1,959.46
9	189.80	−50.11	1,641.54	185.73	−37.78	1,969.68
10	189.80	−46.11	1,615.89	190.46	−37.58	1,976.19
11	189.80	−41.94	1,587.67	195.32	−37.37	1,978.49
12	189.80	−37.60	1,556.64	200.30	−37.15	1,976.04
13	189.80	−33.08	1,522.50	205.41	−36.94	1,968.23
14	189.80	−28.38	1,484.94	210.64	−36.74	1,954.41
15	189.80	−23.49	1,443.64	216.02	−36.56	1,933.84
16	189.80	−18.41	1,398.20	221.52	−36.41	1,905.70
17	189.80	−13.13	1,348.21	227.17	−36.29	1,869.10
18	189.80	−7.66	1,293.23	232.96	−36.24	1,823.04
19	189.80	−1.98	1,232.75	238.90	−36.25	1,766.44
20	189.80	3.91	1,166.22	245.00	−36.34	1,698.09
21	189.80	10.01	1,093.04	251.24	−36.54	1,616.65
22	189.80	16.31	1,012.54	257.65	−36.87	1,520.67
23	189.80	22.84	924.00	264.22	−37.35	1,408.52
24	189.80	29.57	826.00	270.96	−38.02	1,278.41
25	189.80	36.52	719.45	277.87	−38.89	1,128.38
26	189.80	43.68	601.59	284.95	−40.02	956.27
27	189.80	51.04	471.95	292.22	−41.44	759.68
28	189.80	58.61	329.34	299.67	−43.20	535.98
29	189.80	66.38	172.47	307.31	−45.34	282.27
30	189.80	74.34	0.00	310.82	−43.60	0.00

[1]Calculated by adding the net after-tax cash flow ($60 per acre growing 5% a year) and the tax savings on interest payments (28% of the interest paid). From this total subtract the annual amortization payment.

for calculating increasing amortization loan payments are given in the appendix to this chapter.

Delayed payments—Another alternative for reducing the potential cash flow problems associated with debt servicing requirements is to negotiate a contract which allows you to delay payment of principal and interest in case of a crop failure. This creates a cushion in your debt servicing requirements that can be called upon in case of emergency.

In negotiating with a seller for such a "delayed payment" option a buyer should point out that the seller will eventually get all of the money including interest. Also buyers should point out the risky nature of farming and the importance to the buyer of protection against loss of the farm or ranch because of one bad year. Sellers should recognize the need for financial flexibility and may be willing to delay payment of part or all of the principal and interest in a low income year.

A word of caution: If you negotiate for a "delayed payment" in a low income year, be sure that the conditions for a "low income year" are clearly specified. Otherwise, disputes can arise with the seller over what is or is not a "low income year." Such disagreements can be avoided if you negotiate for a "delayed payment" in *any* one year of your choice.

For buyers using conventional financing, it is unlikely that you will be able to negotiate a "delayed payment" option with your lender. However, the lender may provide much the same type of arrangement by offering to refinance or provide a loan extension in cases of financial emergency. Also, the lender may offer a plan whereby you can pay ahead in high income years into a reserve fund that can be used in low income years. Accumulations in such reserves may pay interest at the same rate at which you are paying interest on the loan.

APPENDIX
Computing Payments on an Increasing Amortization Loan

The use of an increasing amortization loan plan can help reduce the "early year" cash flow problems associated with land purchase. There are, of course, an unlimited number of increasing amortization plans that could be developed. In our examples, the plans are based on modifications from equal annual payments.

There are two basic components to the increasing amortization plan: (a) the starting percentage, and (b) the rate of increase in the amount of payment each year. The starting percentage is the ratio between *equal* annual payments and the *first* payment under an increasing amortization plan. For example, assume a loan requires equal annual amortization payments of $100. If the first payment under the increasing amortization plan is $80, the starting percentage is 80%. If the first payment was $90, the starting percentage would be 90%.

The rate of increase in the amount of payment reflects how fast the payment must grow each year to pay off the loan in the same length of time as the equal annual payment amortization plan. The lower the starting percentage, the higher must be the rate of increase to pay off the loan in the same length of time. Table 7.1A illustrates the trade offs involved.

Table 7.1A Multiplication factors for determining increasing annual amortization loan payments.

Length of Loan	Starting Percentage	Interest Rate				
		8%	9%	10%	11%	12%
10 yrs	80	1.05610	1.05710	1.05819	1.05919	1.06029
10 yrs	82	1.04999	1.05100	1.05187	1.05278	1.05379
10 yrs	84	1.04409	1.04487	1.04573	1.04660	1.04742
10 yrs	86	1.03832	1.03901	1.03970	1.04039	1.04121
10 yrs	88	1.03250	1.03312	1.03380	1.03439	1.03501
10 yrs	90	1.02591	1.02740	1.02790	1.02839	1.02899
10 yrs	92	1.02133	1.02180	1.02221	1.02261	1.02301
10 yrs	94	1.01591	1.01620	1.01649	1.01681	1.01710
10 yrs	96	1.01050	1.01079	1.01098	1.01110	1.01139
10 yrs	98	1.00530	1.00541	1.00552	1.00558	1.00569
15 yrs	80	1.03839	1.03941	1.04049	1.04169	1.04282
15 yrs	82	1.03429	1.03521	1.03621	1.03718	1.03830
15 yrs	84	1.03019	1.03112	1.03201	1.03288	1.03377
15 yrs	86	1.02633	1.02700	1.02783	1.02860	1.02938
15 yrs	88	1.02239	1.02299	1.02370	1.02432	1.02500
15 yrs	90	1.01852	1.01900	1.01961	1.02019	1.02072
15 yrs	92	1.01470	1.01510	1.01559	1.01597	1.01651
15 yrs	94	1.01097	1.01132	1.01159	1.01199	1.01229
15 yrs	96	1.00727	1.00751	1.00770	1.00790	1.00816
15 yrs	98	1.00362	1.00370	1.00390	1.00397	1.00410
20 yrs	80	1.03031	1.03141	1.03259	1.03379	1.03499
20 yrs	82	1.02711	1.02812	1.02911	1.03021	1.03134
20 yrs	84	1.02390	1.02480	1.02581	1.02671	1.02770
20 yrs	86	1.02079	1.02160	1.02241	1.02321	1.02410
20 yrs	88	1.01768	1.01839	1.01911	1.01979	1.02061
20 yrs	90	1.01470	1.01519	1.01580	1.01643	1.01709
20 yrs	92	1.01173	1.01209	1.01260	1.01312	1.01362
20 yrs	94	1.00869	1.00909	1.00938	1.00979	1.01010
20 yrs	96	1.00579	1.00598	1.00626	1.00650	1.00669
20 yrs	98	1.00292	1.00301	1.00308	1.00329	1.00341
25 yrs	80	1.02574	1.02691	1.02817	1.02950	1.03090
25 yrs	82	1.02301	1.02411	1.02519	1.02639	1.02760
25 yrs	84	1.02039	1.02128	1.02230	1.02337	1.02450
25 yrs	86	1.01769	1.01857	1.01938	1.02038	1.02131
25 yrs	88	1.01509	1.01580	1.01660	1.01738	1.01821
25 yrs	90	1.01249	1.01310	1.01384	1.01438	1.01511
25 yrs	92	1.01001	1.01050	1.01099	1.01150	1.01199
25 yrs	94	1.00748	1.00779	1.00822	1.00861	1.00900
25 yrs	96	1.00503	1.00520	1.00550	1.00571	1.00599
25 yrs	98	1.00250	1.00259	1.00269	1.00291	1.00299
30 yrs	80	1.02292	1.02425	1.02551	1.02692	1.02841
30 yrs	82	1.02051	1.02173	1.02290	1.02418	1.02552
30 yrs	84	1.01809	1.01919	1.02021	1.02144	1.02260
30 yrs	86	1.01581	1.01672	1.01760	1.01857	1.01971
30 yrs	88	1.01351	1.01430	1.01510	1.01590	1.01680
30 yrs	90	1.01120	1.01180	1.01251	1.01320	1.01397
30 yrs	92	1.00890	1.00938	1.00997	1.01056	1.01109
30 yrs	94	1.00669	1.00709	1.00751	1.00792	1.00829
30 yrs	96	1.00435	1.00467	1.00501	1.00528	1.00558
30 yrs	98	1.00218	1.00240	1.00248	1.00270	1.00279

For example, suppose you borrow $100,000 at 8% interest for 10 years and want to use an increasing annual amortization plan. First calculate the *equal* annual amortization payment. Using equation 6.5 we find the equal annual payment would be $14,903.13. Table 7.1A can be used to calculate the amount of each payment. For example, starting at 80%, the first payment is $11,922.50, i.e. (80% × $14,903.13). To compute the payment in subsequent years multiply the previous year's payment by the factor 1.05610. The payment due at the end of the second year is $12,591.36; (11,922.50 × 1.0561). The payment due at the end of the third year is $13,297.73 (12,591.36 × 1.0561). Other factors in Table 7.1A can be used in a comparable fashion.

Notice that for a given length of loan and starting percentage, the multiplication factor increases as interest rates increase. However, the multiplication factor decreases as the starting percentage or length of loan is increased.

CHAPTER 8
HOW TERMS OF FINANCING INFLUENCE FARMLAND VALUES

The terms of financing for real estate loans can have a significant impact on the price you can pay when purchasing farmland. Terms of financing include the interest rate, length of loan and amount of downpayment. In this chapter we examine how these terms of financing influence the price that can be paid for farm and ranch land.

Interest Rates

The interest rate paid on borrowed funds is one of the more important determinants of the price you can pay for farmland. However, there are many different types of "interest rates" that you may hear or read about. Understanding the differences among these interest rates is important. Four of the most common interest rates are: (a) the contractual rate, (b) the annual percentage rate (APR), (c) the real rate, and (d) the after-tax rate. A brief description of each follows:

Contractual rate—The rate of interest stated on the loan document is referred to as the contractual rate. The contractual rate expressed in percentage terms is the relationship between the amount of interest paid and the amount of principal. Because noninterest costs such as service charges on the loan are not included, this rate may be deceiving.

The problems with the contractual rate can best be illustrated with an example. Suppose you borrow $50,000 for two years at an annual contractual rate of 12%. Whether the service charge is zero, $300 or $3,000, the contractual rate of interest is still 12%. But the dollar cost of borrowing varies substantially with the amount of the service charge.

Annual percentage rate—Recognizing the potential deception associated with the contractual rate, the United States Congress several years ago passed "Truth-in-Lending" legislation. Lenders subject to the regulation are required to inform borrowers of the total amount of finance charges and the annual percentage rate of interest (APR). The APR must be calculated with both interest and noninterest costs taken into account. In the example of a two year loan for $50,000 at 12% contractual rate of interest, if there is a $500 service charge, the APR is approximately 12.75%. Procedures for calculating the APR are given in the appendix to this chapter.

The APR serves as a common method of reporting interest rates and helps avoid deceptive lending practices. However, there are exceptions to Truth-in-Lending legislation which are important to farmers and ranchers. While all farm real estate loans are covered, non-real estate loans over $25,000 are not subject to the legislation. Also, legislation has been proposed in Congress to exempt *all* agricultural loans from Truth-in-Lending.

Real rate—The term "real rate" of interest means different things to different people. Some use the term in reference to the APR. But technically, the "real rate of interest" refers to an interest rate which has been adjusted for inflation. For example, if the APR for borrowing is 10%, but the inflation rate is 7%, the real interest rate is only about 3%.[1] Farmers and ranchers need to recognize that, at times, the real cost of borrowing for the purchase of real estate can be relatively low and could even be negative.

Since 1950, the APR on farm mortgage loans has tended to increase, while the real interest rate has tended to decline. In fact, for several of the years in the late 1970's and in 1980, real interest rates on farm mortgages were negative. Farmers and ranchers who borrow under these conditions are repaying a total dollar amount that has less purchasing power than the dollars received when the loan was originally made.

After-tax interest rates—Interest payments are a tax deductible expense.[2] Therefore, increases in interest payments reduce the amount of income taxes paid by the land owner. For example, suppose that a land owner has a loan with an APR of 10%. Further, assume that the land owner earns enough income so that additional income is taxed at 28%. The after-tax interest rate on the loan is 7.2% (10% × 1 − .28).[3] Notice that the after-tax interest rate is lower for borrowers in high tax brackets. Hence, there may be greater incentive to borrow money if you are in a high tax bracket than if you are in a low tax bracket. A "real after-tax interest rate" would account for both the inflation effect and the tax effect.

Effect of Service Charges on APR

Service charges, also sometimes referred to as points, loan origination fees, or loan fees, are an important part of the cost of borrowing. For example, in early 1980, the Federal Land Bank in one district was

---

[1]The formula for calculating the real interest rate is: $\frac{1+i}{1+P} - 1$ where i is the APR and P is the rate of inflation. In our example the calculation is $\frac{1+.10}{1+.07} - 1 = .028 = 2.8$ percent.

[2]For investors, a maximum limit is imposed on interest deductions. The interest deductions cannot exceed $10,000 plus the taxpayer's net investment income.

[3]A formula for calculating the after-tax interest rate is: $r_t = r(1-t)$, where r_t is the after-tax interest rate, r is the annual percentage rate, and t is the borrower's effective income tax rate.

charging loan fees as high as 8% of the farm mortgage loan amount. For a loan of $100,000 with a service charge of 8%, the borrower would need to pay $8,000 at the time the loan is originated. Such service charges are sometimes paid by borrowing the additional amount needed. Service charges of this nature can have a significant impact on the annual percentage rate of interest (Table 8.1).

Table 8.1 Impact of service charges on the annual percentage rate of interest for a $1,000 loan at a 10% contractual rate of interest.

Service charge as a percent of the loan amount	Length of loan repayment (years)					
	1	3	5	10	20	30
	annual percentage rate (APR)					
0	10.00	10.00	10.00	10.00	10.00	10.00
1	11.11	10.58	10.40	10.24	10.15	10.13
2	12.25	11.16	10.80	10.48	10.31	10.26
3	13.40	11.75	11.21	10.72	10.46	10.39
4	14.59	12.35	11.62	10.96	10.62	10.51
5	15.79	12.97	12.04	11.23	10.77	10.64
6	17.02	13.59	12.47	11.47	10.93	10.77
7	18.28	14.23	12.91	11.73	11.09	10.90
8	19.57	14.88	13.35	11.99	11.27	11.03

As shown in Table 8.1, the APR increases as service charges increase. The shorter the length of loan repayment, the greater the impact on the APR. But for a loan repaid over a 30-year period, the initial service charge has a relatively small impact on the APR. Farm borrowers should recognize the impact of service charges on interest rates and determine under what conditions, if any, the lender is likely to alter the service charge.

Effect of Stock Purchase Requirements on the APR

Stock purchase requirements affect the APR on Federal Land Bank loans. The exact effect depends upon the percentage of stock purchase required, the amount of dividend, if any, paid on the stock, and the timing of stock redemptions. Borrowers must purchase stock in their association in an amount not less than 5% nor more than 10% of the face amount of the loan. The money necessary to purchase the stock normally comes from the proceeds of the loan. In practice, the FLB's seldom pay stock dividends, preferring instead to keep the cost of borrowing as low as possible.

To illustrate the effects of the stock purchase requirement, assume you borrow $100,000 at a 10% contractual rate of interest, a 5% stock purchase requirement, no stock dividend, and no stock redemption until the end of the loan. The APR on this loan is approximately 10.71%. A 10% stock purchase requirement would raise the APR to approximately 11.44%. If the stock is redeemed as the loan is repaid, the interest rate would be somewhat lower.

Effect of Interest Rates on the Net Present Value of Land

In Chapter 6, we introduced the concept of the net present value of land. The interest rate used in those calculations was the APR. In general, the higher the interest rate, other factors constant, the lower is the net present value of an acre of land. That is because the outflows associated with the purchase are larger with more interest being paid. However, the extent of the impact on the net present value of land depends on the discount rate used in the calculations as well as the marginal income tax rate of the borrower.

Table 8.2 illustrates the impact on the net present value of land because of a change from a 30-year loan at 10% interest to a 30-year loan at 12% interest. Panel A shows the net present values of an acre

Table 8.2 Net present values per acre of land for alternative levels of annual net after tax returns, inflation rates of land, discount rates, and interest rates on loans.[1]

If you expect:			Price you can pay per acre to get a return of:		
Annual net after tax cash flow per acre of:	Annual rate of growth in net after tax cash flow of:	Annual rate of inflation in land value of:	10% after income taxes	12% after income taxes	14% after income taxes
			PANEL A (10% Interest Rate)		
$ 20	8%	7%	$ 960	$ 657	$ 506
40	8%	7%	1,921	1,315	1,013
60	8%	7%	2,881	1,972	1,519
80	8%	7%	3,842	2,629	2,026
100	8%	7%	4,802	3,287	2,532
			PANEL B (12% Interest Rate)		
$ 20	8%	7%	819	582	458
40	8%	7%	1,637	1,165	915
60	8%	7%	2,456	1,747	1,373
80	8%	7%	3,275	2,329	1,830
100	8%	7%	4,093	2,912	2,289
			PANEL C (Dollar Reduction in Value Per Acre)		
$ 20	8%	7%	141	75	48
40	8%	7%	284	150	98
60	8%	7%	425	225	146
80	8%	7%	567	300	196
100	8%	7%	709	375	243

[1]Calculations are based upon the assumption that the buyer is in the 28% tax bracket, has a 30-year planning horizon, and finances the purchase with a 30% downpayment and a loan for 30 years. Capital gains taxes are assumed to be 20% of the full gain.

of land calculated under the assumption the loan is at 10% interest. Values in Panel B are calculated under the same set of assumptions except the interest rate is increased to 12%. Panel C reports the dollar reduction in net present value per acre as a result of moving from a 10% to a 12% interest rate on the loan. All values are based upon after-tax calculations.

Dollar reductions in value per acre are over $700 for the highest priced land and less than $50 per acre for the lowest priced land. In percentage terms, the dollar reductions are almost 15% when a 10% discount rate is used. However, as shown in Panel C, dollar reductions per acre are less than 10% when a 14% discount rate is used. In reality, it may be somewhat unusual to expect interest rates to increase without changes in other variables. For example, higher interest rates are typically associated with higher inflation rates. Hence, it is somewhat unrealistic to expect interest rates to increase and remain high over a 30-year period without altering the assumption about expected inflation.

The values in Panel C of Table 8.2 suggest that an increase in interest rates, other things constant, could substantially reduce the value of land. While not shown in Table 8.2, the lower values combined with the higher interest rate would leave the cash deficiency in making loan payments virtually unchanged. Hence, an increase in interest rates which is fully reflected in the price of land has little impact on the cash flow feasibility of purchase. That is, the lower value of land would offset the higher interest charge. However, higher interest rates are often accompanied by tight credit conditions which cause lenders to require more downpayment. A larger downpayment could substantially affect the cash flow feasibility of purchase.

Estimating the Trade-off Between Price and Interest Rate

In land purchases involving seller financing, there are usually opportunities to negotiate tradeoffs between price and interest rate. A seller in a high income tax bracket may find it advantageous to sell for a high price with a low interest rate. That is because only 40% of the gain is taxable while all interest income is taxable.[4] But the question of how much of a trade-off to make between price and interest rate is often not carefully evaluated.

You should recognize that the buyer and seller may have substantially different economic situations. For example, the tax rates and discount rates of the buyer and seller may differ greatly. Therefore, an evaluation of the tradeoffs between price and interest rate usually differ for the buyer and the seller. The following discussion provides such analyses, first from the perspective of the buyer, then from the perspective of the seller.

---

[1]The Internal Revenue Service establishes minimum interest rates for each contractual obligations as land purchases. The current minimum is 9% except for a minimum of 7% on up to $500,000 in sales price for transactions between family members. See chapter 16. When the 9% rate applies, if a rate below 9% is used, it is refigured at ten percent compounded semi-annually.

Buyer's Perspective—No matter what the combination of price and interest rate on the land contract sale, the physical amount of corn, beans, cotton, cattle, or other agricultural products that can be produced on the land is not affected. Therefore, the buyer can evaluate the tradeoff between price and interest rate by assessing the differences in the *cash outflows*. Cash inflows are unaffected. A worksheet for determining the net present value of cash outflows is illustrated in Table 8.3. The worksheet can be completed in the following steps:

Step 1: Record the purchase price, loan amount, interest rate, and length of loan in the top right-hand section of the worksheet.

Step 2: Record the downpayment in Column A.

Step 3: Determine the amount of each annual payment and record in Column B. These payments may be constant, increasing, or decreasing over time and depend upon what you negotiate with the seller.

Step 4: Allocate each payment to principal and interest and record the values in Columns C and D, respectively. Also, determine the loan balance and record in Column E.

Step 5: Determine your marginal tax rate. If you are in an expansion period this rate may increase over time. Record your expected tax rates in Column F. Then multiply Column D × F and record the values in Column G.

Step 6: Add the values in Column A and B and subtract the values in Column G. Record these values in Column H. Multiply Column H by the discount factor (Column I) and record the values in Column J.

Step 7: Sum the entries in Column J and compare with alternative financing plans. The plan with the *lowest* net present value of net cash *outflows* is best.

An example can illustrate the importance of these calculations. Suppose you are offered land for $1,000 per acre, $200 down, balance in 15 equal annual installments with interest at 12%. As an alternative, you offer the seller $1,150 per acre, $200 down, balance in 15 equal annual installments with interest at 8%. Table 8.3 shows the calculations for the original terms of the loan.

As shown in Table 8.3, the net present value of the cash outflows is $935.81 per acre. Table 8.4 presents the calculations for a purchase price of $1,150 per acre, $200 down, 8% interest, and a 15-year loan. The net present value of the cash outflows is $925.57 per acre—$10.24 per acre less than under the original terms of the loan. Hence, the alternate terms are advantageous to the buyer.

There are also several advantages to the buyer which may not be readily apparent from the information in Tables 8.3 and 8.4. Any problems in meeting debt servicing requirements are less severe for the terms as outlined in Table 8.4. Annual payments are $117.46 per

Table 8.3 Buyer's worksheet for determining the trade-off between price and interest rate.

1. Purchase Price: $1,000/acre 3. Interest Rate: 12%
2. Loan Amount: $800   4. Length of Loan: 15 years

End of Year	A Down Payment	B Annual Payments	C Principal	D Interest	E Loan Balance	F Marginal Tax Rate	G Tax Savings D×F	H After-Tax Net Cash Outflow A+B−G	I Discount Factor 12%	J Net Present Value of Outflows H×I
0	200.00				800.00			200.00	1.0	200.00
1		117.46	21.46	96.00	778.54	.28	26.88	90.58	.909	82.34
2		117.46	24.04	93.42	754.50	.28	26.16	91.30	.826	75.41
3		117.46	26.92	90.54	727.58	.28	25.35	92.11	.751	69.17
4		117.46	30.15	87.31	697.43	.28	24.45	93.01	.683	63.53
5		117.46	33.77	83.69	663.66	.28	23.43	94.03	.621	58.39
6		117.46	37.82	79.64	625.84	.28	23.20	94.16	.564	53.11
7		117.46	42.36	75.10	583.48	.28	21.03	96.43	.513	49.47
8		117.46	47.44	70.02	536.04	.28	19.61	97.85	.467	45.70
9		117.46	53.14	64.32	482.09	.28	18.01	99.45	.424	42.17
10		117.46	59.51	57.95	423.39	.28	16.23	101.23	.386	39.07
11		117.46	66.65	50.81	356.74	.28	14.13	103.23	.350	36.13
12		117.46	74.65	42.81	282.09	.28	11.99	105.47	.319	33.64
13		117.46	83.61	33.85	198.48	.28	9.48	107.98	.290	31.31
14		117.46	93.59	23.82	104.87	.28	6.67	110.79	.263	29.14
15		117.46	104.87	12.59	0	.28	3.53	113.93	.239	27.23
16										
17										
18										
19										
20										
TOTAL										$935.81

FARMLAND

Table 8.4 Buyer's worksheet for determining the trade-off between price and interest rate.

1. Purchase Price: $1,150/acre 3. Interest Rate: 8%
2. Loan Amount: $950 4. Length of Loan: 15 years

End of Year	A Down Payment	B Annual Payments	C Principal	D Interest	E Loan Balance	F Marginal Tax Rate	G Tax Savings D×F	H After-Tax Net Cash Outflow A+B−G	I Discount Factor 10%	J Net Present Value of Outflows H×I
0	200.00				950.00			200.00	1.0	200.00
1		110.99	34.99	76.00	915.02	.28	21.28	89.71	.909	81.55
2		110.99	37.79	73.20	877.22	.28	20.50	90.49	.826	74.74
3		110.99	40.81	70.17	836.41	.28	19.65	91.34	.751	68.60
4		110.99	44.08	66.91	792.33	.28	18.73	92.26	.683	63.01
5		110.99	47.60	63.39	744.73	.28	17.75	93.24	.621	57.90
6		110.99	51.41	59.58	693.32	.28	16.69	94.30	.564	53.19
7		110.99	56.52	55.47	637.80	.28	15.53	95.46	.513	48.97
8		110.99	59.97	51.02	577.83	.28	14.29	96.70	.467	45.16
9		110.99	64.76	46.23	513.07	.28	12.94	98.05	.424	41.57
10		110.99	69.94	41.05	443.13	.28	11.49	99.50	.386	38.41
11		110.99	75.54	35.45	367.59	.28	9.93	101.06	.350	35.37
12		110.99	81.58	29.41	286.01	.28	8.23	102.76	.319	32.78
13		110.99	88.11	22.88	197.90	.28	6.41	104.58	.290	30.31
14		110.99	95.16	15.83	102.74	.28	4.43	106.56	.263	28.03
15		110.99	102.74	8.22	0	.28	2.30	108.69	.239	25.98
16										
17										
18										
19										
20										
TOTAL										$925.57

acre for the loan outlined in Table 8.3, but only $110.99 per acre for the loan outlined in Table 8.4. While not accounted for in the calculations, the buyer also benefits from the higher income tax basis ($1150 versus $1000) when the property is sold.

Seller's Perspective—Individuals selling land and providing financing either through a land contract or mortgage may also benefit from a trade-off between price and interest rates. The seller should recognize that a higher price in exchange for a lower interest rate likely increases the amount of capital gains taxes due and decreases the amount of interest income over time. The seller can best evaluate the tradeoffs by computing the net present value of cash inflows under various price-interest rate alternatives. A worksheet for this purpose appears in Table 8.5. The table can be completed in the following steps:

- Step 1: Record the following in the upper-right hand corner of the worksheet: (1) the sales price less tax deductible costs of selling, (2) the loan amount, (3) the interest rate, (4) the length of loan, (5) the basis in the property sold, and (6) the capital gains tax factor which is equal to item 1 minus item 5, that sum divided by item 1 and the result multiplied by .4.
- Step 2: Record the downpayment in Column A, and the annual payments (principal + interest) in Column B. Annual payments may be constant, increasing or decreasing over time.
- Step 3: Allocate each payment to principal and interest and record the values in Columns C and D, respectively. Also determine the loan balance and record in Column E.
- Step 4: Determine the amount of taxable income and record in Column F. The amount of principal payments subject to capital gains tax is found by multiplying the capital gains tax factor by the amount of downpayment (Col. A) and principal (Col. C) received each period. All interest payments (Col. D) are taxable.
- Step 5: Record the expected tax rate for each year of the loan in Column 6. Remember that this rate could be high if you have substantial capital gains. Next, multiply taxable income (Col. F) by the marginal tax rate (Col. G) and record the answer in Column H.
- Step 6: Compute net after-tax income by adding Columns A and B and subtracting Column H. Record the answer in Column I. Then multiply the discount factors in Column J by Column I and record the net present values in Column K.
- Step 7: Sum the entries in Column K and compare with alternative financing plans. The plan with the *highest* net present value of cash *inflows* is best.

FARMLAND

Table 8.5 Seller's worksheet for determining the trade-off between price and interest rate.

1. Sale Price less Cost of Selling: $1,000/acre
2. Loan Amount: $800
3. Interest Rate: 12%
4. Length of Loan: 15 years
5. Basis: 200
6. Capital Gains Tax Factor
$Z = ((1-5)/1) \times .4 = .32$

End of Year	A Down Payment	B Annual Payment	C Principal	D Interest	E Loan Balance	F Taxable Income $((A+C) \times Z) + D$	G Marginal Tax Rate	H Tax Payments	I Net After Tax Income $A+B-H$	J Discount Factor 12%	K Net Present Value of Cash Inflows
0	200				800	64.00	.50	32.00	168.00	1.0	168.00
1		117.46	21.46	96.00	778.54	102.87	.50	51.44	66.02	.909	62.01
2		117.46	24.04	93.42	754.50	101.11	.50	50.56	66.90	.826	55.26
3		117.46	26.92	90.54	727.58	99.15	.50	49.58	67.88	.751	50.98
4		117.46	30.15	87.31	697.43	96.96	.50	48.48	68.98	.683	47.11
5		117.46	33.77	83.69	663.66	94.50	.50	47.25	70.21	.621	43.60
6		117.46	31.82	79.64	625.84	91.74	.50	45.87	71.59	.564	40.38
7		117.46	42.36	75.01	583.48	88.66	.50	44.33	73.13	.513	37.52
8		117.46	47.44	70.02	536.04	85.20	.50	42.60	74.86	.467	34.96
9		117.46	53.14	64.32	482.90	81.32	.50	40.66	76.80	.424	32.56
10		117.46	59.51	57.95	423.39	76.99	.50	38.50	78.96	.386	30.48
11		117.46	66.65	50.81	356.74	72.14	.50	36.07	81.39	.350	28.49
12		117.46	74.65	42.81	282.09	66.70	.50	33.35	84.11	.319	26.83
13		117.46	83.61	33.85	198.48	60.61	.50	30.31	87.15	.290	25.27
14		117.46	93.59	23.82	104.89	53.77	.50	26.89	90.57	.263	23.82
15		117.46	104.87	12.59	0	46.15	.50	23.08	94.38	.239	22.56
16											
17											
18											
19											
20											
TOTAL											727.83

The calculations illustrated in Table 8.5 are for the original terms of the loan proposed by the seller. Table 8.6 illustrates the calculations for the terms proposed by the buyer. A comparison of the net present value of the cash inflows shows that the seller is $2.13 per acre (729.96−727.83) better off under the terms proposed by the buyer. Notice that, in this example, both the buyer and the seller are better off under the alternative financing proposed by the buyer. This illustrates an important point—a modification in terms proposed by the buyer is not necessarily detrimental to the seller.

It should be recognized that the values selected in computing Table 8.3 through 8.6 were selected carefully to illustrate a point. In many circumstances, the sellers may ask for financing terms which are most advantageous to them. Buyers may likely counter-offer with terms which are most advantageous to them. This is all part of negotiating a land sales contract. The worksheets can help you decide what is best for you as a buyer or seller. You can also estimate what, if anything, it would cost the other party to accept the terms you are proposing.

Down Payment

It is well recognized that the amount of downpayment required on the purchase of land has an important bearing on the cash flow feasibility of purchase. Perhaps what is less clearly recognized is that the amount of the downpayment also influences the price one can justify in paying for the land. To illustrate this point, Table 8.7 shows the net present value of land under alternative assumptions concerning the level of downpayment and the discount rate.

Panel A shows the net present value per acre of land when the purchase requires a 10% downpayment. Values reported in Panel B and C are computed under the same set of assumptions except that the downpayment is raised to 30% and 50%, respectively. A comparison of the values quickly reveals that the level of downpayment does have an impact on the price you can pay to get a given return.

If one uses a 6% discount rate, the price you can pay per acre to get a given return *increases* as the amount of the downpayment increases. (For example, find and compare the values $815, $852 and $892 in Panels A, B, and C, respectively.) In contrast, if you use a 9% discount rate, the price you can pay per acre *decreases* as the downpayment increases. For a 7% discount rate, the level of downpayment has little impact on the price you can pay per acre. What explains this phenomenon?

The key here is the relationship between the discount rate—what you can achieve in your next best alternative—and the after-tax interest rate on the loan. In our example, the interest rate on the loan is 10% and the marginal tax bracket is 28%. This gives an after-tax interest rate of 7.2% (i.e. 10% × .72). If your discount rate is *less* than your after-tax interest rate, a larger downpayment *increases* the price you

Table 8.6 Seller's worksheet for determining the trade-off between price and interest rate.

1. Sale Price Less Cost of Selling: $1,150/acre
2. Loan Amount: $950
3. Interest Rate: 8%
4. Length of Loan: 15 years
5. Basis: 200
6. Capital Gains Tax Factor
   $Z = ((1 - 5/1) \times .4 = .33$

End of Year	A Down Payment	B Annual Payment	C Principal	D Interest	E Loan Balance	F Taxable Income $((A+C) \times Z) + D$	G Marginal Tax Rate	H Tax Payments	I Net After Tax Income $A + B - H$	J Discount Factor 10%	K Net Present Value of Cash Inflows
0	200				950.00	66.00	.50	33.00	167.00	1.0	167.00
1		110.99	34.99	76.00	915.01	81.55	.50	43.78	67.21	.909	61.09
2		110.99	37.79	73.20	877.22	85.67	.50	42.84	68.15	.826	56.29
3		110.99	40.81	70.17	836.41	83.63	.50	41.82	69.17	.751	51.95
4		110.99	44.08	66.91	792.33	81.46	.50	40.73	70.26	.683	47.99
5		110.99	47.60	63.39	744.72	79.10	.50	39.55	71.44	.621	44.36
6		110.99	51.41	59.58	693.32	76.55	.50	38.28	72.71	.564	41.00
7		110.99	55.52	55.47	631.80	73.79	.50	36.90	74.09	.513	38.01
8		110.99	59.97	51.02	577.83	70.81	.50	35.41	75.58	.467	35.26
9		110.99	64.76	46.23	523.07	67.60	.50	33.80	77.19	.424	32.73
10		110.99	69.94	41.05	443.13	64.13	.50	32.07	78.92	.386	30.46
11		110.99	75.54	35.45	367.59	60.38	.50	30.19	80.80	.350	28.28
12		110.99	81.58	29.41	286.01	56.33	.50	28.17	82.82	.319	26.42
13		110.99	88.11	22.88	197.90	51.96	.50	25.98	85.01	.290	24.65
14		110.99	95.16	15.83	102.74	47.23	.50	23.62	87.37	.263	22.98
15		110.99	102.74	8.22	0	42.12	.50	21.06	89.93	.239	21.49
16											
17											
18											
19											
20											
TOTAL											729.96

FARMLAND

can pay per acre. That is, you would be better off to borrow less money because the after-tax interest rate is higher than the return you can achieve in your next best alternative. However, if your discount rate is *higher* than your after-tax interest rate, a larger downpayment *decreases* the price you can pay per acre. That is because you now have a better use for your money than to tie it up in a large downpayment. Remember that a lower downpayment could, however, substantially increase your cash flow problems in meeting debt servicing requirements.

Estimating the Trade-off Between Downpayment and Price— The information in Table 8.7 suggests that if, as a buyer, your discount rate is above your after-tax interest rate, then you can gain from a lower downpayment. Under these conditions it may be to your advantage to negotiate with the seller to accept a lower downpayment in exchange for a slightly higher price. The worksheets in Tables 8.3 and 8.5 can be used to evaluate this type of trade-off from both the buyer's and seller's perspective. The following example illustrates the calculations.

Table 8.7 Net present values per acre of land for alternative levels of annual net after tax returns, inflation rates of land, discount rates, and downpayments on loans.[1]

	If you expect:		Price you can pay per acre to get a return of:		
Annual net after tax return per acre of:	Annual rate of growth in net after tax return of:	Annual rate of inflation in land value of:	6% after income taxes	7% after income taxes	9% after income taxes
			PANEL A (10 Percent Downpayment)		
$ 20	5%	4%	$ 815	$ 696	$ 550
40	5%	4%	1,630	1,393	1,100
60	5%	4%	2,445	2,089	1,649
80	5%	4%	3,261	2,785	2,199
100	5%	4%	4,076	3,481	2,749
			PANEL B (30 Percent Downpayment)		
$ 20	5%	4%	$ 852	$ 701	$ 523
40	5%	4%	1,704	1,402	1,047
60	5%	4%	2,556	2,102	1,571
80	5%	4%	3,408	2,803	2,094
100	5%	4%	4,260	3,504	2,618
			PANEL C (50 Percent Downpayment)		
$ 20	5%	4%	$ 892	$ 705	$ 500
40	5%	4%	1,785	1,411	999
60	5%	4%	2,677	2,116	1,498
80	5%	4%	3,570	2,822	2,045
100	5%	4%	4,462	3,527	2,498

[1]Calculations are based upon the assumption that the buyer is in the 28% tax bracket, has a 30-year planning horizon, and finances the purchase with a loan for 30 years, at 10% interest. It also is assumed that capital gains taxes will be 20% of the full gain.

FARMLAND

Suppose that you are offered land for $2,000 per acre, 29% ($580 per acre) down, balance in 15 equal annual installments at 10% interest. You do not have the necessary funds for the downpayment so you counter-offer with $2,100 per acre, $210 per acre down, and the balance in 15 equal annual installments at 10% interest. The net present values of the cash outflows under these two alternatives are shown in Tables 8.8 and 8.9, respectively.

Under the original terms offered by the seller, the net present value of the cash outflows is $1,643.88 per acre. The counter-offer of the buyer results in a net present value of the cash outflow of $1,626.06, or $17.82 per acre less than for the original terms of the loan. The buyer should recognize, however, that the lower downpayment combined with the higher price increases annual payments from $186.69 per acre to $248.49 per acre. Remember that overcoming a downpayment problem by creating a debt servicing problem may not be in the best interest of the buyer.

Tables 8.10 and 8.11 show the net present value of the cash inflows for the seller under the original offer and the counter-offer of the buyer. The counter-offer of the buyer yields a net present value of cash inflows of $1,749.12—$29.58 per acre higher than the original terms offered by the seller. The seller should decide if the higher value is worth the additional risk associated with a lower downpayment. If desired, the seller could even calculate the returns under the counter-offer using a higher discount rate to reflect the added risk.

In comparing trade-offs between price and interest rate, it was noted that the arrangement has more potential to be mutually advantageous if the seller is in a much higher tax bracket than the buyer. For a trade-off between price and downpayment, however, there is greater potential for a mutual advantage if the seller is in a low tax bracket and the seller has a lower discount rate than the buyer.

Length of Loan

The length of loan can have an important bearing on the cash flow feasibility of land purchase. The length of loan also has a bearing on the net present value of land and hence the price you can pay to get a given rate of return.

Net present values of land under different lengths of loans, with all other factors constant, are shown in Table 8.12. Values reported in Panel A are calculated for 10-year loans, while Panels B and C are for 20- and 30-year loans, respectively. Using a 6% discount rate, the price you can pay declines as the length of loan increases. For example, when using an $80 net after-tax return per acre, the price you can pay to get a 6% return drops from $3,746 per acre to $3,546, or $200 per acre. However, the decline is less in going from a 20-year to a 30-year loan.

Using a 9% discount rate, the net present value increases as the length of loan increases. As with downpayment, the key is the rela-

Table 8.8 Buyer's worksheet for determining the trade-off between price and downpayment.

1. Purchase Price: $2,000/acre 3. Interest Rate: 10%
2. Loan Amount: $1,420 4. Length of Loan: 15 years

End of Year	A Down Payment	B Annual Payments	C Principal	D Interest	E Loan Balance	F Marginal Tax Rate	G Tax Savings D × F	H After-Tax Net Cash Outflow A+B−G	I Discount Factor 12%	J Net Present Value of Outflows H × I
0	580.00				1,420			580.00	1.0	580.00
1		186.69	44.69	142.00	1,375.31	.28	39.76	146.93	.893	131.21
2		186.69	49.16	137.53	1,326.15	.28	38.51	148.18	.797	118.10
3		186.69	54.08	132.62	1,272.07	.28	37.13	149.56	.712	106.49
4		186.69	59.48	127.21	1,212.59	.28	35.62	151.07	.636	96.08
5		186.69	65.43	121.26	1,147.16	.28	33.95	152.74	.567	86.60
6		186.69	71.97	114.72	1,075.19	.28	32.12	154.57	.507	78.37
7		186.69	79.17	107.52	996.02	.28	30.10	156.59	.452	70.78
8		186.69	87.09	99.60	908.93	.28	27.89	158.80	.404	64.16
9		186.69	95.80	90.89	813.13	.28	25.45	161.24	.361	58.21
10		186.69	105.38	81.31	707.75	.28	22.77	163.92	.322	52.78
11		186.69	115.92	70.78	591.83	.28	19.82	166.87	.287	47.89
12		186.69	127.51	59.18	464.32	.28	16.57	170.12	.257	43.72
13		186.69	140.26	46.43	324.06	.28	13.00	173.69	.229	39.78
14		186.69	154.28	32.41	0	.28	9.07	177.62	.205	36.41
15		186.69	169.78	16.91		.28	4.73	181.96	.183	33.30
16										
17										
18										
19										
20										
TOTAL										$1,643.88

Table 8.9 Buyer's worksheet for determining the trade-off

1. Purchase Price: $2,100/acre 3. Interest Rate: 10%
2. Loan Amount: $1,890 4. Length of Loan 15 years

End of Year	A Down Payment	B Annual Payments	C Principal	D Interest	E Loan Balance	F Marginal Tax Rate	G Tax Savings D×F	H After-Tax Net Cash Outflow A+B−G	I Discount Factor 12%	J Net Present Value of Outflows H×I
0	210.00				1,890			210.00	1.0	210.00
1		248.49	59.89	189.00	1,830.11	.28	52.92	195.57	.893	174.64
2		248.49	65.48	183.01	1,764.63	.28	51.24	197.25	.797	157.20
3		248.49	72.03	176.46	1,692.60	.28	49.41	199.08	.717	141.74
4		248.49	79.23	169.26	1,613.37	.28	47.39	201.10	.636	127.90
5		248.49	87.15	161.34	1,526.22	.28	45.18	203.31	.567	115.28
6		248.49	95.87	152.62	1,430.45	.28	42.73	205.76	.507	104.32
7		248.49	105.45	143.05	1,325.00	.28	40.05	208.44	.452	94.21
8		248.49	115.99	132.50	1,209.01	.28	37.10	211.39	.404	85.40
9		248.49	127.59	120.90	1,081.42	.28	33.85	214.64	.361	77.49
10		248.49	140.35	108.14	941.07	.28	30.28	218.21	.322	70.26
11		248.49	154.38	94.11	786.69	.28	26.35	222.14	.287	63.75
12		248.49	169.82	78.67	616.87	.28	22.03	226.46	.257	58.20
13		248.49	186.80	61.69	430.07	.28	17.27	231.22	.229	52.95
14		248.49	205.48	43.00	224.59	.28	12.04	236.45	.205	48.47
15		248.49	224.59	23.90	0	.28	6.69	241.80	.183	44.25
16										
17										
18										
19										
20										
TOTAL										1,676.06

FARMLAND

Table 8.10 Seller's worksheet for determining the trade-off between price and downpayment.

1. Sale Price Less Cost of Selling: $2,000/acre
2. Loan Amount: 1,420
3. Interest Rate: 10%
4. Length of Loan: 15 years
5. Basis: 1,200
6. Capital Gains Tax Factor
$Z = ((1-5)/1) \times .4 = .16$

End of Year	A Down Payment	B Annual Payment	C Principal	D Interest	E Loan Balance	F Taxable Income $((A+C) \times Z) + D$	G Marginal Tax Rate	H Tax Payments	I Net After Tax Income $A+B-H$	J Discount Factor 10%	K Net Present Value of Cash Inflows
0	580				1,420					1.0	554.02
1		186.69	44.69	142.00	1,375.31	149.15	.28	25.98	554.02	.909	131.74
2		186.69	49.16	137.53	1,326.15	145.40	.28	41.76	144.93	.826	120.58
3		186.69	54.08	132.62	1,272.07	141.27	.28	40.71	146.98	.751	110.49
4		186.69	59.48	127.21	1,212.59	136.73	.28	39.56	147.13	.683	101.32
5		186.69	65.43	121.26	1,147.16	131.73	.28	38.28	148.41	.621	93.03
6		186.69	71.97	114.72	1,075.19	126.24	.28	36.88	149.81	.564	85.36
7		186.69	79.17	107.52	996.02	130.19	.28	35.35	151.34	.513	78.61
8		186.69	87.09	99.60	908.93	113.53	.28	33.65	153.04	.467	72.34
9		186.69	95.80	90.89	813.13	106.22	.28	31.79	154.90	.424	66.55
10		186.69	105.38	81.31	707.75	96.17	.28	29.74	156.95	.386	61.45
11		186.69	116.92	70.78	591.83	87.33	.28	27.49	159.20	.350	56.59
12		186.69	127.51	59.18	464.32	79.58	.28	25.01	161.68	.319	52.45
13		186.69	140.26	46.43	324.06	68.87	.28	23.28	164.41	.290	48.55
14		186.69	154.28	32.41	169.78	57.09	.28	19.28	167.41	.263	44.89
15		186.69	169.78	16.91	0	44.07	.28	15.99	170.70	.239	41.67
16								12.34	174.35		
17											
18											
19											
20											
TOTAL											$1,719.58

FARMLAND

Table 8.11 Seller's worksheet for determining the trade-off between price and downpayment.

1. Sale Price Less Cost of Selling: $2,100
2. Loan Amount: $1,890
3. Interest Rate: 10%
4. Length of Loan: 15 years
5. Basis: 1,200
6. Capital Gains Tax Factor
   $Z = ((1-5)/1) \times .4 = .17$

End of Year	A Down Payment	B Annual Payment	C Principal	D Interest	E Loan Balance	F Taxable Income $((A+C) \times Z) + D$	G Marginal Tax Rate	H Tax Payments	I Net After Tax Income $A+B-H$	J Discount Factor 12%	K Net Present Value of Cash Inflows
0	210.00									1.0	200.00
1		248.49	59.89	189.00	1,830.11	199.18	.28	55.77	192.72	.909	175.18
2		248.49	65.48	183.01	1,764.63	194.19	.28	54.37	194.12	.826	160.35
3		248.49	72.03	176.46	1,692.60	188.71	.28	52.84	195.65	.751	146.93
4		248.49	79.23	169.26	1,613.37	182.73	.28	51.16	197.33	.683	134.78
5		248.49	87.15	161.34	1,526.22	176.16	.28	49.32	199.17	.621	123.68
6		248.49	95.87	152.62	1,430.35	168.92	.28	47.30	201.19	.564	113.47
7		248.49	105.45	143.05	1,325.00	160.98	.28	45.07	203.42	.513	104.35
8		248.49	115.99	132.50	1,209.01	152.22	.28	42.62	205.87	.467	96.14
9		248.49	127.59	120.90	1,081.42	142.59	.28	39.93	208.56	.424	88.43
10		248.49	140.35	108.14	941.07	132.00	.28	36.96	211.53	.386	81.65
11		248.49	154.38	94.11	786.69	120.35	.28	33.70	214.79	.350	75.18
12		248.49	169.82	78.67	616.87	107.54	.28	30.11	218.38	.319	69.66
13		248.49	186.80	61.69	430.07	93.45	.28	26.17	222.32	.290	64.47
14		248.49	205.48	43.00	224.59	77.93	.28	21.82	226.67	.263	59.61
15		248.49	224.59	23.90		62.08	.28	17.38	231.11	.239	55.24
16											
17											
18											
19											
20											
TOTAL											$1,749.12

Table 8.12 Net present values per acre of land for alternative levels of annual net after tax returns, inflation rates of land, discount rates, and lengths of loans.[1]

If you expect:			Price you can pay per acre to get a return of:		
Annual net after tax return per acre of:	Annual rate of growth in net after tax return of:	Annual rate of inflation in land value of:	6% after income taxes	7% after income taxes	9% after income taxes
PANEL A (10-Year Loan)					
$ 20	5%	4%	$ 937	$ 710	$ 483
40	5%	4%	1,873	1,419	965
60	5%	4%	2,810	2,129	1,448
80	5%	4%	3,746	2,839	1,931
100	5%	4%	4,683	3,548	2,413
PANEL B (20-Year Loan)					
$ 20	5%	4%	$ 886	$ 704	$ 507
40	5%	4%	1,773	1,409	1,013
60	5%	4%	2,659	2,113	1,521
80	5%	4%	3,546	2,818	2,028
100	5%	4%	4,432	3,522	2,534
PANEL C (30-Year Loan)					
$ 20	5%	4%	$ 852	$ 701	$ 524
40	5%	4%	1,704	1,402	1,047
60	5%	4%	2,556	2,102	1,571
80	5%	4%	3,408	2,803	2,094
100	5%	4%	4,260	3,504	2,618

[1]Calculations are based upon the assumption that the buyer is in the 28% tax bracket, has a 30-year planning horizon, and finances the purchase with a loan for 30 years, at 10% interest. It also is assumed that capital gains taxes will be 20% of the full gain.

tionship between the after-tax interest rate (7.2% in our example) and the discount rate. If the after-tax interest rate is higher than the discount rate you should seek a loan with the shortest maturity that is consistent with your cash flow. If the after-tax interest rate is less than the discount rate, you should seek a loan with the longest maturity possible.

Because the length of loan influences the price you can pay per acre, there is potential for negotiating with the seller for a trade-off between price and length of loan or between downpayment and length of loan. In some cases, proposed trade-offs may be advantageous to both the buyer and the seller. The worksheets presented in Tables 8.3 and 8.5 can be used to assess such trade-offs.

APPENDIX

Computing the Annual Percentage Rate of Interest—The procedure for calculating the annual percentage rate of interest (APR) takes into account service charges and other money costs of borrowing. Deceptive practices in quoting interest rates are reduced if the procedure is followed.

The formula for calculating an APR when payments are equal over the entire length of the loan was given in equation 6.5 and is repeated here:

$$A = V_o/(EPIF_{i,n})$$

where:

A = the amount of each equal amortization payment

V_o = the initial amount borrowed

$(EPIF_{i,n})$ = the present value interest factor for an interest rate (i) and number of time periods (n). Appendix A, Table 4 at the end of this book gives the values for different levels of i and n.

Given the values for A, V_o, and n, you can solve the equation for i, the interest rate per period. The APR is then found by expressing i on an *annual* basis. Two examples will be used to illustrate the calculations involved.

Example 1

Problem: You borrow $20,000 at a 9 percent contractual rate of interest. Repayment is to be in 60 equal monthly installments of $415.18. There is a service charge of 6 percent of the original loan amount. What is the APR on this loan?

Solution: From the information given we know that

V_o = $20,000 − ($20,000 × .06) = $18,800
A = $415.18
n = 60

Placing these values in the equation we find:
$18,000 = $415.18 $(EPIF_{i=?,\ n=60})$
18,800/415.18 = $EPIF_{i=?,\ n=60}$
45.282 = $EPIF_{i=?,\ n=60}$

To solve for i, look in Appendix A, Table 4 at the end of this book under n=60 until you find the value 45.882. At 0.75% you find the value 48.172 while at 1% you find the value 44.953. The value 45.282 falls between these two numbers and in fact is much closer to 44.953. Hence the interest rate is close to 1% per period. Interpolation between the numbers reveals that the value for i is .90% per period. Since this is a rate per month you simply multiply by 12 to get the APR. The APR on this loan is 10.8 percent.

Example 2

Problem: You want to borrow $100,000 through yor local Federal

Land Bank Association. The contractual rate of interest is 10%, and the stock purchase requirement is 5%. There is also a 3% service charge. All money to cover the stock purchase plus the service charge is borrowed. The total loan is for $108,425 and is to be repaid in equal annual installments of $12,735,57. The $5,425 worth of stock pays no dividend, but is refunded at the end of the loan. What is the APR on this loan?

Solution: This problem is complicated by the stock purchase requirement. To account for the stock, add the discounted value of the stock redemption to the initial amount received. Therefore

$V_o = \$100,000 + (1+i)^{-20} \times (\$5,425)$
Using the contractual rate of interest as the discount rate
$V_o = \$100,000 + .149(5,425) = \$100,808.32$
Now apply formula 6.5
$100,808.32 = 12,735.57 \, (EPIF_{i=?, \, n=20})$
$100,808/12,735.57 = (EPIF_{i=?, \, n=20})$
$7.9155 = (EPIF_{i=?, \, n=20})$

Looking in Appendix A, Table 4 at the back of this book under n=20 we find the value 8.514 at 10% interest and the value 7.469 at 12% interest. Interpolation between these values gives the value for i as 11.10%. Since this is already an annual rate, the APR on this loan is approximately 11.1%.

CHAPTER 9
FINANCING THE PURCHASE OF LAND

As shown in Chapter 2 purchases of farmland require substantial sums of money. Financing the purchase can be a major challenge. Buyers have three basic alternatives: (1) purchase with cash using accumulated savings, (2) make a cash downpayment and borrow the remainder either from an institutional lender or from the seller, or (3) obtain cash from other "outside" investors who will then have an equity interest in the land being purchased. In today's economy it is difficult to accumulate sufficient savings to purchase farmland with cash. Likewise, the use of "outside" investors in farmland has not been particularly popular. As shown in Chapter 2, 90% of the purchases of farmland are credit financed. Before proceeding to a discussion on sources of funds, we define some of the more common terms associated with farm real estate financing.

Terms Associated With Farm Real Estate Financing

Term of loan—This refers to the length of the loan, i.e. the length of time until the final payment is due. For farm real estate loans, the term is usually stated in years, such as 20 years.

Balloon payment—When the last payment of a loan is substantially larger than any of the other payments, the last payment is called a balloon payment. For example, the purchase of land under a 5-year land contract may require annual principal payments of $10,000 at the end of the first four years, with a balloon payment of $50,000 in principal at the end of the fifth year. Balloon payment loans are commonly used by life insurance companies and by sellers providing financing under a land contract or mortgage. In practice, funds for the balloon payment often come from refinancing the loan.

Refinancing—A loan is refinanced when the original terms (interest rate, term, and amount borrowed) are altered by mutual agreement between the borrower and lender. Refinancing occurs most often when (a) there is a balloon payment due, (b) the borrower wants to obtain additional loan funds and the lender wants to maintain just one loan, or (c) when the borrower has problems meeting scheduled payments and the lender agrees to refinance with terms that better match the cash flow position of the borrower.

Payout—The payout period is the length of time required to pay off the loan at the scheduled rate of payments. This need not be the same as the term of loan. For example, a life insurance company may offer a 5-year balloon payment loan with a 25 year payout. This means that the periodic payments (annual, semi-annual, quarterly or monthly) are sufficient to completely repay the principal and interest over a 25 year period. However, this payment series stops at the end of 5 years, and the entire amount of principal yet unpaid becomes due as a balloon payment.

Payment plans—Payment plans refer to the manner in which the borrower is to pay principal and interest on the loan. Payment plans can require equal payments of principal and interest, equal payments of principal plus interest on the remaining balance, or payments of principal tied to changing economic conditions. Procedures for calculating payments required under various payment plans were given in Chapter 7.

Prepayment privilege—A prepayment privilege clause in a real estate loan agreement allows the borrower to prepay part or all of the loan at any time without penalty. Most, but not all, lenders now provide a prepayment privilege in real estate loan agreements. Without a prepayment privilege, the borrower generally has no right to make prepayments. If prepayments are made, they are usually treated as the last payment(s) due and cannot be used later as a current payment to avoid default. Therefore, a buyer may want to negotiate for a prepayment provision that allows any prepayments to be applied as current payments to avoid default.

Prepayment penalty—A prepayment penalty clause establishes penalties to be paid by the borrower if there is a payment before the due date. For example, some life insurance companies may require the borrower to pay a 2% penalty if more than 20% of the loan is paid off in one year. If financing is provided by the seller, there may also be some desire not to accept early payments since such payments would affect the income tax liabilities of the seller.

First mortgage—A mortgage is a legal document which gives the lender rights in the real estate pledged as security, if the borrower defaults. Since a borrower could mortgage the same real estate more than once, the *first* mortgage holder has first claim on the real estate serving as collateral should the borrower default.

Second mortgage—A second mortgage is sometimes called a "junior mortgage." The second mortgage holder has claims on the pledged real estate of the borrower only after any claims by the first mortgage holder have been satisfied.

Wrap-around mortgage—This is a special case of a second mortgage. An example can best illustrate how it works. Suppose a farm is being sold for $200,000 with an existing first mortgage of

$70,000 at 7%. The buyer agrees to pay $40,000 down, and agrees to pay the seller 9% interest on the remaining $160,000. The seller gets a 9% return on $90,000 plus an interest bonus—the two percentage point spread between the old mortgage rate of 7% and the 9% rate paid by the buyer. The buyer may gain by getting a 9% loan in a 10% market. However, the use of a wrap-around mortgage can be dangerous to the seller, particularly if the first mortgage has a due-on-sale clause which makes the entire mortgage due if the property is sold.

Variable interest rate—A variable interest rate is one that adjusts over time to changing conditions in financial markets. For example, the Federal Land Banks use a variable rate which is tied to their average cost of acquiring funds in the national financial markets.

Fixed interest rate—An interest rate which is unchanged over the entire term of the loan is referred to as a fixed interest rate. Fixed interest rates are commonly used in seller financing.

Participation loan—A participation loan is one in which two or more lenders share in making a loan to one individual or firm. For example, the Federal Land Banks and the Farmers Home Administration might participate in a loan to a borrower with low equity. Participations among commercial banks are also common. Participations are most commonly used when the loan request exceeds the lending limits of the lender, or the lender feels the risks associated with the loan are too great to bear alone.

Land contract—Sometimes called a "contract for deed" the land contract is a conditional sale of real estate. The buyer makes principal and interest payments in exchange for a promise to receive title to the land once all, or a specified percentage, of the principal payments have been made. The seller retains title to the property until the specified payments have been made. While the seller retains title, the buyer takes possession of the property.

Sources of Farm Real Estate Loan Funds

Given the extensive use of credit financing for the purchase of real estate, it is important to recognize the various sources of such financing and the terms and conditions under which it is made available. There are five major sources of debt financing for the purchase of real estate: (1) Federal Land Banks, (2) Farmers Home Administration, (3) life insurance companies, (4) commercial banks, and (5) individuals and others. The distribution of farm real estate debt by lending source is shown in Figure 9.1. Federal Land Banks and individuals are the major sources of farm real estate loans. A detailed description of the terms and conditions of real estate loans from each source follows.

Federal Land Banks

Federal Land Banks are a part of the Farm Credit System—a nationwide borrower cooperative designed to meet the short, intermediate, and long-term credit needs of farmers. There are 12 district Federal

Figure 9.1 Distribution of Farm Real Estate Debt by Lending Source.

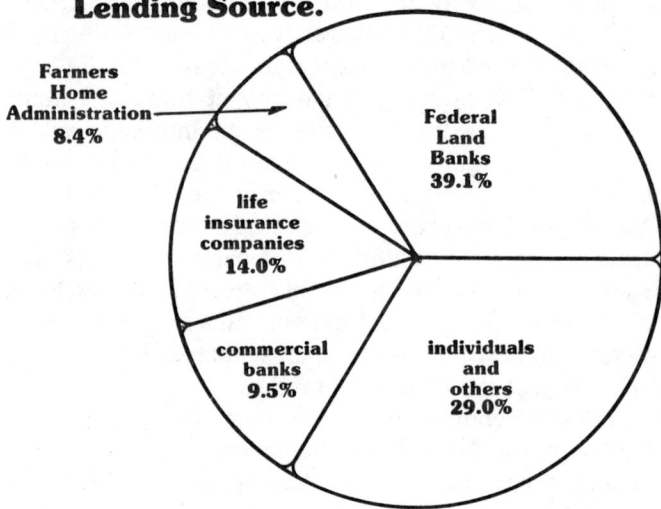

Land Banks (FLB's) and 520 Federal Land Bank Associations (FLBA's). Initially, the government provided some of the capitalization for the System, but by 1947 the money was repaid and the FLB's became completely owned by their member borrowers. Hence, the Federal Land Banks are not a government lending agency as the name implies. However, the Federal Land Banks are supervised by the Farm Credit Administration, an independent agency of the Federal government.

The FLB's make first mortgage farm real estate and rural residence loans of 5 to 40 years. While district FLB's provide the loan funds, borrower contact and servicing of loans are handled by local FLBA's. Each local FLBA serves a number of counties within a state and many have established branch offices. Except in rare cases, the area served by one FLBA does not overlap with that of another. Consequently it is not possible to "shop around" for a FLBA that will best serve your needs.

FLBA's are capitalized in two ways—the earnings which are retained in the association and the sale of stock to the borrowers who must purchase stock or participation certificates in an amount ranging between 5 and 10% of the total loan. Money needed to purchase the stock is normally included in the loan amount requested by the borrower. Since there is typically no dividend paid on the stock, the stock purchase requirement raises the effective interest rate because the borrowers are paying interest on more loan funds than the amount to which they have access. The stock can be refunded as the loan is paid off or at maturity of the loan.

Funds used by the district FLB's in making loans are obtained through the sale of Federal Farm Credit Banks consolidated system-wide bonds and notes. Bonds are sold through a nationwide network of securities dealers. Bonds with a variety of maturities can be issued but most bonds issued for FLB's have maturities ranging from one-to-five years. Notes are issued with a maximum maturity of 270 days, but with most issued in the 5-to-30 day range.

Because the funds used to provide farm mortgage loans are issued with fixed interest rates, and maturities are shorter than most loans on farm real estate, the FLB's need to protect themselves against adverse interest rate changes. For example, suppose a FLB made a 30-year farm loan at a 10% fixed interest rate, and acquired the funds with the sale of a 3-year bond at a fixed rate of 9%. At the end of 3 years the bond is due in full, but only a small fraction of the loan amount will have been repaid by the borrower. To cover their position, the FLB would have to reissue (roll over) the bond. If changes in market conditions have caused rising interest rates, it is possible that the reissued bond would carry a higher interest rate than the interest being paid by the borrower. Problems of this nature have caused FLB's to move to a variable interest rate loan program. Now the cost of funds to borrowers is based upon the average cost of bonds outstanding for their district FLB. For example, assume a district FLB has an average cost of funds of 8.5%. The borrower pays that cost, plus a margin for operating the FLB and FLBA, say 1%. The farmer then pays 9.5% interest. However, the interest rate can be changed as the average cost of funds to the FLB changes. The variable rate charged borrowers protects the FLB against losses of the kind described above.

The variable interest rate program is not without problems. In a period of rapidly escalating costs of issuing bonds, as occurred in early 1980, the *average* cost of bonds outstanding rises, but not nearly as rapidly as the cost of issuing bonds. For example, in early 1980, some bonds were issued by the Farm Credit System for over 17% interest, while the interest rates on Federal Land Bank loans were under 11%! This was made possible by the fact that the *average* cost of all bonds was less than 11%. However, there was concern about possible inequities between "old borrowers" and "new borrowers." Why should old borrowers be forced to pay higher rates caused by new borrowers obtaining funds at below the cost of issuance? To rectify the perceived inequities, several FLB districts moved back to a fixed (and relatively high) interest rate for the first two years of the loan on new loans. In other districts, service charges were raised substantially as a means of better reflecting the current cost of funds to new borrowers.

The Farm Credit Act of 1971 limits FLB loans to an amount not to exceed 85% of the appraised value of the security.[1] This does not imply that a borrower is limited to financing 85% of the purchase price. For one, the borrower may have security to offer other than the land being purchased. Additional security could be used as collateral

---

[1] Amendments to the 1971 Farm Credit Act passed in December 1980 allow the Farm Credit System to make loans up to 97% of the value of real estate in cases where loans are guaranteed by Federal agencies.

as a means of obtaining more than 85% of the purchase price. In addition, the seller, relatives, or the Farmers Home Administration may be willing to provide financing in addition to that offered by the FLB and take a second mortgage on the property.

The appraisal process is important in determining the level of financing a FLB will provide. Appraisals are based on "benchmark" farms and have tended toward market value. However, in some areas where land values have escalated rapidly, FLB appraisals may be below market prices. This varies across FLB districts and over time.

Repayment plans offered by FLB's are flexible so as to accommodate the anticipated cash flows of the borrower. However, there are penalties for delinquency ranging from 1 to 4% depending upon the district. If the delinquency results from adverse yields due to weather or from poor income due to low prices, the FLB's strive to restructure the repayment commitments to allow the borrower to continue in farming. However, the borrower is well advised to discuss such problems with the lender as soon as they become apparent and not wait until delinquency occurs.

In evaluating potential sources of loan funds, it should be recognized that by operating as a borrower cooperative, FLB's are not profit oriented per se. They do attempt to generate profits to enhance the capital base, but the local association is more likely to be judged successful if there is a high growth in loan volume and low loan losses.

Farmers Home Administration

The Farmers Home Administration (FmHA) is a government lending agency operating within the U.S. Department of Agriculture. The agency is designed to provide financial assistance to rural America under a variety of programs. Eligible borrowers are those unable to obtain credit from private or cooperative sources at reasonable rates and terms.

FmHA programs are focused in three major areas: (1) farmer programs, (2) rural housing programs, and (3) business and community programs. The title of the agency is misleading because farmer programs now account for less than one-half the funds provided by FmHA.

FmHA may make either "insured" or "guaranteed" loans. Insured loans are made and serviced through the appropriate county or state FmHA office. In contrast "guaranteed" loans are made and serviced by commercial banks or other lenders. FmHA merely guarantees the loan against default for up to 90% of the loan.

Funds for FmHA insured loans are acquired through a number of "revolving funds." An FmHA revolving fund can be thought of as a checking account in which deposits and withdrawals are made. Principal and interest on insured loans repaid to FmHA are returned to the appropriate revolving fund. Most of the new funds needed to expand loan volume are obtained through the sale of "certificates of beneficial ownership" (CBO's) to the Federal Financing Bank. The

CBO's are fully insured against default. The Federal Financing Bank—an agency of the U.S. Treasury—acquires their funds through the sale of Treasury bills, bonds, and notes.

Because of administrative expenses and the availability of some FmHA loans at subsidized rates, money in the revolving funds could become depleted over time. Congressional appropriations (subsidies) may be needed from time to time to restore the revolving funds. Congress sets limits on FmHA lending authority which in turn limits the amount of appropriations necessary to maintain the revolving funds. Frequently, there is the mistaken impression that all funds for FmHA insured loans come from tax revenues. In recent years less than 10% of cash inflows to the revolving fund for farmer programs have come from congressional appropriations.

There are a variety of farm loan programs available through the FmHA. FmHA loan programs closely associated with the ownership and management of farmland are: (1) farm ownership loans, (2) soil and water loans, and (3) emergency loans.

Farm ownership loans—FmHA farm ownership loans can be used for a variety of purposes, including, but not limited to, purchases of land. To be eligible for a farm ownership loan, the applicant must (1) be a citizen of the United States, (2) have sufficient training or experience to assure success of the proposed operation (3) be or will become an operator of a family farm, and (4) be unable to obtain sufficient loan funds elsewhere to finance actual needs at reasonable rates and terms.

Farm ownership loans may be extended for up to 40 years. In addition, initial payments for beginning farmers or others with few resources may be reduced or deferred. Principal is then amortized over the remaining years of the loan. Maximum lending limits per borrower are $200,000 on an insured loan and $300,000 on a guaranteed loan. However, the FmHA can also participate with other lenders, such as Federal Land Banks in providing real estate loan funds.

Interest rates on insured farm ownership loans are at fixed rates which are based on the cost of funds to the Treasury. On loan participations, however, only the FmHA portion of the loan carries the government rate. The favorable Treasury rate coupled with the ability to purchase land with little equity has resulted in demand for farm ownership loans which frequently exceeds the FmHA's farm ownership lending authority.

FmHA can also make special farm ownership loans for beginning and low-income farm operators. Interest rates on these loans are 5% or less, and the principal repayment schedule is reduced in early years of the loan to allow the farm operator to get established financially.

Soil and water loans—Soil and water loans can be used to finance land and water developments, forestation, drainage of farmland, irrigation, and related land-and-water-use adjustments. To be eligible for such loans, the applicant must be unable to obtain funds elsewhere and must be a farm tenant, owner, partnership, or corporation engaged in farming.

Soil and water loans can be extended for up to 40 years. Loan limits are $200,000 on insured loans and $300,000 on guaranteed loans. Interest rates on insured loans are fixed and are based on the cost of Treasury borrowings.

Emergency loans—Natural disasters, such as flooding or drought, can have a severe economic impact on farm operators in the affected area. When disaster strikes, a need for agricultural credit arises that often cannot be met by commercial lending institutions. To alleviate the financial stress on farm operators affected by natural disasters, the President or Secretary of Agriculture can designate disaster areas eligible for emergency loans from the FmHA.

Under the FmHA's emergency loan programs, loans can normally be extended for the actual amount of losses incurred. Emergency credit can be extended either as insured loans or as loan guarantees. The terms of the loan vary with the type of disaster and the type of losses incurred. Most insured disaster loans are made for 7 years or less and many carry a subsidized, and fixed interest rate. In recent years, greater emphasis has been given to economic emergency loans. Economic emergencies can arise due to either a general tightening of agricultural credit or an unfavorable relationship between production costs and prices received for agricultural commodities. Economic emergency loans can be used to pay installments on existing debts, refinance existing farm debts including farm mortgages, reorganize the operation to make it more viable or pay farm operating expenses. The total amount of an economic emergency loan cannot exceed $400,000 per borrower. Sometimes, proceeds of the loan program have been used to bail-out operators with delinquent farm mortgage loans.

Loan supervision and graduation—One of the eligibility requirements for FmHA loans is that the borrower be unable to obtain adequate loan funds from other lenders on reasonable terms. In addition, the FmHA often makes loans to borrowers with little equity. On farm ownership loans, for example, FmHA can extend up to 100% of the appraised value of the property as long as its lending limits are not exceeded. Consequently, FmHA loans are high-risk loans. To reduce lender risk and to help farmers and others with limited resources become better established, FmHA personnel provide technical assistance and supervision on loans.

The technical assistance provided by FmHA begins at the time of the loan application. The county supervisor helps the applicant develop a "farm and home plan" for a farm ownership loan. This plan includes a projected cash flow statement and other relevant production and financial plans. Copies of lease agreements are checked and, if another lender is participating with FmHA, an application conference would likely be held among the parties involved.

FmHA insured loans are provided under the stipulation that the borrower cannot obtain adequate credit elsewhere at reasonable terms. Once the borrower has gained financial strength to the point where other lenders are willing to provide adequate funds, then the borrower must "graduate" to another lender. That is, the borrower is no longer eligible for FmHA assistance.

Life Insurance Companies
Life insurance companies are organized as corporations under state laws and are designed primarily to provide life insurance and pensions to their clients. The funds available to life insurance companies come from premiums paid by policyholders, accumulations in pension plans, and returns on investments. Life insurance companies accumulate large sums of money. These monies are invested and produce income. Some of the funds must be disbursed to pay annuities and to pay beneficiaries of policyholders.

The investment decisions of life insurance companies are guided by three major objectives: (1) adequate liquidity, (2) yield, and (3) security of investment. Adequate liquidity is needed to pay policy benefits, provide funds for policy loans, and meet investment commitments. The yield on investment is also important since returns on investments are a major source of income. Farm mortgages can compete with other investments only if they generate a return comparable to other investments with the same degree of risk. Security of principal is essential to the continued growth and survival of insurance companies. This prevents insurance companies from making highly speculative investments.

Life insurance companies finance farmland purchases primarily through first mortgages. Second mortgages do not meet the security requirements of life insurance companies. Life insurance companies also provide a significant amount of loan funds to agri-business firms such as grain elevators and machinery manufacturers. The following discussion, however, focuses only on real estate loans for acquisition of farm and ranch land.

Real estate farm loans made by life insurance companies are typically for large amounts. The emphasis on large loans has increased in recent years. Real estate farm loans made by life insurance companies are, on the average, three to four times larger than those made by Federal Land Banks. The desire of life insurance companies to finance larger-scale operations is evident from company lending policies. Most life insurance companies now impose lower limits on the size of real estate farm loans. For example, one company does not normally consider real estate farm loans below $250,000.

Some state laws restrict the size of farm real estate loans made by life insurance companies to 75% of the appraised value of the property. In practice, most loans are limited to less than 75% of the appraised value. In some cases, insurance companies set a maximum loan amount per acre in a given area. For example, in one area of the

Midwest where farmland was selling at between $2,500 and $3,000 per acre, one insurance company had a self-imposed loan limit of $1,500 per acre. Such restrictions are based upon the expected income potential of the property and the security requirements of the life insurance company.

Life insurance company farm real estate loans have typically carried fixed interest rates. This is in contrast to Federal Land Bank loans which have variable interest rates. A liquidity crunch which hit life insurance companies in the early 1980's appears to be reshaping the loan terms offered by life insurance companies. Some companies are beginning to write loans with variable interest rates. Other companies have opted for loans with partial amortization, and with a balloon payment at maturity. The maturity on these loans has been decreased from 15 years to 3–5 years by some companies. By shortening the maturity, the company has the opportunity to adjust the fixed interest rate if the balloon portion of the loan is refinanced. Farm mortgage contracts of this nature usually include language that allows for renewal if the payment record has been satisfactory.

Some insurance companies are investigating the use of shared appreciation mortgages (SAMs). A SAM works in this manner. In exchange for a *fixed* interest rate at below current market rates, the borrower agrees to give a percentage of the capital gains on the land to the lender. For example, if market interest rates are 14%, the lender may agree to a 30-year loan with a fixed interest rate of 11% in exchange for 30% of the capital gains on the asset being financed. If not sold at the end of 10 years, the asset would be appraised and 30% of the gains would be paid to the lender in cash or, more likely, tacked on to the loan balance.

Insurance companies may include some form of prepayment penalty in their loan contracts. For example, borrowers may be required to pay a 2% penalty if they repay more than 20% of the loan in one year. The prepayment penalty is used because if the loan is paid off rapidly, the interest income could be insufficient to cover the cost of initiating the loan.

Real estate loans to prospective land owners are originated for life insurance companies through three basic contacts: (1) field representatives, (2) bank correspondents, and (3) other farm mortgage correspondents. Field representatives are employees of the company. Most field representatives cover a large geographic area and many deal in agribusiness loans as well as farm real estate loans. Field representatives have considerable flexibility in negotiating loan terms, but the final loan agreement must usually be approved by the home office.

Some commercial banks provide contacts for life insurance companies on farm real estate loans. The bank may be unwilling or unable to supply the loan funds needed by one of its customers. In this instance, the bank merely serves as an agent for which it may receive a finder's fee. The life insurance company actually makes and approves the loan.

Life insurance companies also enter into agreements with mortgage investment firms, real estate brokers, accountants, and other persons well acquainted with farm mortgage lending. For a fee, the correspondent originates and services loans for life insurance companies. These correspondents operate much like a field representative except that they may have other business interests and may work on a correspondent basis for several life insurance companies.

Commercial Banks
Commercial banks make both real estate and non-real estate loans to farmers and ranchers. Banks have historically been the largest institutional source of non-real estate farm loan funds. Commercial banks are also an important source of farm real estate loan funds and they often provide referral services that help farm operators acquire real estate loan funds from other lenders.

Commercial banks vary greatly in size and structure of operations. For example, there are a large number of small and medium-size rural banks that provide funds and services to farm operators and local agribusiness firms. In some states, large banks serve farm operators through many branch offices located in rural areas. In other states, large metropolitan banks serve agriculture primarily through correspondent arrangements with smaller rural banks and by lending to agribusiness firms. The commercial banking system constitutes a network of independent institutions serving financial and related service needs of agriculture.

Like life insurance companies, commercial banks have many investment alternatives. Two major investment categories for banks are loans (installment, commercial, farm real estate, farm nonreal estate) and investment securities such as Federal, state and municipal securities, or bonds and notes of agencies and corporations. The investment policy of each bank is determined by liquidity, risk, returns, and needs of the local community.

In many rural areas, the growth in farm loan demand has exceeded the growth in deposits at commercial banks. As a result, loan-to-deposit ratios have been rising. Since loans tend to be more risky and less liquid than government securities, banks have become more concerned about the liquidity and risk position of the bank. Turning away loan requests—including farm real estate loans—is one way to deal with the problem. However, many banks are seeking other more positive approaches to dealing with the problem.

There are several methods available for dealing with a situation where growth in loan demand exceeds deposit growth. For one, banks may enter into participation agreements with correspondent banks to provide participation loans. Banks may also enter into participation agreements with Production Credit Associations. In some cases banks have established subsidiary agricultural credit corporations. Agricultural credit corporations can, in turn, obtain loanable funds through the sale of commercial paper in the financial markets or through discounting with their district Federal Intermediate Credit Bank. Banks can also serve some customers by making FmHA guaranteed loans and then selling off the guaranteed portion to obtain loanable funds.

In the past, rural commercial banks were somewhat isolated from the national financial markets. Deposits were the primary source of funds and savers had limited alternatives for obtaining higher yields on their savings. However, moves to improve returns on savings (money market certificates) and the search for nondeposit sources of funds have removed this isolation from national financial markets. As a result, interest rates paid by borrowers at rural commercial banks are more likely now than in the past to be tied to conditions in national financial markets.

Another problem faced by some rural banks is that legal lending limits per customer are too low to serve the credit demands of larger customers. Many banks have a legal lending limit of under $200,000 per customer. This limit has important implications for the financing of real estate purchases. If a loan request exceeds the legal lending limit per customer, banks can request that correspondent banks accept the "overline" portion of the loan. But if loan limits are restrictive, the bank normally prefers to finance the operating loan and have the farm real estate financed elsewhere.

Banks that do not carry many real estate loans often will assist a potential land owner in getting a mortgage. Various working arrangements with life insurance companies, mortgage companies, the FmHA, and other sources are used to help customers get farm real estate financing.

Real estate loans commit bank funds for a number of years, in contrast to the deposit commitments which may be of short-term nature. Therefore, it is understandable why banks are not a major source of farm real estate loans, as they are for non-real estate farm loans. When a bank does extend a farm real estate loan, the term is usually 15 years or less. A payment program may be established on a 20 to 25 year basis with a balloon payment at loan maturity. The closing costs for such loans are often less than those of other real estate mortgage lenders. Since a bank normally has a close and ongoing working relationship with its clients, individually tailored arrangements are often made.

Individuals

Sellers of farm real estate provide a substantial portion of the financing for the purchase of real estate under a land contract, first mortgage, or second mortgage loan. The popularity of these methods of financing varies substantially among different regions of the country. Nationwide, land contracts account for about one-fourth of all farm real estate financing. First and second mortgages to the seller account for about 10% of the total financing provided. However, the land contract form of financing is far more popular in the upper Midwest, Corn Belt and Western regions of the United States. In contrast, the land contract is less popular in the Eastern States and Delta regions. There is also considerable regional variation in the use of second mortgages to the seller. Second mortgages to the seller are more popular in the Delta States and Pacific region than in other regions of the country.

Seller financing is popular because it may allow the buyer to acquire real estate with a lower downpayment than is possible from institutional lenders. As shown in Chapter 17, there may be tax advantages to an installment sale. And as pointed out in Chapter 8, there can be important trade-offs between price and interest rates when financing is provided by the seller.

Sellers are far more likely to accept second mortgages than are other lending groups. This can be particularly important in a sale involving an existing mortgage at a favorable interest rate. In this circumstance, the seller is much more likely to accept a wrap-around mortgage than are institutional lenders.

From the buyer's perspective, seller financing may reduce the ability to acquire additional financing at a later time. For example, suppose that you have purchased a farm under a 10 year land contract. Suppose further that you have made the first 6 years' payments on time. The reduction in principal plus capital gains has generated substantial equity in the land. Now another farm comes up for sale, but the seller wants all proceeds immediately. You do not have adequate cash to purchase directly, so you approach your local Federal Land Bank Association about financing. Given the relatively small amount of cash you have to put into a downpayment, you would like to use some of the equity in the farm you bought 6 years ago under land contract. However, Federal Land Banks will generally not accept a second mortgage. The Land Bank personnel might suggest that you attempt to get the seller of the first farm to agree to early payment on the land contract, offer the first farm plus the one being purchased as security, and obtain the funds needed to pay off the land contract and buy the second farm on a first mortgage from the Federal Land Bank.

The catch comes in that the seller may not agree to accept early payment because this could create substantial capital gains tax liability in one year. Also, the seller may be upset because the value of the land could have doubled from the time of your purchase. The seller may want to vent his or her anger by foiling your attempts to buy another farm. In contrast, if the financing on the first farm had involved a mortgage with a prepayment privilege, there would be no problems in completing the deal. A growth oriented investor in farmland will want to assure the availability of equity capital by negotiating for a prepayment privilege in a land contract.

Sources of "Outside" Equity Capital

While most of the equity capital required for the purchase of farmland comes from the savings and retained earnings of a single buyer, there are alternatives for attracting "outside" equity capital for the purchase of land, i.e., capital from investors other than relatives or friends. While there are a variety of methods of attracting outside equity capital, the two most common in agriculture are: (1) issuance of stock by a corporation, or (2) issuance of "shares of partnership interest" in a limited partnership.

Corporations—There have been several corporations formed with the goal of attracting equity capital for the purchase of farmland. Investors buy shares of stock in the corporation and the corporation in turn invests in farmland. The primary objective usually is long-term capital gains. Equity capital is supplemented with borrowed funds to acquire more land.

In practice, there appears to be very little equity capital raised in this manner. A problem has been the lack of a good market for such stocks. Investment in land is usually considered a long-term investment, so frequent purchases and sales of farmland are not consistent with the objectives of long-term capital gains. Lacking a sufficient demand for stock in such corporations, stockholders may have problems disposing of their stock at reasonable prices.

Limited Partnerships—The use of limited partnerships to attract equity capital into agriculture has been quite common. The limited partnership is formed by a general partner who issues shares of "partnership interest" to outside investors who are limited partners. Huge sums of outside equity capital were attracted into some types of agriculture, notably cattle feeding, via the formation of limited partnerships in the late 1960's and early 1970's. The attractive feature of limited partnership investments was the substantial tax shelter advantages. Tax legislation in 1976 and 1978 substantially reduced the tax shelter advantages.

Much of the equity capital entering agriculture through sales of partnership interest were in cattle feeding and egg production enterprises. Hence much of the capital was not used to acquire land. However, there was also a significant amount of equity capital invested in citrus production and forestry activities as well as relatively minor amounts for the purchase of cash grain farms. Because changes in tax laws have reduced the tax shelter advantages of limited partnership investment in agriculture, acquisition of equity capital for agriculture through the sale of shares of partnership interests will likely be rather modest during the early 1980's.

CHAPTER 10
GETTING BUYERS AND SELLERS TOGETHER

Sale of any commodity, including farm real estate, requires that a potential seller and potential buyer be brought together. This is typically accomplished through direct contact or through an intermediary like a central market or a broker. For items that are bought and sold frequently, the market place is highly sophisticated and matches buyers and sellers with routine ease. Examples include the supermarket for food and the stock exchange for ownership in American companies. Farm land sales present more challenges. Each tract is immobile and unique with respect to geographic location. Such characteristics preclude a centralized market where potential buyers and sellers can meet.

A potential buyer may be able to define the characteristics of farm land desired, but there is no assurance that such a tract is available for sale. Potential sellers, on the other hand, may struggle with the challenge of locating potential buyers wanting the type of property they have for sale. The most highly developed system for bringing buyers and sellers together is the real estate broker, typically operating as an agent for the seller. The real estate broker is a real estate professional and makes a living by matching buyers and sellers and providing the negotiating link between the two, to find terms and conditions for transfer that are acceptable to both parties.

Before describing the real estate brokerage business, two other arrangements are examined that potential buyers and sellers may use. One way is negotiating a private agreement without the aid of a real estate agent, and the other is sale via a public auction.

Private Agreements
Private agreements are when buyers and sellers do not engage the services of a real estate agent, but rather get together on their own. In such cases the buyer and seller together reach agreement on all terms of sale. To finalize the transaction, an attorney may be engaged to prepare the necessary legal documents. The attorney often provides counsel to be sure all relevant matters are considered. Private agreements are typical for real estate transactions between family members. Here a potential seller and potential buyer are already matched and desire to transfer ownership of the property. Their concerns may

center around the price, and the mechanics and timing of the transfer. Tax matters are often of considerable significance. Legal advice and assistance are often all that is needed to close such transactions.

Private agreements between unrelated parties pose more problems. First is the task of determining a fair market value. The seller may not have sufficient knowledge and expertise to properly value the land. The seller may have a value in mind that is well above or well below the price for which similar land has been selling. In this case a professional appraiser can be employed to help establish a fair price.

A second problem is finding a potential buyer without going through a real estate broker. Exposing a property to serious potential buyers is normally necessary. That typically involves a fairly extensive advertising program. It is possible to advertise in local newspapers, but that does not alert anyone outside the circulation area. As shown in Chapter 2, many purchases of land involve non-local buyers. Even local advertising requires the ability to write an advertisement that will attract interested parties. However, potential buyers of real estate often already live in close proximity to the farm. In such cases, a seller may be able to work out a private agreement to sell the farm.

Negotiating directly between seller and potential buyer can present a further challenge. The buyer may feel uneasy telling the seller directly the problems with the property, but in reality would be willing to buy at a lower price. The two parties may not realize that with discussion and negotiation they could likely arrive at a mutually agreeable price and terms. Legal expertise and counsel may be the turning point in helping two parties reach a private agreement.

Laymen often don't know what steps are required to get from the discussion stage to a written agreement or contract and on to the final transfer of property. In summary, private agreements may be possible, especially with family transactions. However, unless potential sellers have above average knowledge of real estate transactions, and access to some potential buyers, they may want to seriously consider selling through public auction or by employing a real estate agent.

Auctions

In some areas, land is commonly sold at public auction. In broad terms this is accomplished with the aid of an auctioneer who allows interested parties to place their bids orally. The bidding continues for some period of time (several hours perhaps) during which anyone can increase the previously high bid. The property goes to the highest bidder. These public auctions are typically held on the farm itself or at the local county courthouse. Auction sales are either voluntary or involuntary, with the latter situation typically resulting from mortgage foreclosure or enforcement of a lien on the property.

On occasion the "sealed bid" type of auction is used. Here potential buyers are asked to submit their bid in writing. Often the bid is just an "entry ticket" and bids can be raised during the opening.

Normally the highest bidder gets the property. This is a common method of letting non-real estate contracts, on construction projects for example, or for service or sale contracts to the government. But it is not widely used as a method of selling farm real estate. The emphasis in the following discussion is on the public auction with oral bidding.

Involuntary auctions—Sales to foreclose mortgages or enforce other liens generally must be at public auction. Such involuntary sales must comply with requirements for due process of law. The lien may be a judgment, a foreclosure on a delinquent mortgage, a lien resulting from non-payment of taxes, a mechanics lien or other lien. In some instances, the sheriff conducts the sale. In other cases, the sale is conducted by a referee appointed by and responsible to the court.

Many legal requirements and stipulations underlie forced sales of real estate. Potential buyers should understand all stipulations and terms associated with such purchases. At the same time there may be some good buys associated with these stress sales because of the additional legal entanglement and adversary relationships that exist. The sales must be advertised through newspapers and, in some instances, through a posting of the notice in prescribed prominent public places. Terms of the sale are detailed in the advertising. Such auctions are normally held at the county courthouse.

Voluntary auctions—Sellers may conclude that the chances are better for a higher price via the public auction route, rather than listing the property with a real estate broker and dealing with potential buyers one at a time. There are reasons for this. Emotion and the competitive spirit are major considerations. Often times, two or three neighbors may want a property and are willing to bid aggressively for it at a public auction. There is not time to ponder the purchase once the auction has begun—it is now or never. Individuals may have wanted and dreamed about owning a specific tract of land for years. So obviously emotion can run high at a land auction.

Another reason for many auction sales stems from the fact that fiduciaries, such as executors and trustees, may be obligated to sell for the best price. To avoid criticism, the public auction is often used. With larger tracts, the auction allows alternative ways of dividing up and selling the property. Sellers can then choose the method which generated the highest total price. For example, a 640 acre tract in the Corn Belt might be sold either as one unit, or in four tracts—320 acres with buildings and three bare land tracts of 160, 80 and 80 acres each. Neighboring farmers might bid aggressively for the smaller tracts for expansion, but would not bid at all on the 640 acre tract. The farm is sold both ways, with the seller being able to choose the alternative that generates the highest total dollars.

The success of an auction may depend on skill of the auctioneer or other promoter to expose the property widely through an advertising program, and to create serious buying interest. A good auctioneer will likely canvass, rather thoroughly, the immediate surrounding area farmers, to determine potential interest—and to further stimulate interest.

Beyond that, what type advertising program might be used? Several media are possible. Newspapers and magazines are most popular. Newspaper alternatives range from local and regional daily or weekly papers to a national daily like the *Wall Street Journal*. Magazines may be trade-type magazines, such as those going, for example, to dairy or hog farmers, or perhaps national farm magazines like *Farm Journal* or *Successful Farming*. Either display or classified ads may be used. It is essential that the potential target buying audience be defined.

Direct mail, including pamphlets and circular letters, may be valuable for selling farm land. The secret here is development of material that attracts attention and actually arouses sufficient interest that readers will investigate further. An obligation, of course, is distribution of the materials to actual potential buyers. To accomplish this, the purchase and use of mailing lists may be needed.

Billboards, signs, and posters have some use in farm sales—especially the posters that can be widely distributed in a fairly localized geographic area. Once the potential types of buyers are defined it is necessary to design an advertising program that fully exposes a property to such people. In some communities, radio and T.V. advertising are useful in reaching a defined audience.

These advertising alternatives are not unique for use with auction sales, but also are used by real estate brokers to promote listings of real estate for sale. Advertising is costly and the benefits are difficult to measure. It is well to discuss such plans with an auctioneer before committing to the auction—or for that matter, with a broker who proposes to list your property and sell it privately.

For communities familiar with land auctions, this can be a good technique. Creditors and beneficiaries often use protective bidding to establish the minimum they want or need. If their minimum bids are not raised, they of course would get the property. A seller may establish a price under which the land would not be sold, and the auctioneer can announce that price in the beginning. Potential buyers should remain sensitive at an auction to be sure there is no "boosting." This is where fictitious bids are placed by agents of the owner to induce higher bids by others. Such a practice hurts the reputation of auction sales and discourages honest bidders. Ethical auctioneers do not knowingly permit such activity.

There is no guarantee on when an auction will generate the highest sale price, but there is a general feeling that auctions are probably most effective in a strong market and in a community accustomed to having auctions.

Real Estate Brokers
All 50 states and the District of Columbia require that real estate

brokers and salespersons be licensed to engage in the business of handling other peoples' real estate. *Real estate brokerage* is the term describing the business of bringing together persons desiring to make a transaction in real estate. Since much farm land is sold through real estate brokers, this section provides details on their function and operation.

A real estate broker is an individual licensed to buy, sell, exchange or lease real property for others for a fee. Whomever employs the broker is known as *the principal,* and the broker acts as *agent* for the principal, and is compensated with a commission that is contingent upon the broker's successful completion of performing the service for which the individual is employed.

The most common situation is where a seller employs a broker as agent to sell a tract of real estate. The broker-principal relationship is governed by the laws of agency—a major area of law. Real estate brokers often carry out their responsibilities through a "salesperson"— an individual licensed by the state to buy, sell, exchange or lease real property, under the direction and supervision of a broker. Salespeople are responsible only to the broker under whom they are licensed and are allowed to perform only those responsibilities assigned by the broker. This means salespersons cannot act as agents for another person and cannot list or advertise property under their own name.

Salespeople are related to brokers in one of two ways: 1) as employees, or 2) as independent contractors. As an independent contractor, the individual operates somewhat independently in carrying out functions, even though the broker controls to some extent what is done. Buyers do not need to worry about how the salesperson is associated with the broker, but they do need to realize that this individual is acting for the broker and cannot act as an agent for the buyer.

Licensing laws are not uniform throughout the United States. Generally, however, state law requires that individuals pass an examination to demonstrate real estate knowledge and competency. Organized efforts were begun in the early 1970's to upgrade the quality of real estate licensing examinations. There are, typically, additional requirements related to age, education, and apprenticeship—the latter is especially true for licensing as a broker. The basic purpose of license laws is to provide assurance to the general public that brokers and salespersons are properly qualified. Clearly, efforts have been made by the real estate profession and by the state legislatures to enforce high standards for the real estate profession.

As either a seller or buyer of farm land, it is important to understand clearly the agency relationship between the principal and the agent (broker); the principal may be an owner or purchaser, lender or borrower, lessor or lessee—but is most often an owner contracting with the broker to sell property. A contractual relationship is established that defines the responsibility of the agent in dealing with third parties.

The contract is an agreement for employment with the principal and should normally be in writing.

The law prescribes that the agent is to act in a fiduciary capacity and, as such, is required to render faithful service similar in all respects to that imposed upon a trustee. Any and all interests that the agent may have in the property must be disclosed. It is an agency relationship and the broker, acting as agent, may not act for both parties. For example, if the principal is a seller, the agent can not secretly accept any compensation from the buyer. Agents have a duty to act in the interests of their employer, and to exercise care, skill, and integrity in carrying out instructions. The broker-agent in a typical sale is compensated by a commission that is earned by producing a purchaser who is ready, willing and able to purchase on the terms offered by the seller, or on terms the seller is willing to accept.

A broker who is the "procuring cause" of a sale is judged to have earned a commission. Buyers should realize that the broker-agent is representing the seller as a condition of earning the commission. Even though brokers do not represent both parties, they should act in a manner to gain the confidence of the buyer as well as the seller. While that requires high ethics and a somewhat unique relationship, it is clearly different than the agency relationship that exists with the seller.

Functions performed by broker—Functions served by a broker include responsibilities in the following areas: 1) listing; 2) advertising and promoting the property as part of the prospecting activity; 3) showing the property; 4) negotiating terms and conditions of a sale; and 5) closing the transaction.

When the broker is employed by a seller, the agreement between seller and broker is called a *listing agreement*. On occasion a prospective buyer employs a real estate broker to locate land and serve as agent in the purchase of land. In these instances a *finder's agreement* is used to define terms and conditions of the contractual relationship. We concentrate here on the various types of listing agreements. Regardless of which type listing agreement is used, terms of the offer to sell are described.

The agreement is the contract that defines the role of broker as agent for the seller. If the broker finds a buyer, willing and able to pay the price as listed in the listing agreement, the seller is obligated to pay the commission (which is also defined in the agreement) even if the seller should decide against selling under the terms established earlier.

Open listing—With an open listing a seller retains the right to employ any number of brokers as agents, as well as the right to sell the property personally and not be obligated for commission. Only if one of the broker-agents is in some way a procuring cause in the transaction can a commission be earned. Brokers do not like this arrangement because so much is outside their control. Each of the agents may be hesitant to spend much money and time on advertising of the property since potential buyers may go to other brokers.

Exclusive agency listing—This type of listing authorizes only one broker to sell a property but allows the seller to retain the right to sell the property personally without any obligation to pay a commission. Sometimes a seller of farm land knows one or more people in the community who might be interested in buying. It may be important to such a seller to be able to negotiate freely with such prospects.

Exclusive right to sell—Under this arrangement a broker gets a commission regardless of who sells the property. The full responsibility for getting the property sold is placed in the hands of a specific broker. The advantage of this type listing is that the broker can justify considerable expenditure for advertising and promotion, knowing that costs will be recovered at the time of sale. Without an exclusive right to sell, a broker may still try hard to sell a property but may stop short of going all the way on advertising and promotion.

Multiple listing—This type listing agreement is common in urban areas where a large number of brokers in the community are members of a multiple listing organization. In effect, it is like an exclusive right to sell, but with the additional authority granted by the listing broker to distribute the listing to other brokers making up the multiple listing organization. If a broker other than the listing broker makes the sale, the commission is split in a predetermined manner. Multiple listing is much less likely to be encountered in farm land sales, than with sales of urban residences.

Net Listing—Under this arrangement the seller specifies the amount desired for a property, but allows the broker to offer it for a higher price. This type of listing is not recommended and is even outlawed in some states. It opens the door for unethical deals, since the broker, with superior knowledge of land prices, would be able to obtain a listing at a lower price and keep all money above the net listing price.

The length of time the listing agreement will remain in effect is an important consideration. A seller wants to allow sufficient time for a broker to organize and carry out an effective sales campaign; but if too much time is allowed the broker may become complacent and the seller will be unable to switch to a different broker until the listing agreement expires. Theoretically, the agreement can be terminated if the broker fails to spend time on trying to sell the property, but it is very difficult for a potential seller to prove abandonment. Six months is a typical period for a listing contract. For some properties a shorter contract might be sufficient, e.g. where not much advertising or promotion outside the local community is required. It is possible to renew a listing contract following expiration if both the seller and broker agree.

Establishing the price at which to offer property for sale is a major consideration. Sellers often have limited knowledge of current land prices. Good brokers, active in selling a given type of property can be quite helpful with advice and counsel on a proper listing price.

The seller, however, must assume final responsibility for setting that price. Good brokers normally do not want to list a property if the price is substantially exaggerated, knowing that no one will buy, if the price is clearly out of line. On occasion, brokers have been known to take a listing, even though unrealistically high, in hopes that without any action on the property for some period of time it will be possible to persuade the seller that a lower listing price should be used. Recent comparable sales provide useful guidance in establishing the listing price. For sizeable farm land tracts, it may well be worth hiring a professional rural appraiser for an opinion on value. Once the price is established and the listing agreement signed, the broker should begin a program of prospecting for potential buyers.

An advertising and promotion program should be designed especially for the sale property. Brokers have skill in identifying potential buyers and design promotion efforts to reach those individuals. Special circulars are sometimes prepared on farms, providing ample pictures and details. Special mailing lists are then used to reach the target audience. Trade magazines may be useful to reach a homogeneous group. For example, a large intensive hog operation might be advertized in the *National Hog Farmer*. If absentee, executive type individuals are the target, ads in the *Wall Street Journal* may be appropriate. If local farmers are considered as likely prospects, substantial personal canvassing might be the best approach. How good a job a broker does is in part related to the ability to identify correctly the potential buyers and to design and carry out an effective advertising program.

Showing a property involves more than simply driving prospects to the farm. Nearly anyone could do that. Effective farm land brokers should have substantial knowledge about soils and agricultural operations and should have full details on the farm from the ASC and SCS offices. Soil and property maps may be provided to potential buyers. They should be able to project income and expenses accurately for what would be a typical operation, as well as having details verified on the past operation. They should know the sale property very well— all its characteristics, strengths, and weaknesses. A broker is clearly an agent for the seller. But to make sales, brokers must earn credibility with potential buyers and be able to help buyers assess the property in view of the buyer's needs.

Negotiating the final sales price requires considerable skill and is often best handled by a third party—in this case the broker-agent. A potential buyer may be more candid with the broker about concerns and reservations than the individual would in direct discussion with the seller. The broker is able to help the potential buyer formulate an offer to buy. If the offer is for less than the full asking price (which is typical), the agent can interact with the seller about how the buyer is viewing the property. This may prompt the seller to accept the offer or to counter offer at a price in between the offer price and the original listing price.

The agent continues to negotiate the price between the potential buyer and seller until either an agreement or an impasse is reached. The broker's skill and know-how in following through on matters affecting financing, insurance, taxation, accounting, and property management may be valuable in bringing the buyer and seller together on terms.

Once the offer to purchase is signed by both parties, the broker often takes leadership in seeing that the various steps required for closing the transaction are carried out. This may involve working with financial institutions, attorneys representing both the buyer and seller, a title or abstract company, insurance agents, and others. A knowledgeable broker plays an important role in moving the transaction along toward the final closing date, final payout to the seller and transfer of title and possession to the buyer.

Choosing a real estate agent—How can you evaluate and choose a real estate agent? The reputation of a broker in a community should be carefully evaluated. Talk to buyers and sellers who have dealt with the agent. Visit with lenders and others knowledgeable about real estate transactions. One positive measure of real estate professionalism would be membership in the National Association of Realtors® (NAR). The R indicates that the term "Realtor" is registered and may be used only by members of the National Association of Realtors. Membership requires that the individual also belong to a local or regional board of realtors that is affiliated with the NAR.

The NAR started in 1908 and today has grown into a trade organization of major importance. Its most significant single contribution to the furtherance of the real estate business was the drafting, distribution, and enforcement of the Realtor® code of ethics. Being a Realtor® means the individual is a member of the NAR and thus must abide by the code. Those Realtors® who are members of the Farm and Land Institute of the NAR specialize in the sale of all types of agricultural and urban land. This institute grants the designation "Accredited Farm and Land Member" (AFLM) to those members who have met rigid educational standards, and have demonstrated proficiency and expertise with land through examination and training.

Another matter to consider when choosing a real estate agent is the continuity of advertising. Good agents are well-known to the buying community and keep that way through their promotion program. The individual chosen should know land well and examine it closely. The agent should be from the local geographical area and be knowledgeable about the area. Also, a big plus exists for those agents who cooperate with a network of other professional brokers in the state, and, possibly, nationally. Sellers have a lot at stake in making a wise choice on which broker to utilize.

A dilemma may arise on whether to work with only one broker

or with several. One possibility is to give a "cooperative exclusive" listing to two or more brokers working cooperatively. Under such arrangements, each shares in the commission regardless of who sells the property. Combining the talents of several agents may be beneficial. However, avoid having several agents involved, when none of them wants to spend money or time to promote and advertise the property. Each may want to concentrate efforts where there is assurance of an income when the property sells. Just having several brokers "on the alert" for potential buyers is insufficient. It normally takes more than that to get property sold.

Potential buyers and sellers should consider their responsibilities to the broker-agent, and adopt a pattern of behavior that will allow the broker to be of most help. The seller has a contractual relationship, with the terms defined in the listing agreement. For greater assurance of action by the agent, standards of performance may be included in the listing contract.

The extent and type of advertising, use of a mailing list, the use and location of signs, the extent of involvement with multiple listing services and other brokers, and some detail about regular clientele that might be potential buyers, all might be included in the listing agreement. By having some predetermined plan of action, a seller may feel more comfortable allowing time for the agent to carry out the selling program. Sellers need to be realistic on pricing and need to tell the agent what modifications of terms might be acceptable. For example, would a contract sale be possible, or is 100% of the sale price needed at closing? If a contract is to be carried for part of the proceeds, the seller should specify what terms related to interest rates, down payment, and length of contract might be acceptable.

Buyers likewise need to share considerable information with the agent. First, it should be clear what type of property is desired. How much cash does the buyer have available for down payment and what ability is there to get a loan? Brokers attempt to "qualify" prospective buyers by establishing that they can, indeed, come up with the money that would be needed to purchase the type of real estate being considered.

Agents should know what items and considerations are most important to buyers. This makes it possible for the agent to help negotiate agreements between sellers and buyers. A buyer is responsible to an agent on only those properties shown by the agent. For example, an individual may learn about a neighbor's farm being for sale through the broker. It would not be proper for the buyer and seller to get together privately and agree to hold off on the transaction until the listing agreement expires, and then to close the transaction and cut the broker out altogether. The broker is a "procuring cause" in the above example and would be entitled to a commission from the seller.

Summary
Real estate transactions can occur only when a willing buyer and

a willing seller get together and agree on terms of a transaction. There are many ways in which these two parties can be brought together. Private agreements are especially important for family transactions, but may be of limited usefulness beyond that, unless potential buyers and sellers know each other and can negotiate effectively. Auctions are used nation-wide but usage varies widely from one area to another. Forced sales are often by auction. Beyond that it depends heavily as to what people in a given area are accustomed. Real estate brokers play an important role in bringing buyers and sellers together. For the most part, they are trained professionals, devoting their full efforts to a real estate brokerage business. Many communities or regional geographic areas have one or more brokers specializing in the sale of farm and urban land. Since land transactions are normally large relative to other transactions for buyers and sellers, it behooves each to make sure they are dealing with a capable and reliable agent in the purchase and sale of farm land.

Chapter 11
The Land Transaction

When the search for land produces an acceptable tract to purchase, the opportunity may be lost unless a firm commitment to sell is obtained from the seller. Similarly, a seller at some stage in negotiation seeks firm assurances from an interested purchaser that a binding obligation exists to buy the property. Therefore, to avoid frustration and disappointment it is well to anticipate the steps that lead to land purchase and sale as well as the steps involved in the process of acquiring title to the land. This stage of land acquisition often seems highly legalistic yet the rules involved are designed to reduce uncertainty and protect expectations—both those of the seller who is seeking the certainty of knowing that the tract of land has been sold and the certainty of the buyer who wishes to protect the land buying opportunity.

Preliminary Considerations

Before proceeding to the point of making a firm commitment to purchase, several matters of a preliminary nature should be given serious consideration. These points should be checked out well in advance of preparation of an offer to purchase or the signing of a contract to acquire the land.

Zoning and land use restrictions—An important overall consideration is the zoning status or any general land use restrictions on the property under consideration. Such restrictions may arise in various ways.

- The governmental unit, usually the county in the case of farmland, may have enacted a formalized set of zoning restrictions. In some areas, agricultural land is exempt from most zoning limitations but in others constraints are imposed upon some types of agricultural development, notably those involving buildings. Information about the zoning status of the tract in question can be obtained from the local office charged with administering zoning laws. In some instances, a specific office has that responsibility. In others, the information can be obtained from the city or county recorder's office or the office of the administrator for the local governmental unit.
- A second type of limitation on land use is that imposed by previous owners—or the current owner—of the land in ques-

tion. Limitations of various types are occasionally imposed by property owners on subsequent uses—and users—of the land. Such restrictions may preclude types of development thought to be undesirable or may impose specific limitations on improvements placed on the land. Thus, land use restriction might specify the minimum square footage for units in a housing development, a minimum setback distance for dwelling units from streets or highways, restrictions on types of construction material that can be utilized, specification on types of sewage disposal systems, to mention a few of the major types of restrictions in general use.

Some land use restrictions may be viewed as positive in nature in enhancing the value of the property in the future by preventing uses on adjoining tracts that might reduce its value. On the other hand, restrictions may be viewed as a factor likely to depress the value of land by ruling out uses of the land that would add additional value. Thus, if the highest and best use for a tract of land is for high rise apartment units, limiting the tract to single family residential housing would likely limit the value of the land for development purposes.

- A third type of limitation on land use involves easements that have been placed upon the land with the approval of either prior owners or the current owner of the land or that have been acquired by right of eminent domain on the part of governmental agencies or, in the case of utilities, by virtue of authority obtained from governmental agencies. For example, land near airports is often subject to an easement preventing the construction of improvements above a certain elevation. While such a limitation may not be of great significance so long as an agricultural use is contemplated, a limitation of that type could have a substantial effect upon land value if it rendered unacceptable a land use with high potential development value.

Trend of development in the area—For tracts of land near major cities, another preliminary consideration that should be given attention before negotiations have proceeded to the serious stage is the general trend of development in the area. Once a development trend has become established, potential developers and purchasers may acquire expectations that are difficult to change. For example, the emergence of a trend toward commercial or industrial development, even though not conclusive for an area, could make it difficult for purchasers to look upon the area as a potential site for home construction.

Availability of utilities—If development of the land is contemplated in the near term, it is especially important to check, as a preliminary matter, the availability of utilities to the site. The cost for obtaining access to sewer systems (or the cost for constructing a sewage disposal plant), the availability of electricity and the presence of a dependable municipal water supply are all important factors to the development process. Likewise, if development is contemplated, it is important to consider the philosophy and attitude of the governmental unit most likely to be involved in the development process including the matter of costs expected to be borne by the land developer.

Even for land acquired for agricultural purposes, these points may be significant inasmuch as a higher use than agriculture could emerge in a period of a few years especially if the tract in question is near a metropolitan area, a recreational or resort area of some significance or a major transportation or traffic artery which is likely to spawn development.

Loan commitment—If it is anticipated that the land purchase contemplated will be financed by borrowed funds, one should confer with lenders before the search for land becomes serious. The lender may provide guidance on the value of land that can be financed in light of the purchaser's financial situation, and can advise as to the probable availability of funds. Some lenders may be willing to provide a loan commitment in writing indicating a willingness to supply funds up to a specified maximum amount. Such a commitment may facilitate the land search by narrowing the tracts to those within the range of feasibility from the standpoint of financing.

Selecting an attorney—Another important preliminary matter is the selection of an attorney to represent the purchaser in the acquisition process. In general, it is believed important for the purchaser to be represented by his or her own attorney. Although at first glance relying upon the seller's attorney might appear to involve a savings of a few hundred dollars, the attorney for the seller cannot ethically represent both parties to the transaction.

Moreover, questions frequently arise in the course of the transaction that tip the seller's interest against the buyer's interest, thus making it highly advisable for each to have their own attorney. For example, questions may arise as to responsibility for property tax payment as between the seller and buyer. Details must be resolved relating to property insurance coverage and the disposition of proceeds in the event of a loss. Title problems that emerge after the contract is signed must be addressed. And a host of other problems could arise. Only if the buyer is represented by his or her own attorney can the buyer be assured of a source of objective information directed to the buyer's own interest.

Even the attorney representing the lender, if mortgage funds are to be involved, does not represent the buyer. The interest of the attorney for the lender is principally that of assuring that good title is obtained and that other problems do not loom sufficiently large to jeopardize the lender's security interest in the property. Again, that interest may not parallel completely the concerns of the buyer. Ethical considerations preclude an attorney from representing two parties to the same transaction that may have divergent interests.

In many rural areas, a problem of representation by an attorney is complicated by the fact that only a limited number of law firms are available in the community. If that problem exists, it may be necessary to seek legal assistance in a nearby town rather than place a local attorney in a conflict of interest situation. Because it is generally wise for the attorney for the buyer to review the offer to buy or land contract before it is signed, any preliminary work in locating an attorney should be completed before the land search has reached the serious stage. However, if the task of selection of an attorney has not been accomplished previously, it is generally advisable for the buyer to delay signing the contract to permit retention of an attorney to review the contract and advise the buyer of problems likely to arise from the document.

Information from visual scrutiny—Also, in the realm of preliminary consideration, a buyer may find it helpful to give the tract of land under serious consideration careful scrutiny for potential problems of a legal nature. Although an opportunity is generally available later in the process of transfer to raise questions, it may be helpful for the buyer to keep an eye out for problems that could be detected by close visual inspection of the premises. For example, individuals (other than the land owner) may be making use of the property in some fashion. If such use continues for a specified period, typically ten years, the users may have acquired a right to continue the use even over the landowner's objections. Thus, the use of land by neighbors as a passageway regularly for more than ten years, could result in an easement giving those individuals a right to continue using the property for that purpose. Inquiry of knowledgeable persons may be advisable if any indications point to the possibility that such use might have been made in the past.

Another problem area involves the location of boundaries. In many states, an erroneously located boundary that is respected for a period of years, again typically ten years, may become the true boundary if formal objection is not made during the ten year period. Thus, it may be helpful to locate the boundaries and give a quick visual check for obvious indications of how adjacent property owners have been respecting the boundary and also whether buildings or other improvements may be encroaching on the tract of land in question. These do not show up in the local land records and hence do not arise during the course of the regular title search of the property conducted later by the attorney.

Although the seller is generally under a duty to provide good and merchantable title to property, it is well to watch for indications that the property might have been the subject of substantial improvements in the few weeks or months prior to the time the contract was prepared for signature by the potential buyer. In general, suppliers of materials for land improvement have a right for a specified period (typically 60 to 90 days) to file a lien against the property which takes priority even over those who have purchased the property during that period. Thus, as a purchaser you could become subject to such mechanic's liens arising out of improvements made shortly before sale of the property.

In a small percentage of the cases, the poor financial standing of the seller may suggest an even more careful checking of claims already filed against the property including court judgments entered and other amounts which any purchaser would be required to pay. A seller is generally under a duty to satisfy such claims but a buyer may not want to suffer the delays and frustration that may accompany efforts to obtain performance from the seller.

For much of the preliminary checking, neighbors and others living and working in the general locality may be helpful sources of information. Individuals observing the property regularly are often in a good position to know whether much improvement activity has occurred, others are making use of the property, disputes have arisen over location of boundary lines, or other problems have emerged that could bear upon the attractiveness of a purchase of the land in question. In general, it is much easier to extricate one's self from a land transaction before the offer to buy or land contract is signed.

The Offer to Buy or Land Contract

The first document likely to surface in the land transaction is the *offer to buy* **or** *land contract*. The legal effects are essentially the same for the two documents. When signed, an offer to buy becomes a contractual commitment with the same legal functions as a document entitled a "land contract."

The usual practice is for the seller to prepare the offer or land contract. In some instances, the real estate broker representing the seller may have a partially or completely filled in document available as soon as significant interest is demonstrated in the property. It should be kept in mind always, in a land transaction, that the broker is an agent of the seller. Although the broker may provide helpful information to the buyer and the broker's expertise may be of great assistance to the buyer, the buyer should realize that the broker is an agent of the seller, is compensated by the seller, and is retained to sell the property in question.

The broker is not an agent of the buyer. Occasionally, the buyer may be represented by another broker whose allegiance, in that case, is to the buyer and not to the seller. That, however, is a relatively rare circumstance.

The offer to buy or land contract should be reviewed carefully before it is signed. Once signatures are affixed, the document becomes a legal obligation with the expectation that the contractual commitment will be carried out in accordance with its terms.

Offers to buy or land contracts typically contain standard clauses. Although most are basically form documents, offers to buy or land contracts should always be read with care, even the fine print, inasmuch as form contracts are not all identical. As a general rule, form contracts tend to favor the sellers of real property. The reason is that sellers are almost always represented by an attorney in land transactions. Buyers do not always have legal counsel. Therefore, attorneys for sellers have tended over the years to have a disproportionately large influence on the content of land contracts. Eventually, such provisions make their way into form contracts and tend to be used routinely thereafter. For that reason, it may be necessary for the buyer to add provisions or delete or change provisions appearing in the form contract in order to achieve an equitable and fair result. Buyers should not be reluctant to negotiate for contract provisions that depart from those appearing in a form contract.

The following points are usually included in a form contract and should be addressed in any contract involved in the conveyance of land. An example of a form contract is included in Appendix B.

Names of parties—Usually, the first item on a printed contract or offer to buy is a place to enter the names of the seller and the buyer. Usually, the seller or the attorney for the seller enters the correct legal name for the seller. Although an error in listing the seller's name can be corrected later, it is advisable for the seller's name to be entered properly in the interests of avoiding subsequent title changes (if the contract is to be recorded) as well as to assure that the buyer has an enforceable contract against the right party.

The major concern of the buyer, however, is in deciding how the land should be owned by the buyer. Several choices may be open to the buyer.

- The buyer may be purchasing the land in the name of an entity—a corporation, general partnership, limited partnership, or trust. In those instances, care should be exercised in properly listing the name of the entity.
- If the land is to be held in individual ownership, a question may arise as to whether it is to be in the name of the buyer alone, or in the name of the buyer and spouse. If it is to be the latter, it is important to know whether the co-ownership is to be joint tenancy (tenancy by the entirety in some states) or tenancy in common. In the eight community property states of Arizona, California, Idaho, Louisiana, Nevada, New Mexico, Texas and Washington, the same choices exist as to forms of ownership but any property acquired during marriage, except that acquired by gift or inheritance, is generally considered to be owned equally by the spouses regardless of how title is formally held.
- Acquiring title in the name of the buyer alone is the simplest arrangement and enables the buyer to dispose of the land by inheritance, by gift, or by sale. However, any conveyance of real property during life requires the approval of the title holder's spouse even though the spouse's name does not itself appear on the title. Spouses have a "dower" or similar interest

which usually amounts to a one-third interest in all the land owned by the other spouse during marriage. A dower interest translates into a right to claim a portion of the land at death, usually one-third, with the spouse of the title holder also having sufficient interest in the land during the life of the title holder to require that the spouse's signature be obtained on an instrument of conveyance.

- For land held in tenancy in common, each co-owner has an undivided interest in the property and can sell or convey his or her interest. Typically, property held in tenancy in common by husbands and wives is owned one-half by each although a different fraction could be specified in the document of title. Each tenant in common has a right to dispose of his or her own interest during life (with approval by the spouse as noted above) and to dispose of his or her interest by will at death. Typically, a tenancy in common relationship is created by the words "to John Doe and Mary Doe" appearing in the land contract (or deed). In most states, co-ownership of land is assumed to be in tenancy in common unless joint tenancy or tenancy by the entirety is clearly indicated.

  The federal estate tax treatment (and other estate planning implications) of property owned in tenancy in common are discussed in Chapter 18.

- With joint tenancy ownership, there is a right of survivorship with the interest of a deceased joint tenant passing, in effect, to the surviving joint tenant or joint tenants. This right of survivorship distinguishes joint tenancy from tenancy in common ownership. The right of survivorship feature of joint tenancy takes priority over an individual's will for purposes of property disposition at death. Again, the death tax and estate planning implications of joint tenancy co-ownership are discussed in Chapter 18.

In some states a joint tenancy arrangement between husband and wife is termed a tenancy by the entirety. A tenancy by the entirety is similar to joint tenancy except that it is limited to husbands and wives as co-owners and the co-owners do not have the right to demand a partition and sale of the property with the proceeds divided as is possible for joint tenancy.

In general, states do not favor joint tenancy ownership with the result that creation of a joint tenancy must be handled in a very specific and precise fashion or the co-ownership may be considered a tenancy in common. Fairly standard wording for a contract or deed, if joint tenancy is desired, is for the names to appear in approximately the following form—

"To John Doe and Mary Doe as joint tenants with right of survivorship and not as tenants in common."

Each state differs in the proper form for creation of joint tenancy. The use of language less specific than required by state law may produce a different form of ownership or co-ownership.

Although a later opportunity may exist to modify the ownership pattern for the buyer, it is well for the offer to buy or contract to specify the way in which title is to be held. A sizeable down payment on a contract could mean, if a shift is made later in the way title is to be held with the change made in the deed, that a gift could occur at that time. Thus, if funds accumulated for purchase were provided equally by the spouses but with the land contract showing ownership by the husband alone, conceivably a gift occurs at the time the contract is signed of half the payment amount from the wife to the husband. Later, when the deed passes, if title is taken in tenancy in common, it is quite possible that a gift has been made back to the wife by the husband of half the value. Therefore, it is well for the form of title passage to be agreed upon and entered properly in the land contract or offer to buy. For transactions after 1981, any gift to a spouse would be covered by the 100% federal gift tax marital deductions.

Legal description—For the land contract or offer to buy to be effective legally, the document must refer to the land in question unambiguously. Thus, a "unique address" for the tract must appear in the land contract. Although, as a minimum, the legal description must be such as to refer unambiguously to the tract in question, most buyers insist upon a formal legal description for the property. Property passing by will, however, is less specifically described and may pass under a general clause providing for passage of "all of my property, real, personal and mixed, wherever located . . ." or "I hereby give, devise and bequeath my North farm to my son George and my South farm to my daughter Mary." Such descriptions are acceptable if they refer clearly to specific tracts of land without ambiguity.

In general, however, land that is the subject of a sale transaction is referred to formally by one of four methods of legal description.

- In some parts of the United States, notably in the Northeastern states, land outside cities and towns is often described by "metes and bounds." This method of legal description arose prior to the time of formal surveys and makes use of identifiable objects or geographical features on the surface of the land. For example, a legal description might read as follows: "Starting at the center of the intersection of the Bardstown Road and Route 26, proceeding in a southerly direction to the old oak tree, thence northwesterly to the granite boulder, thence due north to the middle of the flowing brook, thence northeasterly to the point of beginning, containing 6.2 acres, more or less." Such legal descriptions are highly practical in nature, the boundaries can usually be identified readily on the ground, and such descriptions pose few problems unless the objects involved are moved or cease to exist. Thus, if the old oak tree dies, someone moves the granite boulder, or the flowing brook changes course slightly, disputes could arise over boundaries with adjacent owners. Metes and bounds-type descriptions are still used in some parts of the United States.

- Courses and distances, as a method of legal description, refers to the method of describing tracts of land by use of survey techniques. Thus, a surveyor is asked to start at a known location and survey the tract producing a direction and a distance for each side of the land parcel. Such surveys are normally accompanied by a plat drawn to scale which shows the direction in degrees, minutes, and seconds and the distance in feet or hundredths of feet. This method of legal description is commonly used to describe irregular tracts of land in all parts of the country.
- For most farmland, the method of legal description used is that based on the Congressional survey. During the nineteenth Century, much of the United States that had not been developed up to that time was surveyed using a unique grid system. Imaginary east-west lines, known as base lines, provide a base for locating tracts. Similarly, imaginary north-south lines, known as principal meridians, provide a reference for locating tracts on a north-south plane. Lines drawn parallel to base lines, six miles apart, create townships north and south from the base line. Likewise, lines drawn parallel to principal meridians, six miles apart, are known as range lines. Together, the intersection of range lines and township lines produce Congressional townships which are generally six miles on a side and contain 36 square miles. Because of the curvature of the earth, townships are generally not square but carry adjustments caused by the convergence of range and township lines. In the United States, corrections are generally made along the westerly and northerly sides of the township.

Within a township, the 36 sections, each containing one square mile (except for corrections caused by the curvature of the earth) are laid out as shown in Figure 11.1. Conventionally, the number commences with the northeast corner of the township, moves across the northwest corner of the township and proceeds row by row with section 36 in the lower right-hand corner. A section containing one square mile is 640 acres in size.

Each section is divided into quarter sections with each quarter section further divided into quarters or other fractional interests that are readily identifiable.

A typical example of a legal description based on Congressional survey is as follows:

The Southeast One-Quarter (SE ¼) of Section 16, Township 67 North, Range 19 West of the 5th Principal Meridian.

That legal description indicates that the land is located approximately 400 miles north of the base line and approximately 114 miles west of the Fifth Principal Meridian. The tract should contain 160 acres more or less. Legal descriptions in contracts (and deeds) should always be read with great care inasmuch as the validity of the document depends upon a proper legal description for the property. Although it may be possible to correct legal descriptions that are in error, it is much easier to make changes before the contract or deed is signed than after the document becomes effective.

- For land located within or near cities and towns, conveyance of land parcels may be in accordance with a system of plats. Thus, a developer may survey the land to be developed with the tract divided into lots and blocks. Once the survey has been completed and filed with the appropriate local office, transfer of lots thereafter may be made by reference to the lot number and block number without the requirement that the entire amount of survey information be included with each transfer. This simplifies the process of conveying land in urban areas. For example a tract of land might be described as follows:

    Lot 3, Block 16, in the Original Town of East Ipswitch, Iowa.

Purchase price—An offer to buy or a contract for the sale of land should include a clear specification of the total purchase price, the amount to be paid as earnest money on signing of the offer to buy or contract and the amount to be paid on the closing of the transaction, or payment specifications if the purchase price is to be paid in installments. In the latter case, the initial document signed may itself be termed an installment land contract or the initial document may specify that an installment contract is to be later completed and signed.

Although relatively few problems arise with respect to specification of purchase price, the buyer should note carefully whether the purchase price is computed at a specified price per acre or whether the purchase price is in the form of a lump sum. An agreement to purchase 160 acres at $3,000 per acre might be interpreted by the seller as equivalent to a purchase price of $480,000. A buyer, however, may discover that the tract of land contains only 155 acres in which case the expectation might be that the purchase price should be $465,000. Tracts of agricultural land rarely contain the precise acreage implicit in the legal description to the land. The initial document signed should make it very clear whether the price is calculated on a per acre basis for the number of acres actually existing within the tract or whether the transaction involves a lump sum amount.

For offers to buy or contracts for sale of land, the purchase price is usually expressed in terms of an amount which can be determined with certainty. Occasionally, in an installment land contract, the amount of each payment or even the amount of payment overall may be related to price or yield indices. These are discussed in a later section.

If the buyer has not obtained assured financing for the purchase, the buyer may wish to include within the contract or offer to buy, a specification that the agreement is dependent upon the buyer obtaining adequate financing at reasonable rates and terms. Although such provisions may not be welcomed by sellers, who are seeking an assurance that the land has in fact been sold, buyers may wish to protect themselves from consequences of inability to obtain adequate loan financing. This is a matter to be negotiated by the parties.

Property tax payment—An offer to buy or contract for the sale of land typically contains a provision allocating the responsibility for paying property taxes. In general, property taxes are assessed one year before property tax payments are made. For that reason, sellers conventionally pay one year's property taxes after the property has been sold.

Example: A tract of land is sold with possession obtained January 1, 1981. In that state, property taxes are levied on a calendar year basis. It would be expected that the seller would pay all of the 1980 property taxes payable in 1981. The buyer would make the first property tax payment in 1982 and would be responsible for all payments for that year and all later years.

For farmland, property taxes are often allocated on the basis of the crop year. Therefore, if possession of farmland is obtained on March 1, 1981, the seller would normally pay all of the 1980 taxes payable in 1981 with the buyer commencing with property tax payment

Figure 11.1 Designation of sections in a Congressional township.

6	5	4	3	2	1
7	8	9	10	11	12
18	17	16	15	14	13
19	20	21	22	23	24
30	29	28	27	26	25
31	32	33	34	35	36

in 1982. In general, buyers in that situation would pay the full amount of 1981 property taxes payable in 1982 even though the buyer had possession for only 10 months out of the year. The fact that the buyer had the land for the crop season generally results in a full year's property taxes paid by the purchaser.

Some states assess property taxes on a fiscal year basis. This complicates the allocation problem but the principle is still basically the same—the property taxes are paid one year after they are assessed.

In some areas, property tax payment responsibility is customarily handled on a current basis. In those instances, a buyer obtaining possession of a farm on January 1, 1981, would pay all of the property taxes due and payable during 1981.

Even in states where property taxes are typically allocated between sellers and buyers in a specific fashion, the parties are always free to include a different property tax payment procedure in the land contract. Therefore, the offer to buy or contract for the sale of land should contain a careful and detailed procedure for property tax payment responsibility. Several thousand dollars may be at stake in the allocation formula agreed upon.

Risk of loss—The offer to buy or contract for the sale of land should contain a specification about risk of loss as to buildings or other improvements damaged or destroyed during the terms of the contract. Also, a specification about who is to keep property insurance in force, should appear in the contract. Moreover, there should be included a specification of who would receive the insurance proceeds in the event of loss.

In some states, the risk of loss passes to the buyer when the contract is signed. In that event, it is especially important that the buyer understand the risk assumed and proceed to cover that risk in some fashion.

The most common procedure is for the seller to add the buyer's name to the present insurance policy on the buildings and other improvements so that the proceeds are then made payable to the parties as their interests appear.

With a signed contract to purchase the land, the buyer has an insurable interest in the buildings and other improvements and could take out his or her own insurance.

A third possibility is for the contract to specify that the risk of loss remains with the seller until the buyer obtains possession with the buyer typically agreeing to accept the proceeds in the event of loss or damage before the closing of the transaction takes place. Some states place the risk of loss on the seller until the buyer obtains possession. Buyers should check state law and ascertain clearly the risk of loss assumed, if any, when the contract is signed.

For a longer term installment contract, the question of who receives the insurance proceeds in the event of loss can take on added importance. If, for example, buildings are destroyed five years after involves variance in names of the previous property owners. Thus, if the property was acquired in 1918 by James S. Smith and in 1947 was sold by J. S. Smith, it is not clear from the record whether those were the same person. For this reason, it is suggested that those owning real property adopt a form of signature and use it without variation.

Unreleased mortgages—Mortgages placed on the land continue to be a cloud on the title until released. If the mortgage has not been paid off, the problem may be fairly serious. In the event the mortgage has been paid off but has not been released on the record, the defect is less serious but still a significant problem to be taken care of unless rendered unimportant by a statutory "curative act" under state law.

A spouse's interest—Failure of a spouse of a land owner to sign the deed on conveyance is a defect in title. In general, the spouse of a title holder to land has an interest (usually one-third interest) in all land acquired during marriage. That is the reason why the spouse of a landowner must join in the deed—to give up that interest in the property. If not done, it is entirely possible that a spouse could come back later and insist on a portion of the value of the property (usually one-third).

Death of the landowner—At the death of a property owner, several steps must be taken to assure good title to the land in the hands of heirs or those taking under the deceased's will. Notice must be given to the heirs, federal estate and state death tax returns filed, income tax returns prepared, property taxes paid, and the estate properly closed, to mention some of the major steps in the process. A significant departure from the specified procedure for estate settlement could result in a defect in title. Thus, the settlement of an estate of a landowner should be handled with substantial care.

Judgments—A court judgment against a landowner generally creates title problems with respect to the land owned by that individual. In many states, a judgment against a landowner is a lien against that person's land for a period (often 10 years) and the judgment may be valid against the individual for even a longer period (often as long as 20 years). Thus, the attorney for the buyer in preparing the title opinion watches for court judgments against the current owner as well as previous owners.

Mechanics' liens—Mechanics' liens that have been filed and thus have become a matter of record are likewise of concern to the buyer because of the priority claim the individual or firm filing a mechanic's lien may have against the property. Various other liens are possible under state law including, in many states, liens for the care of livestock.

Easements on record—Easements that have been filed as a matter of record become part of the title to the property and thus are binding against subsequent owners of the land. These include easements permitting pipelines to be built and maintained across a tract of land, electrical transmission lines to be constructed and maintained, air rights acquired by an airport authority, rights permitting the interruption of the natural flow of a watercourse (the so called flowage easement), to mention the major possibilities. All easements should be checked carefully by the attorney in the preparation of the title opinion.

Easements because of usage—Easements existing by virtue of usage may also constitute claims against the property but do not appear on the abstract of title. Thus, such easements may not be known to the attorney preparing the title opinion. As noted earlier, the buyer should make an independent observation of the property for signs that others may have been using the property such as crossing over the property to reach an isolated, land-locked tract.

Property taxes—Unpaid property taxes also constitute a claim against the property and normally result in sale of the property at a tax sale if the property taxes are not paid. Thus, it is unusual to observe property taxes that have been unpaid for more than a few months.

Quality of title: clearing up defects.
To clear up defects in title to land, several steps may be taken by the attorney for the seller. For minor items that can be resolved by explanation, such as differences in names, an explanatory affidavit could be prepared and filed as a matter of public record. The affidavit would then be included in the next continuation of the abstract and would be available for all subsequent purchasers to review. In some instances, such as where individuals have a specific claim against the land, use could be made of a quit claim deed with that individual asked to sign such a deed to transfer the interest to the current owner. This type of procedure is often used where a spouse of a previous owner failed to sign the deed.

Another procedure for clearing up defects in title is to make use of state "curative acts." Such acts automatically render some defects unimportant. In some states, as noted above, defects arising more than a specified number of years previously are not considered important. Several states have enacted 40 year marketable title acts which render unimportant most claims arising more than 40 years previously. Most states have several curative acts of a highly specific nature that apply to old mortgages, defects in deeds and other types of specific title problems.

For the more serious title problems, that cannot be cleared up by an explanatory affidavit, a curative act, or a quit claim deed, the seller may have to bring a "suit to quiet title." Since this is an expensive procedure, with a law suit filed in court to cut off the rights of potential claimants to the property, it is undertaken only if the defects are relatively serious in nature and cannot be resolved in a less expensive manner.

Types of deeds
With good title assured to the buyer and with financing in hand

for the transaction, the parties are in a position to close the transaction. Usually, the original contract for the sale of land specifies the tentative date for the closing of the transaction. On the occasion of the closing of the transaction, the seller provides a properly executed deed to the purchaser. At the same time, the purchaser makes payment to the seller for the remaining purchase price, either in cash or by executing a mortgage or land contract to the seller if the seller is to be the financier for the transaction.

The type of deed used in the transaction can have considerable significance to the buyer. The highest quality deed, and the one preferred by buyers, is the *general warranty deed*. Such a deed contains promises to the buyer that the buyer is receiving good title to the property. Thus, if the seller has good financial standing, such promises can provide important assurances to the buyer that good title is being obtained. Such promises about good title extend not only over the period during which the seller owned the property, but also over all previous periods of ownership as well.

A *special warranty deed* contains fewer promises than a general warranty deed with the promises of good title limited generally to the period of time during which the seller owned the property. Such deeds typically do not contain assurances about the quality of title during any previous period of land ownership.

A *quit claim deed* is the lowest quality deed of all and contains few, if any, assurances of good title. In fact, a quit claim deed typically promises the buyer that the seller is conveying whatever title the seller had, but no commitment is made about the quality of the seller's title. Thus, if the seller had good title the buyer would receive the same kind of title; if the seller had no title at all that is precisely what the buyer would receive. One who is inclined to sell the Brooklyn Bridge would be advised to use a quit claim deed in the process.

The major use of quit claim deeds is to clear up defects in title. In some states, quit claim deeds are of such low quality that their use creates a defect in title in itself.

Financing the Transaction

Unless the buyer is in a position to pay cash for the land purchased, the period prior to the closing of the transaction may be devoted to securing the necessary capital and preparation of the necessary financing documents. The basic choices are (1) the mortgage transaction and (2) the installment land contract.

Mortgage financing—Financing acquisition of real property through a mortgage generally involves two parties: the *mortgagor* is the land owner (or potential land owner) seeking financing; the *mortgagee* is the person or firm supplying the financing.

Two documents are involved, typically: (1) the note, and (2) the mortgage. The note establishes the promise to pay on the part of the mortgagor and specifies the amount to be repaid, the schedule of repayment and the interest rate charged. The mortgage creates a security interest in the land for the mortgagee so that if there is a default later in making payments, the mortgagee can seek satisfaction from the land which stands as collateral for the transaction. The mortgagee need not rely upon the status of general creditor as would otherwise be necessary.

Copies of a typical deed and mortgage appear as Appendices D and E.

If the transaction involves taking over a mortgage already on the property, additional considerations are involved. The seller may be willing to sell the equity interest in the property with the buyer *assuming* or taking *subject to* the existing mortgage. With a mortgage assumption, the new owner agrees to make payments on the mortgage and to be personally liable for any deficiency that might arise on satisfaction through liquidation of the land and application of the proceeds to the amount due under the mortgage. If the purchaser takes subject to the existing mortgage, the buyer agrees to make payments on the mortgage and could lose the property in satisfaction of the amount still owing but would not be personally liable for any deficiency.

Normally, the original mortgagor continues to be liable on the note until the obligation is discharged or otherwise satisfied. That is the rule even though the property may have been sold to someone who either assumed or took subject to the mortgage. Unless satisfaction of a mortgage occurs through payment of the amount due, the only other avenue for terminating the original mortgagor's liability would be through a *novation*. A novation is a three-party arrangement by which the mortgagee agrees to release the original mortgagor from liability on the obligation and to look solely to the new land owner in satisfaction of the mortgage amount that remains unpaid. A novation, if acceptable to the parties, may be used as a substitute for refinancing the transaction.

Upon default, for mortgage transactions, the mortgagee's remedy is that of *foreclosure*. Foreclosure is a process, conducted under court supervision, by which the property is sold and the proceeds are generally applied first to pay the costs of sale, then to pay any amount due under the mortgage and third, to pay any remaining amount to the mortgagor. Typically, the mortgagor has a period after the sale during which the mortgagor can pay the amount in default plus interest and costs and continue with ownership of the property. In many states, this is a one-year period or a one-year *right to redeem*. In some states, that period of right to redeem can be shortened for some types of property (often limited to residential property) if the mortgagee agrees not to seek any deficiency from the property owner (if the property fails to sell for enough at the foreclosure sale to cover the amount remaining under the mortgage).

In light of the right to redeem for mortgagors, a purchaser of land at a foreclosure sale essentially is only an investor for the period of time during which the right to redeem exists. In the event the mortgagor obtains the necessary funding to redeem the property, the purchaser at the foreclosure sale receives with interest the amount paid at the foreclosure sale. If the mortgagor does not redeem, the purchaser at the foreclosure sale ends up with the property.

Installment land contract—As a long-range financing device, the installment land contract is used if the seller of the property is willing to provide the financing for the transaction and be repaid over a period of several years. Installment contracts may run from two or three years to 35 years or even more, although most installment contracts are from 10 to 25 years in duration.

Installment contracts differ sharply from the note and mortgage as a financing device in that title to the property in the case of installment contracts normally remains with the seller until all or a substantial portion of the payments have been made on the contract. Retention of the title by the seller makes possible the remedy of forfeiture which is unique to installment contracts. With a mortgage transaction, title is with the mortgagor and forfeiture is not available to the mortgagee. A sample installment contract form appears as Appendix F.

With forfeiture, a seller gives notice to a defaulting buyer of an intent to forfeit the contract. The minimum period of notice required for forfeiture is set by state law. Many states permit forfeiture with as little as 30 days notice to a defaulting buyer. At the conclusion of the forfeiture period, if the amounts in default remain unpaid, a seller may recover the land, keep any downpayments or other payments made, and retain any improvements made by the buyer to the property. Forfeiture as a remedy assures return of the land to the seller in a far shorter time than would be required for foreclosure of a mortgage. This is one reason why sellers like installment land contracts. The other major reason is the opportunity to spread the income tax liability from sale of the land over the term of the contract with gain reported as payments are received. The income tax aspects of installment reporting of gain are discussed in Chapter 18.

Forfeiture is often viewed as a relatively harsh remedy against the buyer. Depending upon the amount of payments that have been made, improvements made by the purchaser and change in value of the property itself, forfeiting the buyer's interest with a relatively short notice can lead to a substantial loss on the part of the purchaser if financing arrangements cannot be made to continue contract payments.

For that reason, a purchaser as part of the original negotiations with respect to the installment contract might wish to seek some modification of the traditional forfeiture rules. One possibility would be to ask for a greater period of notice than the minimum required by state law. Therefore, a purchaser could ask for a 60 or 90-day notice before forfeiture could be instituted. Another possibility would be for the purchaser who has paid a specified proportion of the principal amount due to be able to obtain a deed from the seller and give a mortgage for the remaining amount due under the same terms and conditions as the original contract. Such conversions of contracts to a deed and mortgage apparently do not jeopardize the seller's income tax treatment of the gain. Where used, such conversions typically require that the seller have paid at least 50% of the principal balance due under the contract.

Another approach to softening the harsh effects of forfeiture would be for the purchaser to be able to request compensation (on an appropriate amortized schedule) for improvements made on the property by the buyer prior to initiation of the forfeiture action. Sellers might wish to limit such a right to those improvements specifically approved by the seller.

Three other remedies are available to sellers of land under installment contract. A seller could proceed with foreclosure in essentially the same manner as for default of a mortgage. The seller could bring an action and obtain a judgment for the amount in default and then seek to collect that amount by attaching and selling the buyer's property. Finally, the seller could seek to rescind the contract with the land returned to the seller, and the down payments and other payments returned to the buyer. With rescission, the parties are returned to their original positions insofar as possible. A rescission requires mutual consent and thus is rarely used inasmuch as one party typically feels that another remedy would be more advantageous.

The Closing of the Transaction

Upon completion of financing arrangements and resolution of all problems relating to title, the parties are ready to close the transaction. Usually, the tentative date for the closing of the transaction is set in the original contract.

Concerns about the seller's creditors—If the buyer has reason to believe that creditors of the seller might be on the verge of filing additional liens or obtaining judgments that could affect the property, the buyer might wish to take two additional precautions.

One precaution involves holding the closing of the transaction at a location in reasonable proximity to the county office where deeds and installment land contracts are recorded. The objective, of course, is to reduce the time between receipt of the deed by the buyer and recordation of the deed (or installment land contract) to establish the priority position of the purchaser. This gives less chance for creditors of the seller to establish a position ahead of the seller. Occasionally, the closing of the transaction may even occur in or in close proximity to the office where claims against property are filed to facilitate the late checking of filings involving the property.

The second precaution would be to request an escrow amount to be set aside out of the amount otherwise due from the seller to be used to cover any obligations that arise during the time immediately following the closing of the transaction that could become a priority against the buyer. For example, in many states, mechanics' liens can be filed within 60 or 90-days after materials are provided and take priority against the property even though the land has been conveyed to a new owner in the interim. Some form contracts include a provision enabling a purchaser who has reasonable cause to believe that liens might be filed during that period to request creation of an escrow account, reasonable in amount, to cover any such liens with the remaining balance eventually paid to the seller.

As an alternative to the creation of an escrow account, a seller could request lien waivers from the seller. This approach may be acceptable if the buyer knows with some certainty the identities of the firms that have provided materials during the period preceding the closing of the transaction. Thus, if substantial improvements have been made to the property, and the identities of the general contractor and subcontractors are known, the buyer might feel adequately protected if lien waivers were provided, signed by those firms or individuals. However, if the identities of the creditors are not known with certainty, the lien waiver approach alone may not be an acceptable alternative.

Settlement Sheet—In advance of the closing, a sheet should be prepared showing in detail the amount due the seller, any adjustments for unpaid property taxes that are the responsibility of the seller, any other obligations of the seller to be paid from the purchase price, the amount of any mortgage or other indebtedness to be taken over by the buyer, any closing costs to be paid by the seller and, finally, the amount to be paid by the buyer. The informational sheet prepared prior to the closing should be made available to both the buyer and the seller and their attorneys for review well in advance of the date set for the closing of the transaction. A sample form appears as Appendix G.

Recording documents—Immediately following the closing of the transaction, it is generally advisable for the buyer to record promptly the deed or installment land contract received. Recordation establishes the position of the buyer as an owner of the property both with respect to others to whom an unscrupulous seller might convey the property and also with respect to the creditors of the seller who might be seeking to establish a priority position. If a mortgage is involved, the mortgagee usually wants the mortgage recorded promptly to establish a position of priority for the mortgagee.

At the time of recordation of a deed, some states impose a transfer tax. That tax, which formerly was a federal tax at the rate of 55¢ per $500 of consideration, is now a state or local tax, where it is imposed at all. Some states follow the tax rate formerly imposed at the federal level of 55¢ per $500 of consideration. Others impose the tax with a different rate structure. Although the tax is generally considered to be the liability of the seller, the payment of the tax can be a matter of negotiation in the contract.

Additional costs, typically modest, are incurred to record the deed, installment land contract, or mortgage. Although these amounts are generally considered to be the responsibility of the party wishing to record the document (the purchaser, in the case of the deed; the mortgagee, in the case of a mortgage; and the purchaser, in the case of an installment land contract) the responsibility for paying those amounts also is subject to negotiation in the original contract.

Escrow arrangements—For contract purchases, where the deed is not passed until a later time, the practice in some areas is to execute a deed and place the document in escrow with instructions to the escrow agent to deliver the deed on proof that the buyer has performed in accordance with the terms of the contract. The creation of such escrow arrangements should be in conformity with state law. Parties should also reach agreement on who will pay the costs, if any, for the escrow arrangement. In some instances, banks and other financial institutions serve as escrow agents as a matter of accommodation. In others, a charge may be involved.

Additional points for discussion—At the time of the closing of the transaction, several other items should be discussed unless covered previously.

- Any arrangements existing informally or as a matter of formal fenceline agreement relative to division of responsibility for construction and maintenance of partition fences should be discussed. If fenceline responsibility has been established by formal fenceline agreement, a summary of that agreement should have appeared in the abstract and already have come to the attention of the buyer. Informal agreements, existing by virtue of convention or tradition in the community should be communicated to the buyer even though such arrangements may not be legally enforceable. Any special arrangements for the maintenance of flood gaps or any other difficult to fence areas should also be discussed.
- The buyer should also obtain information from the seller as to the history of fertilizer and limestone application and the use of insecticides and herbicides. Herbicide usage is especially critical because of the possibility of carryover into the next crop year. Prior usage of such carryover chemicals should be brought to the attention of the buyer.
- The buyer may also wish to make inquiry as to any disease problems on the premises. Some disease organisms may be viable for a period of several months or even years after the last outbreak of the disease. Those diseases, especially, should be discussed with the buyer.

    In most states, the legal liability of landowners for the spread of disease depends heavily upon knowledge that the disease existed. Therefore, if a farm tenant brings clean livestock on the premises that have previously been infected with a contagious disease, the landowner is generally not liable unless the farm tenant was misinformed by the landowner as

to the status of disease on the premises. Thus, if the tenant failed to raise the question, the landowner would generally not be liable for spread of the disease to the tenant's livestock.
- Information about prior cropping patterns may be useful information for the buyer. Although much of that information may be on record at the local Agricultural Stabilization and Conservation Service (ASCS) office, that may not be the case if the owner was not participating in government programs.
- If the seller has any tile maps or information about the location of tile lines including dates of installation, size of tile, type of tile and location of outlets, that would be very helpful information for the buyer to have. Repairs on tile lines may be expedited if it is known with certainty where the mains and laterals are located. In addition, information about the tile system may be helpful in claiming investment tax credit and in establishing the initial depreciation schedule in terms of useful life or recovery period and depreciable value.
- The buyer may also find it helpful to have information on the age and condition of plumbing systems and waterwells used for domestic, livestock or irrigation purposes. Similarly, it is also helpful to have information on the age and condition of electrical wiring unless condition can be readily ascertained by inspection.

CHAPTER 12
MANAGEMENT ARRANGEMENTS AND INFORMATION

Upon finalizing a farm land purchase, an owner immediately faces the challenge of how best to manage and operate the newly acquired resource. Quite likely, preliminary management alternatives will have been considered as part of the decision to buy. Before plans can be fully developed, a new owner should review these objectives and goals that initially prompted the land purchase. The goals normally include earning a fair return on the investment and protecting and possibly enhancing the value of the land.

Most important for the owner within this set of objectives is choosing a role to assume in the farm's operation and management. Such a choice determines the organization and future planning for the property and perhaps the amount of return realized from the investment.

An owner's role can vary from being active in day-to-day farm operations and decisions, to being completely passive and allowing someone else to operate the farm and make the decisions. Most landowners can be classified into one of three broad categories:
1) the farm owner-operator
2) the owner who custom hires all operations
3) the owner-lessor

While there are specific advantages or disadvantages associated with each, that does not imply that any one role is more appropriate than another. Only the owner can determine which is most appropriate for the unique situation.

The Farm Owner-Operator
Most who decide to operate their own farm are already established in farming and the purchase of new land is but an extension of their existing business. For such owners, the new land may create economies of size. With a larger land base the operator may be better able to utilize resources and can recombine them in a way that will create higher returns. The farmer may also realize additional economies through larger input purchases and product sales. Still other forces motivate farmers to expand, such as intangible goals and personal

motivations. Whatever the composite of reasons for expansion, the owner-operator is buying land to combine with an existing operation, and will operate the additional tract. No further elaboration is needed on this alternative.

Direct operations—Some investors may want to manage and operate a farm but not provide the labor. For such owners, "direct operation" is an alternative. In this case the owner functions as overall manager of the farm but hires all labor and may delegate much of the decision making to a manager placed on the farm as a direct representative of the owner. The owner is responsible for the purchase of all inputs and for selling all products and owns all equipment. This arrangement provides substantial control over the investment. Success of the arrangement depends on the quality and reliability of personnel employed to carry out the farming operations. Skilled laborers are often difficult to find.

Some direct operations are created by vertically integrated corporate investors. An example is vegetable canning companies who control the timing of the crop production and other management decisions and who accomplish all work with hired labor on owned land. Another common approach by the canning companies is contract production with farmers. This is not direct operation since the farmer owns the land and manages the labor, even though specific guidelines must be followed in carrying out the production and harvest operations.

The direct operation fits well for the individual or company who manages to acquire a sizeable operation (not necessarily all in one geographic location) and wants to remain active in management. In such cases, there is usually adequate capital to own fully the machinery and equipment as well as all operating inputs including breeding and feeder livestock. Such an individual might be a business executive or professional person with mostly weekend time for involvement in the operation. The key ingredient is a desire to have considerable input into management decisions.

Custom Hiring

A landowner who has some knowledge of farm management but not enough experience or capital for completing all of the farming tasks may wish to custom hire some operations. That may include every task or only specified operations. This approach involves hiring an individual to bring equipment on the farm and complete a task. Note that both the machine and a skilled operator are employed. Most common are the tasks that require expensive, specialized equipment. Average rates for hiring custom operations, based on studies from the state land grant university, can usually be obtained from your county extension office.

While custom work is increasing across the nation, as costs of machinery rise, custom operators are not always easy to find. Nearby farmers often perform custom work for additional income but they may not be willing to do so until they have finished their own work. Thus, problems arise, e.g. harvesting conditions may not be at an optimum when the custom operator arrives. In addition, most farmers perform custom work for only one or a very small number of tasks. If you desire to have most of your operations done by custom work, several different farmers may need to be employed. However, some communities have individuals making their living primarily from providing custom hire.

Most custom operations are tailored to field work while few livestock operations present an opportunity for custom hiring.

On the plus side, the owner who desires custom hiring maintains a large amount of control over the farm and has no debt or repair commitments associated with machinery. The owner can also receive the benefits of new technology that might not otherwise be affordable. Additional labor is also obtained and is usually of high quality.

When custom agreements are made, one should be sure to involve some form of legal commitment that insures the completion of the task in a satisfactory manner. A simple written contract signed by both parties, stating the farm location, scope of the task at hand, period of service, type of service and payment agreement is often adequate. Special terms should be made for compensation of losses in the event of any breach of contract.

County extension agents can often help assist in locating good custom operators. Local "word of mouth" and newspaper ads are also helpful in addition to direct contact with nearby farmers. Machinery dealers are also a possible source of assistance.

It should be re-emphasized that the owner who custom hires should possess full knowledge in farm management. The owner is entirely responsible for all decisions made—the custom operator merely carries out those decisions.

Leasing the Farm

Most landowners, excepting owner-operators, prefer not to devote their full time to managing a farm. People employed in other businesses, heirs, absentee landowners, widows or others are often too busy or lack the experience or capital to operate a farm. The previously mentioned alternatives may not be feasible for such individuals.

At the same time, many operating farmers have need for additional land, but cannot afford a purchase of their own. These experienced farmers are often willing and eager to lease land as a strategy for expanding their operating unit. Beginning farmers who are enthusiastic and capable may view this arrangement as a "way in."

With land prices increasing, leasing arrangements are popular in most rural communities. About half of all farmland is involved in some type of lease or rental arrangement.

There are several reasons for using a rental arrangement, ranging from the high cost of owning land to a desire for a high level of management. Many of the reasons emerge as the chapter progresses. We expect lease arrangements to become even more popular in the future as it becomes more difficult for existing operators to expand their land base via the ownership route.

What is a farm lease agreement?—Simply stated a farm lease is an agreement transferring certain land rights from the owner (lessor) to a tenant (lessee). It shows how the revenue and expenditures are to be shared by the two parties. Ideally the lease contains an agreement on all matters that concern either the lessor or lessee in operation of the farm. It provides a future reference when questions arise, and helps avoid disagreements.

From a legal standpoint the following five conditions should be included in a written lease agreement: 1) the property to be leased should be accurately described in the agreement; 2) the time period over which the lease extends should be stated; 3) terms concerning amount of payment (in cash and in kind) and place where payment is to be made should be included; 4) names of both the lessor and lessee should be designated; and 5) both lessor and lessee must sign the document. While these conditions meet the legal requirements, it takes far more details to ensure a satisfactory working lease arrangement.

Good farm leases include terms that ensure a fair sharing of revenue and expenses by both parties. It should also include a statement of the farm objectives. The lease should, insofar as possible, assure that both parties' objectives will be met. For the landlord this may mean preservation of the quality and value of the property. For the tenant, objectives may include assurance of continuation of the lease. In any event, all terms should be put into writing for the protection of both parties. A written lease protects heirs if one party dies; it provides a basis for farm practices and, in the event either party disputes these practices, the legal document can be used to help settle disputes. In addition, a written lease provides assurance of the rules that will be applied in the event of a dispute. Lastly, it also specifies how the lease is to be terminated or provisions required to continue the lease beyond the length of term originally specified.

Oral agreements are considered by some to be a sign of good faith in the other party. Indeed, oral leases are binding in most states for terms not exceeding one year. However, the practice of leasing land under an oral agreement is not only unbusinesslike, it can lead to unfortunate circumstances and is not suggested. Unwritten leases leave both parties vulnerable to error, misjudgment, and misunderstanding. Legal requirements of lease termination are less certain with an oral lease in some states. Appendix H contains several lease forms developed for the north central states. You can also check with your county extension agent.

Which type of lease?—Several different lease or rental arrangements are possible. The most appropriate lease depends upon the farm's unique situation, the desired role as a landlord, the tenant's capabilities and the amount of available resources such as capital. Several considerations should govern the choice of a lease plan. One is the degree to which each party is involved in providing improvements and non-real estate capital. There are differences on who will make decisions concerning the organization and operation of the business. Substantial differences exist on degree of risk or uncertainty that each party will assume relative to physical production and price received.

The goals related to ownership of the property should underlie the choice of a lease arrangement. "Material participation" in the farm operation has become a focal issue on how farm owners and their heirs are treated by IRS on several issues, such as "use valuation" of farmland, installment payment of federal estate taxes, and social security payments. The type of lease agreement can be the determining factor. Thus "Material participation" is covered in Chapter 18.

Several types of leases have evolved to meet the needs of landlords and tenants. Most common among these are the:
1) cash lease (or contract cash rent)
2) crop-share lease
3) livestock-share lease
4) labor-share lease
5) standing rent lease
6) combination leases

Cash lease—One of the simplest methods of leasing farmland is that of cash leasing. Cash leases represent an agreement whereby the tenant pays the landowner a fixed amount per acre or a lump sum for the entire acreage for use of the landowner's farm. In return the tenant receives all income produced irrespective of yields and prices. Normally the tenant is responsible for all expenses associated with the land's use with the exception of real property taxes, insurance and certain building repairs. The latter expenses are paid by the landlord out of the cash rent payment. Such a lease frees landlords from most management responsibilities but, above all, the landowner no longer needs to be concerned about the risks associated with agricultural production. The owner receives a definite, fixed income with this type of arrangement.

Risks related to production and commodity prices are passed to the tenant. There is little chance for argument or dispute about production techniques with the cash lease since the terms of the contract are clearly understandable to both parties. An exception to this would be if the landlord desires to specify production practices beyond requirements listed in the cash lease. While the cash lease affords some advantages it also poses disadvantages which should not be ignored. The landowner assumes fewer risks with this lease and theoretically should receive a lower rate of return on the investment.

Historically, landlords have received a rental payment approximating ⅓ of the crop values on good soils and ¼ on less productive soils. It is important to recognize that considerable variation exists on rental rates, with returns in the immediately preceding crop year having considerable impact on the expectations of both parties. The amount of responsibility assumed by the landlord should relate directly to the amount of income received.

A concern of most owners is that of tenant exploitation of the land—primarily fertility depletion and unnecessary erosion caused by a lack of concern for the land on the part of the tenant. Wise landowners should enter into cash lease agreements with the stipulation that certain levels of fertility will be maintained and erosion controlled by the tenant. Such terms should be clearly outlined in the lease agreement. Assuring the tenant a longer occupancy period tends to diminish problems of this nature as the tenant becomes concerned about the longer run productivity of the land if the tenant must depend on it for income.

In general, the cash lease is most advantageous for landowners who seek minimal management responsibilities, a fixed and stable amount of income—known in advance, and who own land that cannot be easily harmed through mismanagement.

Crop-share lease—The crop-share lease is the most popular lease arrangement and can be adapted to a wide range of cropping patterns. The basic elements of a crop share lease consist of a rental payment which usually consists of a share of the grain produced on the farm. The landlord has the opportunity to share in management questions concerning land use and management decisions relating to seed, fertilizer and chemical use, and shares a portion of the expenses associated with these items. As with the cash lease, the owner assumes responsibility for real property taxes, insurance and building improvements. In return, the operator contributes labor and most or all of the machinery needed.

The proportion of inputs contributed and returns that each party is to receive vary. Questions often arise as to what percentage is most equitable. Typically, landlords have received either ½, ⅖ or ⅓ of the value of crops produced as a rental payment. The specific proportion can best be determined by relating the proportion to the type of soil involved. As can be seen in a study done by Franklin J. Reiss at the University of Illinois, the better, more highly productive soils involve a larger rent share (see Figure 12.1). The tenant's share in absolute dollars remains roughly the same despite the level of soil productivity and type of rent share. This can best be explained by considering the economic view of land. Land is always a residual claimant to income and is rewarded by the landlord's share. Thus, the landlord's share decreases as soil productivity decreases, which is normally associated with lower land prices on the poorer soils.

Determining the appropriate rental share is a task which may require additional information from local sources, such as productivity ratings of soils and typical rental arrangements. Determining equitable

Figure 12.1 Gross value of crop production and landlord and tenant shares on one-person grain farm under crop-share leases.

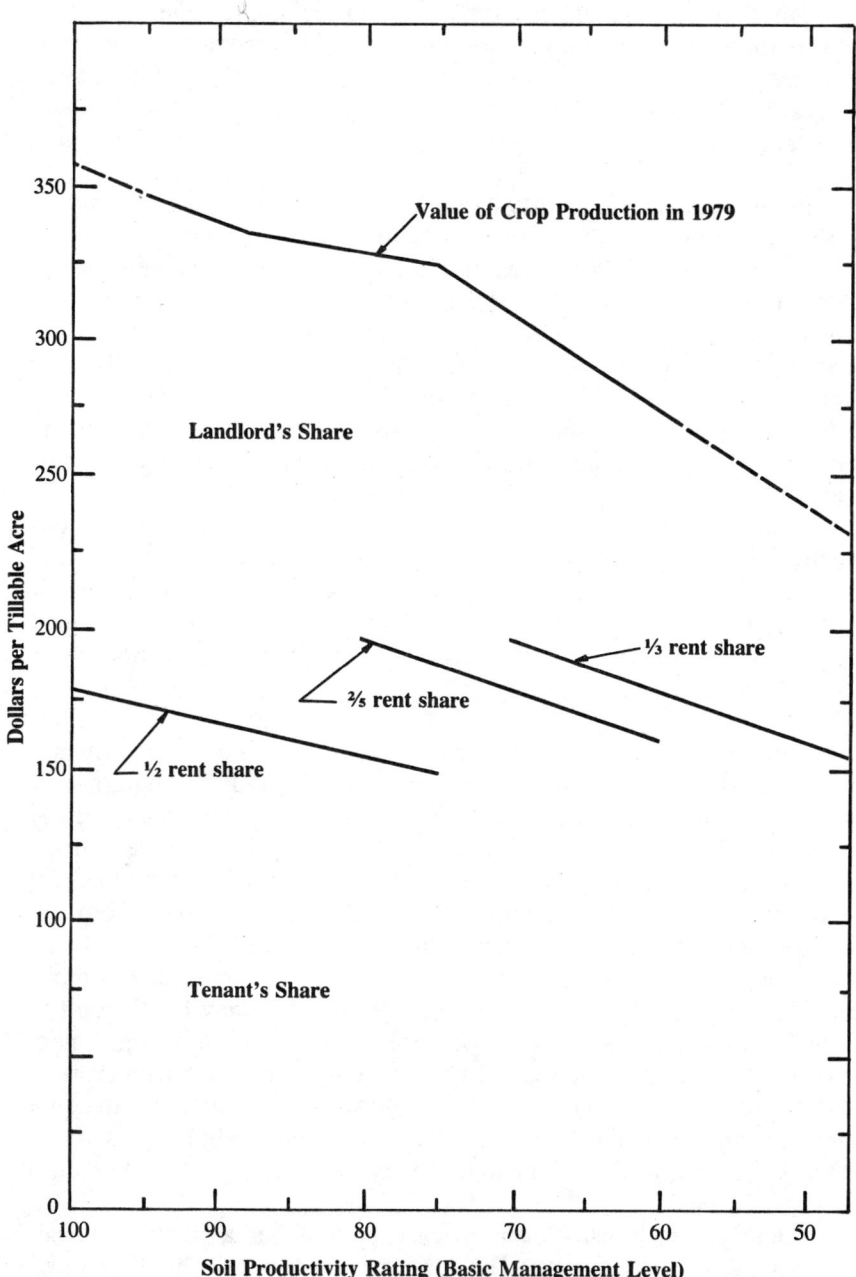

SOURCE: Reiss, F. J., *Landlord and Tenant Shares*, AERR-180, 1979, Department of Agricultural Economics, University of Illinois, Urbana, IL., Aug. 1980.

FARMLAND

rental arrangements is more complex than it might appear at first glance. Local tradition in sharing crop income tends to be an important factor in formulating tenants' expectations. Professional help may be needed to determine which share is most suitable.

Another problem arises from changing technology. New production methods and changing costs linked to those methods may result in a need to change lease terms concerning input proportions of the parties. For example, many recent technological changes have involved labor replacement by larger and more efficient machinery. This may alter the relative contributions. Terms of crop-share lease agreements should be reviewed frequently to ensure that both parties are supplying fair amounts of inputs with respect to the returns received.

Overall, crop-share leases give the owner more management opportunities. The owner can help choose the inputs in the production process and can also choose how to market the landlord's share of the product. The landlord bears some of the downside risks but also has a chance to share in high yields and high prices, when they occur. Studies have generally shown that landowner returns are higher in the long run with crop share arrangements than with cash rental agreements.

With a crop share lease, both parties should be able to communicate well, since such a lease requires many joint decisions. As a landowner, the crop share lease demands that you keep abreast of modern farm management techniques as well as maintain an open mind. If this is accomplished, a crop-share lease can be advantageous for both parties.

Livestock-share lease—In this arrangement, the landowner furnishes the land and buildings. The tenant furnishes labor and most equipment. Animals, feed inventory and special livestock equipment are typically shared equally by both parties, as are other expenses and all the returns. The potential problems and benefits are similar to those described for crop share leasing. Risks are shared, including the risks from the livestock enterprises, as are many management decisions.

It is possible for livestock share leases to become quite complicated. Equitable shares are based on inputs which may be difficult to value. For example, when livestock volume increases it becomes necessary for the tenant to devote a larger amount of labor to the enterprise. In order for the lease to be fair, the landlord should contribute an equal share of capital which helps offset the additional labor needs. How much capital should be contributed may be difficult to answer on a generalized basis.

Clearly, the livestock share lease approaches a partnership between landlord and tenant. The tenant is not merely another hired worker, but instead, a co-manager who frequently works with the landlord. In most states livestock share leases are not considered partnerships. Partnership status generally requires a sharing of net income.

With a livestock share lease, the landowner and tenant each has some deductions. Courts are also influenced by the way the parties hold themselves out to the public. Therefore, landlords and tenants under a livestock share lease should not refer to themselves as partners.

Under a livestock share lease, the landlord should keep abreast of technological changes, have considerable capital, and be able to withstand losses from time to time. Alternative livestock production methods can lead to differences of opinions. Thus good communications between landlord and tenant are crucial. This type of lease imposes a great deal of responsibility on a landowner.

Often the owner of a livestock operation has had prior experience with livestock and may have accumulated a large amount of capital. But for various reasons the landowner may want someone else to provide the labor and daily management needed. Some owners have built up reputable breeding herds that they want continued beyond their own retirement. A livestock share lease is convenient for these situations, also.

For those who do have livestock management abilities, this lease can often be the most profitable of all leasing arrangements and offers the greatest opportunities for further investment in the farm business.

Labor share lease—Unlike the previously mentioned leases, the labor share lease is often utilized where the operator-manager has little, if any, capital to invest. Typically, the landowner owns all capital items (including machinery) but desires that someone else provide the labor and management for the farm's operation. This type of lease is especially suited for parents who have children returning to the farm or for those who wish to allow a relative to enter farming. The young operator contributes labor and management for a proportionate share of the returns. Over time, capital may gradually be transferred to the operator by purchase or gift.

The owner assumes all of the risks involved in the farming operation so it is necessary to maintain some control over the farm business. In some instances this control is gained by having the tenant sign a formal assurance in the form of a promissory note or bond, guaranteeing payment to the landlord should default take place. The guaranty amount covers damages likely to be incurred by the landowner.

The labor share lease is not recommended to the less experienced landowner for it is difficult to find individuals with the expertise to manage a farm wisely in the initial stage of the lease. The operator may need the help of someone with more experience even though the person is well educated in all aspects of farming.

Standing rent lease—Because of rapidly changing price and production risks, various types of "hybird" leases have evolved to provide additional options for landlords and tenants. Most prominent in this development is the concept of standing rent.

Standing rent is based on the idea that landlords should be able to participate in profits from exceptionally high grain prices, but in all other aspects the lease is similar to regular cash rent. Normally, both parties review the historic yields for the past several years on the farm to determine the average yield. Then, a certain portion of the yield is determined to be the landlord's share for payment. The rent is stated, not in dollar terms, but in physical quantities which, when valued at current price levels, provide the landlord's share.

Once the standing rent is determined in terms of bushels or other units of production, prices are recorded at several pre-chosen times during the year and averaged at year-end. The average is then applied to the standing rent amount in terms of units of production to determine the lease payment.

Usually the standing rent allows some margin for automatic adjustment. However, should a disaster affect only a small area around the farm the standing rent may have to be altered. When no historical yield records are available, county-wide averages can be used as an estimate for determining the standing rent amount on initiation of this type lease.

Substantial effort is required to establish this type of lease. The rent must be reduced to a number of bushels of grain per year (using corn, for example, in the corn belt) and then a pricing pattern established to generate a precise technique for determining how the dollar value will be determined each year. Total rent each year involves multiplying the agreed-upon bushels per acre times the calculated price per bushel. Like cash rent, there would seldom be variation from the contractual arrangement, regardless of whether the actual yields are above or below the average yield used in the contract. Only price would vary each year. As with any lease, the end result should be equitable to both parties in terms of contributions made and risks taken.

The proportion of yield that a landlord should obtain normally equals or exceeds the dollar amount the landlord would have received through cash rent, since the landlord is exposed to greater risks with standing rent. To compensate for the increased risk the landlord should receive a higher potential return than in straight cash rent (remember—the landlord may lose also if price levels decrease more than normal). In some cases the owner has received up to 10% more for assuming the added risk. The measure of risk is sometimes difficult to quantify in terms of additional payment. Local county extension advisers can offer more detailed measures applicable for your area.

With this lease there is as much incentive to the operator as with ordinary cash leasing and all other conditions are similar. The tenant bears less risk of lower income due to low prices, but continues to bear all risk associated with variable yields.

The standing rent type lease is growing in importance, especially among landowners who desire a semi-fixed income but are willing to take price risks with the tenant. This type of lease often provides for a lease term of four to five years to permit the tenant to average out crop yields and prices over a term beyond one year. Tenants may be unwilling to enter into such an agreement with a lesser term, since the short-term risks may be viewed as too great.

Combination leases—A large number of leases have been developed which cater to specific or unusual situations. Some leases contain flexible rent clauses which go into effect whenever gross income per acre exceeds a specified level. Others represent combinations of cash leases and crop or livestock share leases. One example is that of a cash lease on buildings combined with a crop-share lease on the crop land. The possibilities are infinite, but as long as both parties are satisfied, such arrangements are entirely acceptable. Unfortunately, parties who depart from the more common leasing arrangements often create more areas for disagreement and greater complexity in determining rental amounts, which may ultimately lead to problems between landlord and tenant. The fact remains that there are such leases.

After choosing an appropriate type of lease, the next step is to become more specific in adapting it to suit the particular farm. Special attention should be given to such areas as the following:

1) the time of making the lease
2) the term of the lease
3) specifics on division of expenses
4) rent adjustments
5) tenant participation in improvement costs
6) management responsibilities of both parties
7) records and reports
8) termination and settlement

Rental arrangements on most farms are completed in early fall for a lease period starting in the following calendar year. Landlords should be responsible for early review of the lease and should ideally notify a tenant six months to a year in advance if a new tenant is desired. This allows the present tenant to plan cropping systems and affords time to find another farm if necessary. This helps to prevent ill feelings or unnecessary resentment. The timing of the lease is often more appropriate if it falls in line with the farm's production schedules. For example, the tenant who is assured of planting and harvesting the same crop will obviously seek to maximize the production of that crop. Many states require that notice to terminate a farm lease be given on or before a specified date.

Unless otherwise specified, most states assume the duration of a lease to be one year. Therefore, if longer terms are desired they should be specified in the lease agreement, including provisions for automatic renewal if that is desired. A tenant normally strives to do a better job if there is some assurance of a lengthy tenure. However, this is not to imply that all tenants will do better jobs if they have a long lease. It is the landowner's responsibility to review the tenant's success every year and to negotiate the duration of subsequent leases with the tenant.

Expenses and revenue are usually shared in the same proportion but variations are limitless. Each situation tends to be unique.

Lease adjustments from year to year vary according to the type of lease. Cash leases may be changed annually, while others are usually adjusted less often. All of the farm business records should be reviewed annually to detect any needed changes in lease terms. Most leases need adjustment at least once every five years and many should be adjusted every three years to maintain fairness.

Management responsibilities and the types of records desired are not identical for any two farms. Nevertheless, all leases should clearly set forth conditions concerning these matters. Early determinations of these matters prevent disputes later. In addition, the records kept may be used to review the tenant's progress towards achieving the owner's objectives. Their importance cannot be stressed too much. As a minimum, records should be kept of all income and expenses, production reports, inventories, cash flow, changes in financial position, soil fertility levels, depreciation, and other pertinent information. Year-end financial statements are also advised. The owner should receive periodic reports of the operator's work. Such reports serve to prevent small problems from becoming major concerns.

If tenants are to have tenure longer than one year the owner may want the tenant to contribute to improvement costs. Upon leaving, the tenant may be allowed to remove the improvement or be reimbursed for the remaining share of the investment based on the remaining useful life. For example a tenant raising livestock on a cash lease farm may wish to build or rebuild fences. If the tenant provides some of the improvements, the desire to remain on the farm may increase and the farm becomes more productive.

Some of the uncertainty of lease termination can be reduced if conditions for renewal or termination are clearly stated in the lease. A tenant should be guaranteed termination notice at least six months prior to expiration, and periods of a year or more can be very beneficial for both parties; the tenant needs time to relocate and the landowner needs time to find another tenant. Conversely, similar terms should specify the notice a tenant must give the landlord to terminate a lease. Again, some states require that notice to terminate a farm lease be given by a date specified by state law.

Only the basics related to lease agreements have been touched upon in this chapter. Professional help may be consulted for more information.

Being a landlord requires an open mind, patience, a desire to accumulate agricultural knowledge and the ability to communicate well with others. Unfortunately not every landlord possesses these qualities, but they serve as a guide for those wanting to be successful landowners.

Professional Farm Managers
An increasing number of landowners find that dealing with tenants

and keeping abreast of the new technology changes takes too much time away from their other activities and work. Such individuals can place their farm in the care of a professional farm manager.

Professional farm managers typically have considerable expertise in managing farms. Most are familiar with the production methods that work best and often can increase the income more than the fee charged for their services.

Farm management services are provided by independent management firms and by commercial banks. Some banks have trust departments which contain farm managers. In addition to expertise in farm matters, managers often possess or have close access to legal and financial expertise in farm matters.

Choosing a farm manager is not easy if you lack first hand knowledge on abilities of potential managers. Membership by a prospective farm manager in the American Society of Farm Managers and Rural Appraisers is a sign that the individual may be active professionally and is one indication that the individual is keeping abreast of changes in farming. Education and experience requirements are imposed for membership. Moreover, members are bound by a code of ethics.

Even more assurance is provided through members who are "Accredited Farm Managers," a title awarded by the American Society of Farm Managers and Rural Appraisers. The title is awarded only to those members who have met high standards of qualification by completing certain additional prescribed educational requirements plus passage of a two-day examination and several years of successful farm management activity. A directory of accredited farm managers and accredited rural appraisers can be obtained from the American Society of Farm Managers and Rural Appraisers, Inc., P.O. Box 6857, 360 South Monroe Street, Denver, Colorado 80206.

After initial contact of an owner with a manager, the manager typically visits the farm and assesses the existing conditions. Following that, the manager usually meets with the owner to determine objectives for the future. The manager then develops both a long and short term plan that seeks to meet the objectives. Alternatives are presented that the owner can review. Such plans often involve aspects of the leasing arrangement, property improvements, cropping systems and soil fertility improvements.

Once a plan is suggested by the farm manager, both owner and tenant can review it and, if acceptable, the plan can be implemented. With a farm manager, there is seldom a need for the owner to confer directly with the tenant—the manager is an agent for the owner and thus has contractual powers to represent the owner in operational decisions related to the farm.

After the plan is initiated, the manager works very closely with the operator to assure that desired objectives are met. The manager continually supervises the farmer in the decision making processes—in purchasing and marketing products, as well as in all phases of production.

The manager is responsible for periodic farm visits and makes reports to the owner, routinely including an analysis of the farm operation and current concerns. The manager also keeps records and financial statements updated for the owner and makes recommendations for future change. Such recommendations include cropping systems, fertilizer applications, changes in buying or marketing strategies and cash flow improvements.

As an added service, many managers provide advice on tax planning. The manager can check insurance programs for the farm and represent the owner in adjusting for losses. The manager also attends hearings or special events that might affect the owner's interests.

Special accounting services are often available, as are crop reporting or other specialized tasks, such as farm appraisal. Professional farm managers typically charge owners a percentage of gross or net income realized from the farm as the fee for management services. They may charge flat fees for consultation or one-time specialized services. A client does not necessarily need full service to benefit from a professional farm manager. Most managers are willing to engage in special problem solving that might be needed on the farm.

Where to go for More Information

As stated throughout the chapter, the nearest source of information is the county cooperative extension office associated with the state land grant university. The county advisers have direct access to information from state universities and are available free of charge to help solve problems. If colleges or high schools at the local level have agricultural educators, their help can be obtained.

Another major source of information is farm management or agricultural consulting firms. Agricultural lenders often have expertise in farm matters and can suggest individuals for further information, or can themselves offer valuable assistance. There are also many agricultural magazine and newsletter services, many of which can more than justify the cost.

The most successful landowners are those who are informed and who work hard at putting their knowledge to work. The task of being a landowner appears at first glance to be enormous but the rewards can be substantial.

CHAPTER 13
LIABILITY CONSIDERATIONS

The ownership and management of farmland can lead to liability claims based upon injury to a person or damage to another's property. The risks involved include injury or damage from events such as ordinary tillage or harvesting operations, which may be resolved under traditional negligence principles. Extra-ordinarily hazardous activities such as the use of blasting materials to create drainageways, the impoundment of water behind a dam or the aerial spraying or dusting of crops may be subject to more demanding strict liability or absolute liability rules.

Injury to a person while on the premises is another common risk with liability in some states dependent upon the status of the visitor. Injury to employees, a risk of increasing concern, may be resolved under workers' compensation, as a "no-fault" system, or it may be handled as a matter of employer fault, depending upon state law.

Anticipating liability claims is essentially a matter of *risk management*. That may involve a careful assessment of liability insurance needs and acquisition of needed coverage and efforts to meet legal standards of care to avoid liability where that can be done.

A Litigious Era
The potential for financial settlements for injuries suffered or damages incurred appears to be greater than a generation ago because of several developments in the law of "tort" liability as noted below. In addition, the chance of being sued appears to be greater, also. The latter half of the Twentieth Century may go down as the period with the greatest amount of litigation in history. And there seem to be several reasons—
- For one, individuals appear to have fewer inhibitions about settling disagreements in court. A generation ago, short of a serious affront to one's economic being, it wasn't exactly the "in" thing to be in court settling differences. In fact, one could easily be labeled a troublemaker for filing a lawsuit to recover for losses suffered. That seems much less true today.
- Second, if injury is suffered or damages are incurred, there's now less of a question of "who was at fault" and more of a

question of "how can I recover what I've lost?" To someone who has suffered permanent injury from a surgical procedure gone wrong, it seems relatively unimportant whether the surgeon was careful or careless. The crucial point is that something went wrong—and the individual suffered.

The attitude of expecting recovery regardless of fault has boosted the "no fault" idea. Indications point to continued emphasis on recovery for injury or damage regardless of fault. Essentially, that calls for spreading the costs of injury or damage over a large group. In the early part of the Twentieth Century, that led to replacement of the fault system of recovery with the workers' compensation concept for injuries or illnesses suffered by employees. Emergence of no-fault auto insurance in the 1960s stemmed from much the same set of forces.

- In many respects, the stakes are higher today than they were four decades ago. An error in selection of herbicide, for example, may affect hundreds or even thousands of acres of crop in a single operation. Contamination of a batch of animal feed can lead to massive economic consequences. Modern technology adds to that aspect of the liability problem.

The importance of status—The liability of a farm manager—or landowner—depends heavily on status. As a general rule, the greater the control over the action that led to the damage or injury, the greater the chance of liability. Likewise, as a general rule, the greater the economic benefit to the individual, the greater the chance of liability. The fault or "tort" system tries to match costs and benefits in approximate fashion. Thus, the role played by the person responsible for injury or damage is an important factor in liability.

Where farm liability is involved, with a farm manager as part of the operation, there are at least five different and distinct roles being played.

1) A farm tenant bears the usual responsibility for his or her own actions. But the farm tenant's acts rarely create liability problems for others. The tenant is generally considered neither an agent nor employee of the farmland owner or of the farm manager.
2) The employee of a farm tenant is responsible for his or her own acts and, in addition, may commit the tenant as employer to liability. This falls under the rule of *respondeat superior*— let the employer respond in damages for acts of the employee causing harm to others. Whatever an employee does while acting within the scope and course of employment—basically, doing what the employee was hired to do—can make the employer liable. That's one good reason why it is wise to hire employees with care.
3) The farm manager is, of course, responsible for his or her own acts. A farm manager may, depending upon the nature of the

relationship involved, commit the landowner to liability. That is based on the principle that the farm manager is an agent of the landowner. More on that in a later paragraph.
4) For an employer of a farm manager, such as a farm management firm, liability normally exists for the acts of employees. That assumes, of course, that the employee was acting within the scope and course of employment.
5) The landowner, represented by a farm manager, bears only limited responsibility to others for defects with respect to the premises. Conditions caused by deterioration or lack of repair during the tenancy are usually the responsibility of the tenant. The general rule has been that a landowner, in the absence of a contractual commitment or a statutory duty to repair, is under no obligation to make repairs. A landowner may be liable to the lessee for latent defects in the property known to the lessor but concealed from the tenant. But a landowner is generally not liable to third parties for injuries or damages suffered if the premises were in good repair at the time the lease was entered into and were under the control of the tenant.

The trend toward greater landowner responsibility—Historically, landowners have shouldered little responsibility for injuries suffered on the premises. But that situation is gradually changing—

- Occupancy of the premises by the tenant does not relieve the landowner of the consequences of his or her own negligence, such as from repairs negligently made by the landowner.
    Example: After a severe wind storm, causing damage to the electrical distribution system, the farm manager contracts with an inexperienced local firm to repair the damage. It was done carelessly and the tenant's hired man was electrocuted. The landowner, farm manager and the local firm doing the work could be liable for the loss.
- A nuisance created on the premises may give rise to liability on the landowner for creating, and on the tenant for maintaining, the nuisance.
    Example: A landowner under a livestock share lease constructs three lagoons for livestock waste disposal. Odors from the lagoon are highly offensive to a neighbor. Both the landowner and tenant could possibly be held liable for the nuisance involved.
- Some state courts have recognized an implied warranty of habitability in the oral or written lease of a dwelling. In general, the landowner is held to an implied promise that there are no latent defects in the facilities and utilities vital to the use of the premises for residential purposes and that these essential features will remain during the lease term in such a condition as to maintain habitability of the dwelling. Although most of the cases involve urban dwellings, the rule could be applied to

farm houses as well.
- Some states have adopted the "Uniform Residential Landlord and Tenant Act" although occupancy primarily for agricultural purposes is excluded from coverage in some of the jurisdictions.

Negligence

Liability for injury or damage depends upon the concept of negligence more than on any other legal rule. Negligence is the applicable legal concept for determining liability unless some other concept such as strict liability, absolute liability or workers' compensation provides the governing rules for determining who pays for injuries or damages suffered.

To be liable for negligence, four conditions must exist:
1) The conduct of the individual responsible for the act leading to injury or damage must have fallen below the standard of the reasonable and prudent person under the circumstances.
   Example: An automobile accident occurs while B, a farm manager, is driving from farm to farm. For purposes of B's liability—and B's employer, the ABC Farm Management Co.—the standard would be that of a reasonable and prudent *person* under the circumstances.
   Example: F, a farm manager, arranges for purchase of a soil insecticide in bulk for use on several farms and forgets to caution a new tenant who has difficulty in reading labels about precautions that should be taken in application of that particular insecticide. F's conduct would probably be measured by what a reasonable and prudent *farm manager* would have done under the circumstances. And that would be more demanding than what a reasonable and prudent *person* would be expected to do.
   Example: T, an owner of 320 acres of farmland, hires a high school student, M, during the summer months to paint the buildings. In the process of painting a building, M carelessly knocks over a ladder, injuring the tenant's young daughter and a neighbor child. Both T and M could be liable for the injuries suffered.
2) Second, there must have been damage suffered. In general, recovery is limited to actual damages.
3) A causal relationship must have existed between the defendant's act and the complaining party's injuries or damages. Unless the injury or damage was reasonably foreseeable as a result of the careless act, an important link in the chain of liability is missing and recovery would be denied.
4) Finally, the negligence of the complaining party may limit or bar recovery. It is called *contributory* negligence in some states. Other jurisdictions view carelessness on the part of the complaining party as *comparative* negligence which can reduce

the amount of recovery.

Liability to persons entering the premises—One of the greatest liability concerns of landowners, farm managers and tenants is the responsibility owed those who come onto the premises and are injured. Historically, liability has been heavily influenced by status—whether the injured person was a trespasser, licensee (person with permission to be there) or an invitee (business guest). Those classifications still have some validity. But several states have moved toward a general duty of care to those who would reasonably be expected to come onto the premises, and away from reliance on the status categories.

But the following categories continue to be significant as liability concepts in many states—

- A trespasser is one who enters land without consent. The land occupier—owner or tenant—is generally not obligated to make the property safe or to warn of dangers but the land occupier cannot intentionally, willfully or wantonly injure a trespasser. Spring guns, loaded shotguns wired to door knobs, electrified fences set to shock trespassers or use of dangerous guard dogs can all create problems of liability. One can use reasonable force to protect property—but cannot inflict great bodily injury or take a life in defense of property.

  By contrast, one can use reasonable force in defense of one's own person that can, if necessary under the circumstances, include the taking of life or the infliction of great bodily injury. Note that, in general, if attacked away from the home, there is a duty to "retreat to the wall"—attempt to extricate oneself from the fray—before going so far as to inflict great bodily injury or take a life in self defense.

  A trespasser who is discovered on the premises has a right to be warned of dangers, including hidden dangers, that may exist on the land.

  In general, trespassing is a civil matter—a dispute between the trespasser and the landowner or tenant. But many states make a knowing trespass a crime.

- Under the "attractive nuisance" doctrine, a child who is attracted to the premises by a dangerous instrumentality—construction site, unattended equipment, abandoned vehicles, to mention a few possibilities—is entitled to a higher standard of care than an adult trespasser. Children by nature lack mature judgment and generally cannot appreciate dangers inherent in the land on which they are trespassing. For that reason, a higher duty of care is imposed on landowners and occupiers where children should reasonably be anticipated. It is especially important to watch the potential for liability for injury to children for land located near a school, playground or other area where children gather.

  Some states have rejected the attractive nuisance doctrine

as it applies to ponds, reservoirs or other artificial bodies of water, at least in sparsely populated areas.
- A licensee is one who comes onto the premises with permission—or at least acquiescence—of the owner or tenant. But the licensee is there for his or her own purposes and not to confer a benefit on the owner or tenant.

    There is no duty to keep the premises safe for licensees—such as hunters. But care must be used to avoid injury to the licensee. And a licensee is entitled to a warning of hidden dangers.
- A social guest generally occupies a position higher than that of a licensee but short of a business guest or invitee in terms of duty owed by the host. A social guest might be able to recover from a fall on a highly waxed floor, a faulty step or a poorly lighted stairway, for example.
- An invitee is a person on the premises for a business purpose or for mutual advantage rather than solely for the benefit of the person entering the property. The owner or tenant is not an insurer of safety of invitees, but a duty is owed to make and keep the premises reasonably safe and warn invitees of existing dangers. An important part of invitee liability is the duty to search out and correct defects in equipment or facilities that could lead to injury or damage to an invitee.

    This category includes those delivering feed or picking up milk, the veterinarian, employees of the rendering works, and salesmen making regular calls.
- Some states have enacted legislation removing much of the risk of liability for letting others use one's premises for recreational purposes without charge. Typically, it takes gross negligence or willful or malicious conduct for an injured individual to recover. And that is difficult for an injured party to prove. As a result, recovery by an injured visitor using the premises without charge for recreational purposes is much more difficult in a state with such a statute.

Special situations

Over the years, states have gradually modified the traditional liability rules based largely on negligence where the outcome of the traditional approach was viewed as unacceptable by the state legislature or the courts.
- In general, there is no legal duty to aid another in peril although it might be the humanitarian thing to do. But if one renders aid, and does so carelessly, that person could be held liable.

    Several states have adopted "Good Samaritan" laws that excuse all but acts or omissions constituting recklessness if a person undertakes in good faith to render emergency care or assistance without compensation.
- In states having a "guest statute," an owner or operator of a

motor vehicle is typically excused from liability for injuries suffered by nonpaying guests riding along unless the driver was intoxicated or reckless. A few states have held such statutes unconstitutional.

- Historically, a manufacturer or seller of defective items was not liable to injured consumers unless a direct contractual relationship existed between the parties.

  However, in a landmark New York case decided in 1916, a different rule was firmly established in the United States. As presently applied, the rule states that a manufacturer who fails to exercise reasonable care in the manufacture of goods which, unless carefully made, constitute an unreasonable risk of bodily injury to users, may be liable for harm caused.

  The most recent trend has been toward making recovery easier in product liability actions. Several states now follow *strict liability* rules with the complaining party only required to show that the defendant sold the product, it was defective and unreasonably dangerous to the user, the product reached the user without substantial change in condition and the defect was the proximate cause of the injury. This is an easier burden than proving that the manufacturer was at fault which is required in a negligence action.

- In activities viewed as inherently dangerous, the party responsible for the activity is responsible for injuries or damages regardless of fault.

  Examples: Keeping wild animals, keeping peculiarly vicious domestic animals, blasting operations, impounding water behind a dam and aerial spraying or dusting of crops (in some states).

The basic idea is that if someone undertakes an extra-ordinarily hazardous activity, with full knowledge of the potential for harm to others, the person choosing such a hazardous approach should respond by paying damages if injury or damage occurs to another or another's property. In those situations it is no defense to say that the landowner or tenant hired an independent contractor to do the work. The normal insulation from liability for acts of an independent contractor is not applicable if the activity was extra hazardous in nature.

Example: Knowing that having an employee dynamite stumps to clear land would lead to liability if rocks and debris were to fall onto neighboring property, the landowner contracts with a firm to do the blasting. In most states, the landowner's liability would not be diminished merely because an independent contractor—rather than an employee—did the blasting.

Employee or independent contractor—In general, as noted in the preceding section, acts of an employee make the employer liable—

if done within the scope and course of employment. But injury or damage caused by an independent contractor is not the worry of the landowner or tenant—except for the inherently hazardous class of activities as noted above.

Example: A farm manager hires an experienced painting contractor to paint the buildings on a farm. If a scaffold falls and injures the tenant's employee, in general that is the painting contractor's worry—not the landowner (or farm manager) who hired the work done.

If the landowner's own employee were painting the buildings when someone was injured, that could result in employer liability.

The distinction between *employee* status and that of an *independent contractor* is an important one. And the line separating the two is not entirely clear. In general, it is a matter of the right to control "the manner and means of performance." So if someone is hired to paint the buildings, using his or her own equipment and working according to their own schedule without close employer supervision, it is likely to be considered an independent contractor relationship. And the landowner is not likely to be held liable unless the operation was considered inherently dangerous—and painting usually is not classified that way. But if the landowner's own employees do the job, and injuries result, employee status generally makes the landowner liable if negligence could be proved.

Insurance Coverage

The best safeguard against liability is for everyone to begin every day with the same resolve—to act always as the reasonable and prudent person under the circumstances. That takes care of 90% of the problem. The other 10% involves concepts of strict liability or absolute liability. In those situations, usually the highly dangerous activities, the amount of care exercised generally does not affect liability.

In all situations, it is well to carry an adequate amount of liability insurance and to be especially watchful for inherently dangerous activities. Many farm liability policies exclude from coverage activities that are extra hazardous in nature. For example, damages from aerial spraying or dusting of crops are excluded from coverage under many liability policies unless special coverage is secured. Liability policies should be reviewed carefully to note exclusions from coverage and any special limitations for designated liability situations.

Some landowners and tenants use "umbrella" policies to boost liability coverage at a modest cost. That involves maintaining designated levels of coverage (typically at a level of $300,000 per event) for motor vehicles and general farm liability. The umbrella policy elevates coverage, usually for all types of liabilities, to a level of $1 million or more for a modest additional premium charge.

CHAPTER 14
FEDERAL AND STATE PROGRAMS AND REGULATIONS

Numerous Federal and state programs and regulations affect the ownership and management of farmland. Knowledge of available programs can provide opportunities for improving and conserving the land, reducing water and air pollution, improving the income flows from the land, and obtaining credit to purchase and operate the land. State and Federal regulations also influence the ownership of land, as well as production and marketing practices. Knowledge of existing regulations is necessary to successfully own and use farmland within the scope of the law. Failure to abide by existing regulations may result in costly legal problems.

Federal Programs

Most Federal programs which have a direct bearing on the ownership of farmland are administered through agencies of the U.S. Department of Agriculture. Federal programs designed to improve and conserve farmland and to improve the income flows from land are enacted by Congress and carried out primarily through the Agricultural Stabilization and Conservation Service (ASCS) and the Soil Conservation Service (SCS) of the U.S. Department of Agriculture.

Agricultural Conservation Program—The Agricultural Conservation Program (ACP), first enacted in 1936, has been updated and continued to date. The goals of the ACP program are the protection, restoration, and preservation of agricultural land. Congress has allocated funds for cost-sharing programs for selected conservation practices. The rationale for cost-sharing is that there are some conservation practices in the public interest which are not economical for the individual land owner. Cost-sharing is intended to encourage conservation practices which would not otherwise be carried out. In many cases cost-sharing is on a 50-50 basis.

At the national level there are over 65 conservation practices which are eligible for cost-sharing. The relative importance of the major practices, as measured by the proportion of total cost-sharing expenditures, is illustrated in Table 14.1.

Table 14.1 Expenditures for cost-sharing conservation practices, U.S. 1978[1]

Practice name	Amount	Percent of U.S. total
	(dollars)	(percent)
Permanent vegetative cover establishment	30,503,695	17.73
Irrigation water conservation	22,760,511	13.22
Permanent vegetative cover improvement	20,027,624	11.63
Water impoundment reservoirs	11,970,804	6.95
Terrace systems	11,956,572	6.94
Underground drainage systems	9,864,236	5.73
Sod waterways	9,531,260	5.54
All other practices	55,554,499	32.26
U.S. Total	172,169,201	100.00

[1] Source: U.S. Dept. of Agric. "The Agricultural Conservation Program" 1978 Program Year.

The ACP is administered at three levels—national, state, and local. The primary focus of the ACP is at the local level where county elected ASCS committees administer the program to cope with local conservation needs. Because of the local orientation, the conservation practices eligible for cost-sharing vary from county to county and from state to state. You can obtain information on conservation practices available for cost-sharing in your area by contacting the local ASCS office.

The percentage of farmers and ranchers participating in the ACP is highly variable throughout the country (Figure 14.1). Participation rates in the Midwest and Western regions of the U.S. are much lower than in other regions of the country. But no matter what region of the U.S. you own farmland or ranchland, it is to your benefit to check carefully the possibilities for cost-sharing conservation practices for your farm or ranch. The benefits to you in terms of improved productivity and increased sale value may well be worth the effort and expense involved.

Commodity programs—There is a long history of farm commodity programs designed to stabilize and enhance the income of farm owners and operators. Products covered by such programs include wheat, feed grains, cotton, rice, sugar, wool and mohair, tobacco, peanuts, sugar, and dairy products. The programs have acted to shift some of the production and price risk from farmers to the general taxpaying public. However, consumers do receive benefits from the programs in the form of adequate food supplies at reasonable prices.

Financial assistance to farm operators under commodity programs may take the form of: (1) disaster payments, (2) deficiency payments, (3) nonrecourse loans, and (4) recourse loans. Features of the com-

Figure 14.1 Percent of total number of farms participating in the Agricultural Conservation Program, 1978

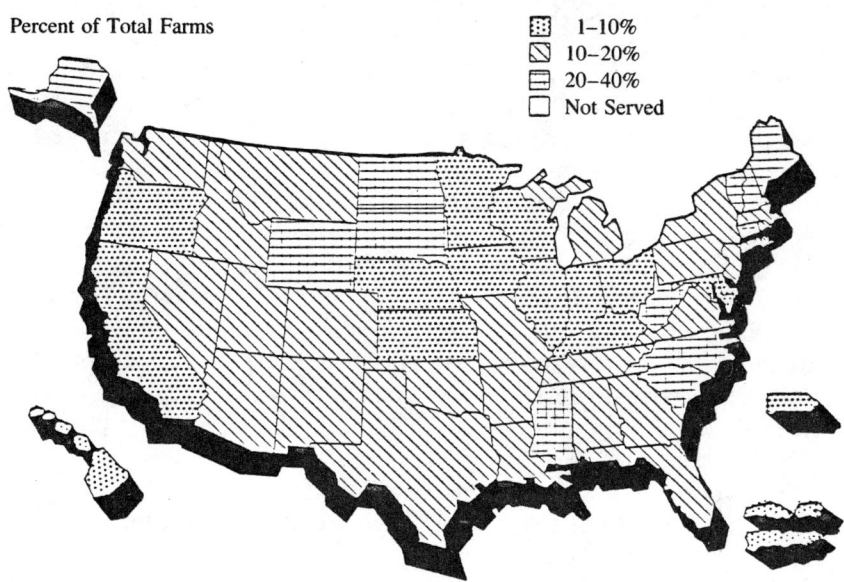

Source: U.S. Dept. of Agric. "The Agricultural Conservation Program" 1978 Program Year.

modity programs can change quickly as a result of changing economic conditions in agriculture. To keep abreast of these changes and to obtain current information, you should contact your county ASCS office.

In the past, Congress has authorized disaster payments to be made to producers of wheat, feed grains, cotton, and rice in the event that the crop could not be planted or if the quantity harvested was substantially less than normal because of drought, flood, or other natural disaster. To be eligible for such payments, producers must have participated in the set-aside program—a program for removing acreage from production. Disaster payments provide a form of crop insurance to participating farmers. Legislation in early 1980 extended the disaster payments concept through the 1980 crop year. Legislation was enacted in 1980 to provide a comprehensive crop insurance program.

Farm operators who participate in set-aside programs are also eligible for Commodity Credit Corporation (CCC) nonrecourse loans.[1] Nonrecourse loans allow farm operators to pledge eligible commodities as collateral on loans. At the option of the borrower, the pledged

---

[1] Because of depressed farm prices in early 1980, CCC loan programs were made available to nonparticipants in set-aside programs. Continuation of eligibility for non-participants seems unlikely.

FARMLAND

commodity can be turned over to the CCC and serve as *full* payment of principal and interest on the nonrecourse loan. For example, assume the loan rate on corn is $2.10 per bushel and you obtain a loan for that amount. Assume the market price for corn when the loan matures is only $2.00. Rather than sell the corn at $2.00 you can turn the commodity over to the CCC as full payment of principal and interest. If the price of corn is $3.00 per bushel you can sell the corn and repay the loan. Some farmers use CCC loans even when they expect the market price to be well above the loan rate. That is because interest rates on CCC loans have often been low relative to rates charged by other lending institutions.

Deficiency payments can be made to farmers when the national weighted average market price of a commodity falls below the "target" price of the commodity. For example, assume the target price on corn is $2.20 per bushel. If the market price is $2.00 per bushel and you have obtained a loan at $2.10 per bushel, you are also eligible for a deficiency payment of up to 10 cents per bushel. The loan program in conjunction with deficiency payments provides a "floor" on the prices of eligible commodities. Provisions for deficiency payments have been extended through 1981.

CCC *recourse* loans are also available for the purchase, construction, or installation of storage and drying equipment on farms. These loans cannot be repaid by turning over commodities to the CCC. Eligibility for such loans does not require participation in set-aside programs. Additional information on CCC loans can be obtained from county ASCS offices.

The net effect of commodity programs on farmland values has been the subject of much research. The general consensus is that the benefits of such programs are quickly capitalized into the value of land. As a potential purchaser of land, this means that the price you pay for the land will likely reflect the value of such programs. As an owner of land, it means that any improvements in the terms of government commodity programs enhances the value of your property.

Emergency Conservation Measures Program—This program is administered by ASCS and is designed to provide financial assistance when new conservation problems have been created on farmland as a result of natural disasters. The need for assistance is determined on a farm-by-farm basis after a careful assessment of the damages and the farmer's capacity to repair the damaged farmland. The objective is to restore agricultural lands to their original productive capacity.

Emergency Livestock Feed Program—This program provides financial assistance to enable livestock owners to purchase feed in emergency conditions. The program is designed to prevent the forced liquidation of livestock due to the lack of feed resulting from floods, droughts, hurricanes, tornadoes, earthquakes, diseases, or other natural disasters. Because such natural disasters can strike in a relatively small area, an individual farmer or group of farmers can be designated as being eligible for such assistance. In areas affected by disasters, ASCS personnel have the responsibility of notifying eligible producers of the assistance available. However, absentee owners may need to check on the availability of such assistance following a disaster.

Forestry Incentives Program—The Forestry Incentives Program (FIP) is designed to encourage landowners to plant trees on suitable open lands and to improve existing timber stands. Since developing and improving forest areas is a long-term investment which may be uneconomical for current landowners, cost-sharing is available. FIP agreements are limited to eligible owners with at least 10 but not more than 1000 acres. Cost sharing assistance cannot exceed 75% of costs and payments per owner per year are limited to $10,000.

Water Bank Program—The Water Bank Program (WBP) is designed primarily to preserve and improve the habitat of migratory waterfowl and to preserve and improve wetlands. The WBP operates primarily in the northern part of the Mississippi River flyways—the migratory routes used by waterfowl. Participants in the program can enter into 10-year agreements and receive annual payments for the conservation of water and wetlands. Such land cannot be used for agricultural purposes. Investors acquiring or owning wetlands should check with their county ASCS office to determine eligibility for such payments.

Government loan programs—The Federal government provides loan funds to agriculture through the Farmers Home Administration (FmHA), the Commodity Credit Corporation (CCC) and the Small Business Administration (SBA). Loan programs of the FmHA that have a direct bearing on land ownership were identified in Chapter 9. CCC and SBA loans are directed more toward the financing of operating inputs and therefore have only an indirect impact on land ownership.

State Programs

State programs which have an impact on farmland are many and varied. A detailed listing is beyond the scope of this book. However, some of the programs which you, as a landowner, may find useful to enhance the value of your property are identified in the following paragraphs.

Cost-sharing programs—A number of states including Iowa, Wisconsin, Minnesota, Nebraska, Idaho and Ohio have developed state or local cost sharing programs for a variety of conservation and pollution abatement practices. In most cases where such programs are available, the state pays 50% of the total cost of approved program practices. State funds for such programs come from general revenues or from special funds. For example, in Iowa part of the funds for wind erosion control come from state road use tax revenues.

The conservation practices eligible for cost sharing vary among states. In Iowa, approved practices include critical area planting, pasture and hayland planting. Minnesota and Nebraska have a less extensive list of approved practices. Both Ohio and Nebraska have cost sharing programs for animal facilities.

The development of state and local cost-sharing programs is in the infancy stage. Existing programs are relatively new and a number of other states are in the process of developing such programs. In most cases, the programs are administered by the State Soil Conservation Service and additional information can be obtained from that source.

Farm Loan Progams—A number of states provide financial assistance to farmers in the form of special credit programs. The programs range from financial assistance for 4-H projects to loans for the purchase of farms. However, only a few states have loan programs designed for the purchase of land.[2] The major sources of farm real estate loan funds were identified in Chapter 9.

Federal Regulations

Federal laws regulate some production practices and marketing activities of farm firms. As an owner or manager of farmland you should be aware of the general nature of such regulations. However, a complete enumeration of all regulations affecting the ownership of land is beyond the scope of this book. Rather, a brief summary of the nature of such regulations is provided. An indepth treatment of each topic can be found in *Agricultural Law* by Neil E. Harl, volumes 9–12.

Environmental regulation—Federal environmental regulations are designed to protect and improve the quality of the environment. Regulations focus on controlling air and water pollution.

The United States Congress has enacted legislation such as the Clean Air Act which is designed to prevent air pollution and improve air quality. The Act makes the Federal government responsible for establishing air quality standards, with the states implementing Federal standards. Perhaps the most common form of air pollution from agricultural operations is odor emissions from feedlots. However, since the Clean Air Act requires objective measurement of pollutants, it has been ineffective in controlling air pollution in the form of odor.

Legal challenges to feedlot odors most commonly take the form of nuisance suits. Location of the feedlots, severity of the problem and the history of development in the area appear to be key factors in the outcomes of such suits.

In an attempt to maintain and improve water quality, Congress enacted the Federal Water Pollution Control Amendments in 1972. Two sources of pollution were identified: point and non-point. Point pollution comes from a clearly identifiable source such as a pipe, ditch, or concentrated animal feedlot. Non-point pollution comes from a diffuse source such as run-off from an open field. The enforcement

---

[1]A detailed listing of state credit programs can be found in: "The Cooperative Extension Service Should Provide Farmers With More Information on Farm Credit Sources," U.S. General Accounting Office CED-80-45, Feb., 1980.

of Federal water quality regulations is carried out by the Environmental Protection Agency (EPA). Much of the EPA concern over agricultural point source pollution has focused on controlling feedlot run-off. Nonpoint source pollution control measures for agriculture have been focused toward getting landowners and managers to adopt the conservation practices eligible for cost-sharing as identified in this chapter.

Labor regulation—The Fair Labor Standards Act (FLSA) was amended in 1966 and again in 1974 to include minimum wage and equal pay provisions for some agricultural workers. However, most agricultural workers are still excluded from the minimum wage standards and all agricultural workers are outside the overtime pay provisions of the Act.

Excluded from minimum wage standards are farm workers on farms where not more than 500 man days of agricultural labor were used in any calendar quarter of the preceding calendar year. Also excluded are immediate members of a farm family, workers engaged in range production of livestock, and certain migrant workers.

The Occupational Safety and Health Act (OSHA) was enacted to improve worker safety standards. Because of various amendments, most agricultural workers are excluded from coverage. For example, OSHA regulations are not applicable to agricultural workers employed on farms having ten or fewer employees. A major recent thrust of OSHA regulations on covered agricultural operations has been to promote worker safety against the hazards of chemicals. Field reentry standards following application of certain chemicals have been proposed. Machinery safety devices, such as "slow moving vehicle" signs have also been mandated by OSHA regulations as have standards for storage and handling of anhydrous ammonia, temporary labor camps, pulpwood logging, roll-over protection devices for tractors and safety guards for agricultural equipment.

Marketing regulations—There are many Federal regulations which establish grades, labeling and packaging standards, and provide for inspection of agricultural products. These are essentially consumer oriented regulations to ensure food safety and to better inform consumers of the quality and standards of food and fiber products. For agricultural producers not engaged in direct marketing, the effect of such regulations on the daily operations of the firm is minimal.

Farm firms are perhaps more directly affected by Federal marketing orders. Marketing orders are issued by the Secretary of Agriculture. If approved by a required percentage of producers the orders are binding on all producers and handlers of the product in the specified production or marketing area. Marketing orders are designed to establish and maintain orderly market conditions. The objective is to assure equitable returns to producers while providing adequate supplies at more stable prices to producers. Marketing orders have a direct bearing on the stability of income and hence the cash flows associated with land ownership. Investors seeking stability in returns to land should consider investment in land used to produce commodities covered by marketing orders.

There is a long history of Federal involvement in milk marketing orders. There are also marketing orders in effect for fruits, vegetables, hops, tobacco and nuts. However, cotton, rice, wheat, corn, livestock, small grains and other agricultural commodities are not covered by marketing orders.

State and Local Regulations

Federal environmental, labor, and marketing regulations are often supplemented (and in some cases implemented) by state and local regulatory agencies. Because the nature of such state and local regulations is fairly comparable to Federal regulations, further discussion is not included here.

Alien and corporate ownership of land—Numerous states have enacted statutes which restrict alien ownership of land. Restrictions vary from outright prohibition to those with only minor practical impact. States with major restrictions include: Arkansas, Connecticut, Illinois, Indiana, Iowa, Kentucky, Minnesota, Mississippi, Missouri, Nebraska, New Hampshire, North Dakota, Oklahoma, Pennsylvania, South Dakota and Wisconsin. States with minor restrictions include: Alaska, Arizona, California, Hawaii, Georgia, Idaho, Kansas, Maryland, Montana, Nevada, New Jersey, North Carolina, Oregon, South Carolina, Virginia and Wyoming. Because of the current legislative interest in the topic, state statutes should be consulted for up-to-date information.[3]

A number of states including Iowa, Kansas, Minnesota, Missouri, Nebraska, North Dakota, Oklahoma, South Dakota, Texas, West Virginia, and Wisconsin have state statutes restricting farming activities by certain corporations.[4] The stated objective of such statutes is the preservation of the family farm against the perceived threat of large corporations. All states with such statutes have exceptions that permit "family farm corporations" to own agricultural land and engage in farming. A family farm corporation is typically defined as one having: a) a limited number of shareholders (usually not more than 10), b) one or more stockholders who live on and farm the land, and c) a majority of its income from the sale of agricultural products.

Statutes limiting corporate land holdings have been enacted despite the fact that the involvement of large corporations in agriculture has been minimal except in California, Florida, Hawaii, and Arizona. Some states have statutes applicable to other types of business organizations, such as limited partnerships.

---

[1]For a brief discussion of laws in each state see: Harl, *Agricultural Law*, Volume 13, Section 123.02; Chapter 3, "State Controls and Reporting Requirements," in *Monitoring Foreign Ownership of Real Estate*, Report to the U.S. Congress, 1979.

[2]A discussion of these restrictions can be found in Harl, *Agricultural Law*, Volume 7, Section 51.04.

Zoning—State and local governmental units can control the use of land through zoning ordinances. Zoning was developed originally in urban areas to accomplish a variety of social goals. In more recent times, zoning ordinances have been utilized in rural areas as well.

Rural zoning has been used for a variety of purposes such as to control location of residences in isolated areas, promote conservation, prevent location of residences in disaster prone areas, preclude businesses from expanding and limit population inflows. In many instances, zoning has been used as a defensive mechanism to prevent what is deemed to be "undesirable" development. In some instances, zoning ordinances have been declared invalid by the courts because they are arbitrary and unreasonable. As an owner or manager of land, you should determine the zoning ordinances which are applicable for your locality. Zoning officers for your county or state can provide information on zoning ordinances.

A total of 18 states have "right to farm" laws which limit the right of rural residents to complain about odors and other nuisances.

CHAPTER 15
PROPERTY TAXES

Except for exempt organizations, the ownership of real property involves the burden of taxation. For real property held until death, federal estate taxes (and state inheritance or state estate tax in many states) are imposed as discussed in Chapter 18. Real property transferred during life by sale or other taxable exchange may generate income tax liability as discussed in Chapter 17. Transfers by gift may be subject to federal gift tax (and state gift tax in 13 states). Gift transfers are discussed in Chapter 19.

In this chapter, attention is focused on the property tax aspects of real property ownership with attention to the taxing power and its limitations, the different types of property or property-related taxes and the practical side of property tax administration at the local level.

The taxing power and its limitations—In the United States, the taxing power rests with the legislative branch of government and cannot be delegated to other governmental agencies or bodies. The federal government has only the taxing powers delegated by the United States Constitution. State and local governments have the taxing powers not prohibited by the United States Constitution and not precluded or limited by state constitutional provisions. The basic legal framework for the property tax is thus a matter of statutory law in each state.

Every state legislature is free to tax the property within its jurisdiction so long as the tax is levied in an equitable and reasonable manner and for public purposes. State legislatures may develop classifications of property for purposes of taxation but the classification must be reasonable and not arbitrary and must rest upon differences having a fair and substantial relationship to the objects of the legislation. All persons similarly situated must be treated alike. State constitutions also, typically, specify that all properties of the same class within a taxing district must be taxed uniformly at the same millage rate and with the same assessment-valuation ratio.

Types of Property Taxes
Real estate or real property is subject to several different property taxes. Each of the major categories of tax on real property is discussed in the paragraphs following.

Ad valorem property tax—Ad valorem (according to value) property taxes are imposed universally in the United States on real property and in many states on personal property. Where different tax rates apply to real property and personal property, or where personal property is not subject to property tax, the distinction between the two types of property can be quite important. In a farm setting, some items of property may be classified as personalty under state law such as some components of grain drying and storage facilities, and equipment that is not attached to the land. State laws vary as to the rules defining personal property and should be checked carefully, especially the year property is first entered on the tax rolls.

In imposing the property tax, all property that is not exempt from the tax is assessed at a valuation that is typically a specified percentage of fair market value. The assessed values are multiplied by a tax levy that is uniform within the taxing district to determine the gross tax. Various deductions and credits are allowed by state law to produce the net tax due. If not paid, the tax becomes delinquent with the property eventually sold and the proceeds applied to the property tax due or the property passes to the governmental body levying the tax. A more detailed discussion of property tax administration appears in a later section.

Special assessments—In the United States, special assessments are used to finance the construction of improvements that benefit the area taxed such as sewers, water lines, sidewalks, and streets. In addition, special assessments are widely used to finance drainage, levee, irrigation, fire protection and similar projects. In the latter cases, the special assessment levy is usually made by a special taxing district authorized by state law.

Special assessments are generally assessed and collected along with general ad valorem property taxes with the same default procedures available if the assessment is not paid. Especially with levies such as to finance land drainage, special assessments may involve complex formulas for levying the tax in relation to benefits or anticipated benefits from the project. In other instances, special assessment formulas may be relatively simple with assessment made on a front foot basis (for sidewalks or streets), for example. In general, project costs in the form of special assessments are allocated in proportion to benefits expected.

For purchases of real property, it is important to know—(1) what special assessments have already been levied and would be an enforceable claim against the property, (2) any special assessments that are in the process of being levied and are expected to become a claim against the property in the near future and (3) those projects in the proposal stage that could materialize and result in special assessments. It is noted that some special assessments imposed on farm land, such as for drainage, levee, or irrigation purposes, may have immediate and substantial benefits to the land. In those instances, additional income to pay the special assessment should be forthcoming. In other instances, such as for street, sewer or water improvements installed in areas near cities and towns, special assessments may result in increased value for the land and possible long-term benefits as the land is sold, but short-term benefits are not sufficient to pay the special assessments. This aspect should be checked carefully as part of the general evaluation and review carried out prior to purchase.

Severance taxes—In several states, a *severance* tax is imposed in lieu of, or in addition to, the regular ad valorem property tax. Severance taxes are levied on the mining, or removal, of coal, petroleum, natural gas, ores and sulphur. Some states levy severance taxes on timber.

Typically, severance taxes involve a fixed amount of tax per unit of mineral or timber removed or a percentage tax against gross receipts or market value. For the landowner, a severance tax may be preferable to an annual property tax because severance taxes are levied at a time when funds should be available to pay the tax. Annual taxation of timber or mineral resources may make it necessary to liquidate the resource to pay the tax even though the removal and sale might not otherwise be viewed as timely management of the resource.

Severance taxes are also justified on the basis that the total quantity of the resource is difficult to estimate while in place. This is especially true for mineral deposits.

Prospective land purchasers should check the severance taxes that would be applicable to the land in question and become familiar with the effective rates and taxing procedures.

Recording taxes—Many states impose a documentary or stamp tax on the recordation of deeds. The tax was a federal levy until 1965. Several states impose the tax at rates similar to the federal obligation at the time of repeal of the tax which was 55 cents per $500 of value. A number of states permit the amount of any mortgage or other encumbrance to be deducted before the tax is levied; others do not.

In many areas, sellers of land are responsible for the amount of the tax although the sales contract can specify who has responsibility for payment. In some parts of the country, it is the practice of buyers to pay the tax. As a practical matter, unless agreed upon otherwise in the purchase contract, the buyer may end up paying the tax in order to record the deed. Usually, payment of the tax is required before the deed may be recorded.

Administration of the Property Tax

Because the property tax typically constitutes a significant economic cost, owners of real property should have a general awareness of the procedures by which the tax is assessed and paid. The amount of property tax paid may be influenced by timely objections to changes in valuation, for example, or by timely application for tax exemptions and credits.

Property assessment—Every tract of real property must be appraised for taxation purposes. Although the valuation formula typically sets the assessed value as a percentage of fair market value, the more important question from the standpoint of property tax paid is the value of a tract of real property compared to other taxable properties. With a common millage or tax rate applied within a taxing district, the crucial variable establishing relative property tax burdens is the *comparable* value rather than the absolute value used. Absolute values may have significance in terms of allocations of state property tax relief funds, which are often handled on a formula basis. But otherwise, relative values are the important factor.

In some states, the value of real property used for agricultural purposes is established under special procedures. Some jurisdictions specify that the value of agricultural land is to be set in part (or in total) on the basis of productivity. That usually involves capitalization of expected income from the property at a designated capitalization rate. A number of states provide for special "use" valuation for agricultural land in areas subject to metropolitan or resort area influence with property tax assessments based on the value of the land for farming purposes.

Because relative property values, once established, tend to be continued from year to year unless improvements are made or buildings are damaged by fire, storm or some other casualty, any objection to the established valuation should be made promptly. In each state, a designated time (usually relatively brief) is set aside for filing objections to property valuations. In general, objections to property values require proof that the property in question is over valued relative to other comparable tracts. Most property tax jurisdictions (typically counties) provide forms for filing formal objections to property tax valuations.

The property owner is generally given the opportunity to meet with the local property tax board (known as the board of review in many states) to protest the assessment. The board has the power to correct errors of assessment and to adjust property valuations to bring values into line with what is viewed as equitable as compared with other comparable properties in the taxing district. Before an objection is made to property assessment, careful consideration should be given to the consequences of possible valuation adjustments. For example, an increase in valuation for other properties in the neighborhood could be viewed less than favorably by the other owners.

Further adjustments in property values may occur upon review by the state tax office charged with review of property assessed values. Equalization orders may be entered to assure that assessments are equitable on a relative basis across the boundary lines of local taxing districts. The equalization function is especially important where taxing bodies (such as local school districts) levy a common tax rate in two or more local taxing districts. Unless property assessed values are comparable, inequitable tax burdens may result from such tax levies.

State law usually specifies when all real property is to be revalued.

That usually occurs every two to four years and is another step in maintaining comparable property assessed values. However, even in years that do not involve revaluation, improvements to the land may be added at the direction of the local assessor.

Millage or tax rate—The tax rate imposed on real property historically was expressed in mills (dollars per thousand dollars of assessed valuation). Thus, a millage rate of 80 was $80 per $1,000 of assessed value for taxable property. Some states have abandoned the millage concept and now apply the tax rate as a percentage of assessed valuation.

The tax rate actually applied to a tract of real property is the consolidated rate approved by the local body with tax certifying power. That body, usually functioning at the county level, obtains tax rate requests from local groups entitled to obtain property tax revenues. That usually includes school districts; municipalities; the county itself, for support of county government; area or community colleges; and special groups that have been given legislative authority to raise revenue through the property tax.

Of special interest to agriculture is the fact that county extension districts, in many states, may levy a property tax. Also, a number of states permit property tax levies to support programs to indemnify farmers for loss of livestock because of certain disease problems.

Property tax requests are usually the subject of an open budget hearing scheduled each year by the local governmental body with responsibility for approving the consolidated tax levy. Property owners and others wishing to provide commentary about budget requests or to protest proposed property tax levies have an opportunity to do so at the public hearing.

Once the tax levies have been approved, a computation is made of the actual property tax due for each tract of real property. Typically, the computation is made by 40 acre (or smaller) tracts. A tax statement is mailed to each property owner showing the legal description and assessed value for each tract of real property owned, the tax rates in effect, the date when the net amount of tax is due and the period for payment of the tax without penalty.

Tax collection—State law typically specifies a date when the property tax is due and becomes a lien on the real property. If the tax is paid, the lien is discharged. The usual pattern is for property tax to be paid in two installments each year. Some states operate on a calendar year basis with the first payment for the tax year due in the spring with the final payment due in the autumn. Other jurisdictions operate on a fiscal year basis with the first payment for each tax year made in the autumn with the final payment in the spring.

In the event the tax is not paid by the date for the second payment, a process is set in motion to have the tax collected from the property. From that point forward, the tax lien is discharged only after payment of the property tax due plus interest and any additional costs imposed.

Eventually, if the tax is not paid, the real property is sold at a tax sale with the proceeds used to pay the costs and amounts of tax in default.

Deductions, exemptions and credits—Each state allows property owners to deduct amounts from assessed values for specified reasons and permits designated amounts to be subtracted from the calculated tax. Exemptions for military service during specified periods are allowed in many states. Formal application may be needed to take advantage of this reduction in property tax burden.

Deductions or credits are allowed by many states for residences occupied by the homeowner. An increasing number of states permit a deduction for elderly, disabled or low income taxpayers. A few states authorize a deduction or credit for property tax on lands used for agricultural purposes.

CHAPTER 16
INCOME TAX ASPECTS OF ACQUIRING LAND

Whether farmland is acquired by purchase, by gift or by inheritance, income tax aspects are an important part of the transaction. Initially, the income tax *basis* must be determined for each asset in order to start the depreciation schedule for the depreciable components of the acquisition, such as buildings, fences, tile lines and other depreciable improvements. In addition, some parts of the acquisition may be eligible for investment tax credit. Income tax aspects relating to acquisition of farmland are discussed in this chapter; the income tax considerations on disposition of land are discussed in Chapter 17.

Income Tax Basis

Upon acquisition of a tract of land, an early task involves the allocation of the total income tax basis among the various items of depreciable and nondepreciable property. Income tax basis represents the portion of value of an asset that has already been subject to income tax and need not be taxed again on disposition. Depreciation is limited to the owner's income tax basis in an asset. Income tax basis is, therefore, an important concept both on acquisition of land and on disposition.

Acquisition by purchase—For farmland (or other assets) acquired by purchase, the income tax basis is the purchase price plus such costs as attorney's fees, nondeductible closing costs, and other expenses of acquiring the property. Thus, the initial depreciation schedule is based upon the purchase price. The total purchase price is allocated among the various assets in accordance with their relative fair market values.

>**Example:** A 320 acre farm is acquired for $640,000. The purchaser estimates that the buildings and other improvements represent approximately 10% of the value of the purchase with the land representing the remaining value. In that example, $64,000 of the total $640,000 purchase price would be allocated to the buildings and other depreciable property including fences and tile lines. That allocation must be reasonable based upon the relative values of each component of the purchase.

The initial income tax basis may change. Improvements to the land increase the overall basis by the amount of the improvement. Likewise, depreciation claimed affects the basis with a reduction for the amount of depreciation taken.

Example: Returning to the above illustration of a $640,000 farm, with $64,000 allocated to the depreciable improvements, assume a $30,000 investment was made in grain handling and storage facilities. That would raise the income tax basis to $670,000. If depreciation of $10,000 is claimed in the first year of ownership of the farm, the income tax basis for the entire farm would be $660,000 at the end of the year. Sale of the farm for a greater amount would result in taxable gain.

Upon acquisition of a tract of farmland, it is the purchaser's responsibility to allocate the purchase price (or other income tax basis amount) among the depreciable and nondepreciable components of the acquisition. The amount to be allocated to each asset depends upon relative values. There is no dependable "rule of thumb" for making the allocation. The amounts assigned to each component must be reasonable. Of course, the economic motivation is to allocate as much as possible to the depreciable components so that advantage can be taken of depreciation deductions (referred to as "cost recovery" in the Economic Recovery Tax Act of 1981) and also investment tax credit for eligible property. The allocation of values among the components of the acquisition is entered on the first depreciation schedule.

Acquisition by gift—For farmland acquired by gift, the recipient's or donee's income tax basis carries over from the income tax basis of the giver or donor. Actually, to determine the income tax basis of property acquired by gift, you must know—(1) the income tax basis of the property in the hands of the donor, (2) the fair market value of the property at the time of the gift, and (3) the amount of gift tax, if any, paid.

- If the fair market value of the property at the time of the gift was less than the donor's adjusted basis, the income tax basis for purposes of depreciation is the same as the donor's basis. A key point—when the fair market value of the property is less than the donor's basis, the donee's basis for purposes of determining loss is the fair market value of the property at the time of the gift.

Example: B received 160 acres of farmland by gift from his mother. The farm had been recently purchased by his mother for $150,000. That was her income tax basis and became the income tax basis in B's hands. At the time of the gift, the farmland had declined slightly in value and was valued at $140,000. For purposes of depreciation, B would use the $150,000 figure as the income tax basis. If B later sold the property for $155,000 the income tax basis would be $150,000 adjusted for any improvements made after the gift and for any depreciation claimed. If B later sold the property for $135,000, however, the income tax basis for purposes of determining loss would be the fair market

value at the time of the gift or $140,000. A sale between $140,000 and $150,000 would produce neither gain nor loss.
- If the fair market value of property is greater than the donor's income tax basis, the donor's basis carries over to the donee for all purposes—for computing gain or loss on sale and for purposes of figuring depreciation.

Example: C received from her father 80 acres of land by gift. The fair market value of the tract at the time of the gift was $240,000. The income tax basis of the tract in the hands of C's father was $35,000. For purposes of figuring depreciation, the $35,000 figure would become C's income tax basis. Likewise, if C sold the property for any amount in excess of $35,000 the excess would be taxable gain.

- If any federal gift tax is paid on a gift transaction, the portion of gift tax attributable to the net increase in value of the property may be added to the income tax basis. This is the rule for gifts made after 1976. For gifts made before 1977, the entire amount of federal gift tax paid could be added to the income tax basis of the property.

Example: D acquired 80 acres of land by gift from a parent in 1978. The property had a fair market value of $50,000 and an income tax basis of $20,000. Assume further that federal gift tax of $9,000 was paid on the gift. Of the gift tax paid, $5,400 would be added to the donee's income tax basis. The amount of gift tax to be added to the basis would be determined as follows:

Fair market value	$50,000
Less income tax basis	20,000
Net appreciation	$30,000
Gift tax paid	$ 9,000

    Multiplied by
    $30,000 ÷ $50,000
    = 3/5
    Federal gift tax attributable to net appreciation
    = 3/5 × 9,000
    = 5,400
    Income tax basis of property at the time of gift
    = 20,000
    Income tax basis to recipient
    = 20,000 + 5,400
    = 25,400

As can be seen from the above examples, the giver's or donor's income tax basis carries over to the recipient and becomes the starting point for figuring depreciation and calculating gain (and usually loss) on the transaction. Therefore, recipients of property by gift essentially shoulder the donor's income tax liability.

    Acquisition by inheritance—For property acquired by inheritance, a different rule applies. The starting point, for figuring income tax basis, is usually the value placed on the property in the estate of the person from whom the property was inherited. The value placed on the property for federal estate tax purposes becomes the initial income tax basis for the property.

> **Example:** Grandfather T died August 15, 1980 owning 320 acres of farmland that had been purchased in 1922 for $160,000. The land was valued in T's estate for federal estate tax purposes at $960,000. If the land were sold by the estate six months later for $985,000, there would be $25,000 of gain on the sale.

The rule for determining income tax basis for inherited property means that, at the time of death, all potential gain in the property is eliminated with no tax cost to anyone. Property held until death is, of course, subject to federal estate tax and, in many states, a state death tax. However, part or all of the *income tax* liability is eliminated at the time of death. This constitutes a substantial advantage to retaining property, like land, that has appreciated substantially in value, until death.

In 1976, the Congress moved to eliminate the new income tax basis at death under what became known as the "carryover basis" rules. Those rules would have meant more income tax liability for the heirs on the sale of inherited property. However, the concept of carryover basis was highly controversial, was postponed for three years by legislation in 1978 and in 1980 the carryover basis concept was repealed altogether. Actually, the carryover basis concept could be used on an optional basis for deaths during the period of January 1, 1977 through November 6, 1978 although it is thought that relatively few made use of the concept for deaths during that period.

Under a special rule enacted in 1981, for deaths after 1981 with respect to property acquired after 1981, gifts of appreciated property within one year of death do not receive a new income tax basis at death if the property is acquired from the decedent by the donor of the property or the spouse of the donor. The same treatment applies if the property is sold by the decedent's estate if the donor or the donor's spouse is entitled to the proceeds from sale of the property.

As noted above, the income tax basis becomes equal to the value used for federal estate tax purposes in the estate. In the event that the estate elects to utilize "use" valuation at death, that value also becomes the income tax basis. Therefore, the advantage of use valuation in reducing federal estate tax results in more income tax liability on later sale of the property, if the property is sold, and less depreciation claimable on depreciable assets after death.

Investment Tax Credit
Upon acquisition of farmland, investment tax credit may be available on some components of the purchase. A substantial amount of what would otherwise be viewed as real estate or real property qualifies

for the 10% investment tax credit (a maximum of 11½% for a taxpayer that has an Employee Stock Ownership Plan). The 10% rate, which is the rate available to most taxpayers, is a major tax saver. Investment tax credit may be claimed on acquisition by purchase but not acquisition by gift or inheritance. Effective in 1981, investment tax credit has been amended relative to Accelerated Cost Recovery System (ACRS) property. Among other changes, eligible property in the category of "three year" recovery property is eligible for a 6% investment tax credit with 5, 10, and 15-year eligible property eligible for the regular 10% rate. The amendment bases investment tax credit on the ACRS recovery period rather than on the useful life of the asset. The changes in the rules applicable to property placed in service after 1980 are discussed beginning on page 227.

Property eligible—The investment tax credit was enacted, effective in 1962, to encourage investment in capital assets. The eligibility rules for investment tax credit reflect that Congressional intent. Thus, amounts spent for land itself are not eligible. However, the portion of the purchase price allocable to storage facilities (for the storage of fungible agricultural commodities), feeding floors, paved drives, fences (to keep livestock in or out but not decorative fences), silos, tile lines, outside power and light systems and fruit trees acquired for income producing purposes are all eligible.

In addition, the credit may be claimed on "single purpose agricultural or horticultural structures." That rule, added in 1978, was specifically designed to include greenhouses and livestock confinement facilities. However, the rules are not so limited and extend to facilities "specifically designed, constructed and used" for a particular kind of livestock. There seems little doubt that confinement livestock units are eligible. In addition, other facilities may be eligible if designed, constructed and used for a particular kind of livestock. In one case, a structure used for hay storage and cattle feeding met the test where the storage of hay was incidental to the use of the structure as a feeding facility. The presence of workspace in a facility does not disqualify the unit if related to the livestock operation carried on within the facility. The eligibility of general purpose workspace, such as an office to control the entire farming operation, for investment tax credit may be questioned.

Investment tax credit may also be claimed for rehabilitation expenditures after October 31, 1978, for buildings that have been in use for at least 20 years (30 years after 1981), except for structures used for lodging. To be eligible, at least 75 percent of the walls must remain standing. A building that had been previously rehabilitated is not eligible for the credit until 20 years have elapsed since the prior rehabilitation (30 years beginning in 1982). To be eligible, there must be substantial rehabilitation of the building. The qualifying expenditures over the current year and the preceding year must exceed the greater of the income tax basis of the property or $5,000. The expenditures must be capitalized and must be made for real property with a 15-year recovery period for depreciation. The investment tax credit percentage is 15% for 30-year buildings 20% for 40-year buildings and 25% for certified historic structures. The income tax basis of the property must be reduced by the investment tax credit claimed for 30 and 40-year structures but not for certified historic structures. The investment tax credit on rehabilitation expenditures may not be claimed unless straight line cost recovery is used in depreciating the property.

Example: During 1982, John Anderson spent $15,000 to convert a 60-year old horse barn into a swine farrowing and nursery facility. The conversion included tearing out partitions, replacing the doors, installing ventilating facilities and laying concrete. The useful life given to the expenditures was 15 years. The expenditures should qualify for investment tax credit as rehabilitation expenditures.

Livestock are also eligible for investment tax credit if held for draft, dairy, breeding or sporting purposes (other than horses). Thus, if on acquisition of a farm a beef cow herd is also acquired in the deal, the investment tax credit claimable would be increased. In effect, the availability of a 10% investment tax credit produces a 10% reduction in cost for the asset inasmuch as the investment tax credit is reduced from calculated tax dollar for dollar.

Among other items eligible for investment tax credit are pollution control facilities. In recent years, the full value of certified pollution control facilities has been eligible for investment tax credit even though the fast five year amortization has been used for depreciation purposes.

How the credit is calculated—Investment tax credit may be claimed on the "qualified investment" of eligible property. The qualified investment is a percentage of "cost" and is determined by the useful life of the property or, for property placed in service after 1980, the ACRS recovery period.

The amount of investment tax credit claimable depends also upon the useful life of the asset and how the property was acquired—purchased outright or obtained in an exchange—and whether the item is new or used. Keep in mind that all property acquired on the acquisition of land would be considered used property to the purchaser.

Useful life—For investment tax credit property with a useful life of seven years or more, that is not recovery property, the full amount of investment tax credit is claimable. (See earlier discussion for special rules on pollution control facilities and rehabilitation expenditures.) For most taxpayers, that would be 10% of the "qualified investment" in the asset. Most property placed in service after 1980 is recovery property. Investment tax credit for recovery property is discussed in the next section.

Example: On purchase of a farm, a steel grain bin built several years ago was valued on purchase at $7,000 with a 10-year useful life. The investment tax credit for the grain bin would be $700.

Recovery property—For property placed in service after 1980, investment tax credit is claimed under the ACRS system. For property with a three year recovery peiod. 60% of the investment qualifies for investment tax credit. In effect, the investment tax credit is six percent of the qualified investment in the property. For eligible 5, 10, or 15-year recovery period property, all of the investment qualifies for the regular investment tax credit.

How acquired—As noted above, the amount of investment tax credit claimable depends also upon how the property was acquired—whether obtained by purchase or acquired in a trade. Four rules summarize the procedure for calculating investment tax credit from the standpoint of how the property was acquired.

- For new property acquired outright, the full purchase price would be eligible for investment tax credit. There's an exception for three year recovery property where only 60% of the qualified investment is eligible for investment tax credit.
- On a trade of used items for new, the investment tax credit can be claimed on the undepreciated amount of the old asset traded in plus the cash paid.

Example: A used grain dryer with an undepreciated value of $1,200 is traded for a new dryer with $4,000 paid in boot. Investment tax credit could be claimed on the $5,200 "cost" of the new dryer ($1,200 undepreciated basis for the dryer traded in plus the $4,000 paid in cash).

- For used eligible property acquired by purchase, the full purchase price is eligible for investment tax credit. However, a special limit is imposed on the amount of investment tax credit claimable for used property. That limit has been $100,000 of qualified investment per year ($50,000 on a separate return). The limitation for used property eligible for investment tax credit has been increased to $125,000 in 1981 and to $150,000 in 1985. In the year of a major land acquisition, with substantial improvements on the land eligible for investment tax credit, the limit could be reached. In that event, the taxpayer makes a decision as to which property to select for investment tax credit. That decision should be made primarily on the grounds of (1) the useful life of the eligible item of property and (2) the likelihood that the item might be disposed of before the expiration of the recovery period selected.
- For used property acquired in a trade, generally only the cash boot paid is eligible for investment tax credit. It is not permissible to add the undepreciated amount of the old item traded. This constitutes a disadvantage in "used for used" trades compared to "used for new" trades. For that reason, there is a tendency for individuals wishing to acquire a used item in what would otherwise be a trade situation to sell the old asset and purchase the new one in a separate transaction. That motivation

has resulted in a special rule providing that if old property is sold within 60 days before or after similar used equipment is acquired, investment tax credit can be claimed only on the difference between the undepreciated value of the old property traded in and the price of the used replacement. Thus, only the cash boot paid is eligible for investment tax credit.

The special wash sale rule for livestock should be noted: if livestock eligible for investment tax credit is disposed of and within a period commencing six months before and ending six months after that date substantially identical replacement livestock is acquired, the amount of investment tax credit claimable is reduced.

In the event that investment tax credit is recaptured on the used property traded in, however, investment tax credit may be claimed on the undepreciated basis of the item traded in plus the cash boot paid.

Limit on used property—As noted above, used property is eligible for investment tax credit only to the extent of the first $125,000 of qualified investment. New property does not face a comparable limitation. Thus, it can be quite important whether investment tax credit property is new or used. This may be especially true in a year of acquisition of a farm which may encompass substantial amounts of investment tax credit property, all of which may be used.

Thus, in the year of purchase of a farm, it may be important to note the status of items such as breeding stock.

Example: In the same year that T acquired a 640 acre farm, T purchased 100 head of stock cows at a dispersion sale in the community. The 100 head were mostly four and five year old cows, all of which had produced a calf, although 27 head were bred heifers and 15 head were yearling heifers. The cows that have produced a calf would be considered used investment tax credit property, but the two year old bred heifers and yearlings would be considered new property.

The general rule is that an animal becomes used on the giving of milk or the birth of young.

Property not eligible—Property acquired in some types of transactions is not eligible for investment tax credit even though all of the other eligibility requirements are met. In general, these involve acquisitions from a related person, acquisitions involving prior use of the property or acquisitions from a controlled entity.

- Investment tax credit cannot be claimed on a used asset acquired from a spouse, ancestor, or lineal descendant. Acquisitions of property by children from parents thus are ineligible for investment tax credit. Therefore, if a child purchases a farm, breeding stock or machinery from a parent, no investment tax credit is claimable. Essentially, that rule places a 10 percent premium on purchasing such assets from a parent and a 10 percent price break on purchasing from a stranger.

That is the outcome whether the acquisition is by purchase, gift or inheritance. The same holds true for acquisitions involving the taxpayer and a corporation more than 50% owned, directly or indirectly, by the taxpayer. And comparable rules apply to acquisitions involving trustees and beneficiaries.

The Internal Revenue Service has insisted that property acquired by a partner from the partnership is not eligible nor is property acquired by a partnership from a partner eligible for the credit. Use of property by a partnership has been viewed as constituting use also by the individual partners so that investment tax credit could not be claimed by the partners on acquisition. However, the credit has been allowed in court cases including one where a 50% partner bought out the other partner's interest and the partnership was terminated. Income tax regulations have been proposed by IRS that would allow investment tax credit on acquisitions by a partnership from a partner who owns a 50% or less interest in the partnership and when such a partner acquires otherwise eligible property from the partnership.

- If investment tax credit property is leased by a taxpayer who later purchases the property, the property is not eligible for investment tax credit unless the use before purchase was only "casual."

Example: Thomas Smith, a young farmer, has the opportunity to rent a half section of farmland a mile down the road from a retiring farmer. Because the retiring farmer has no farming heirs, Smith hopes to be able to purchase the land, perhaps at the death of the land owner. If Smith does rent the half section, and later has the opportunity to purchase the tract, *no investment tax credit could be claimed on the purchase*. Prior use of the assets by rental bars investment tax credit for that individual as a purchaser.

If a building site is involved, and the land owner plans to retain substantial control over the building site in any event, it might be well to exclude the building site from the terms of the lease so that those investment tax credit items within the building site would still be eligible for the credit if later purchased by the lessee. Apparently, *any* prior use disqualifies the property from investment tax credit on acquisition even though a period has elapsed from the time of rental and the date of purchase.

Claiming investment tax credit as a lessor—For those purchasing farmland to be leased to tenants, a question may arise as to whether investment tax credit can be claimed on leased assets. There was no problem in claiming investment tax credit as a lessor until an amendment in 1971 created a barrier for investment tax credit by "noncorporate" lessors. The rules added in 1971 apply to individuals, partnerships, trusts, estates and Subchapter S corporations. The limitations do not apply to lessors functioning as regularly taxed corporations.

For a noncorporate lessor to be able to claim investment tax credit

on leased assets, one of two conditions must be met—
1. The property in question must have been manufactured or produced by the lessor in the ordinary course of business, or
2. The term of the lease (including options to renew) must be less than one-half of the estimated useful life of the property, *and* for the first 12 months after transfer of the property to the lessee the sum of deductions allowable to the lessor must exceed 15 percent of the rental income. The "useful life" is the Asset Depreciation Range (ADR) class life as of January 1, 1981.

The tests are further complicated by the fact that for purposes of the 15% rule, the expenditures for depreciation, depletion, interest and taxes cannot be taken into account. That leaves only insurance, repairs and other miscellaneous deductions to count toward the 15% requirement. For many noncorporate lessors, neither of the two tests can be met. The result may be that investment tax credit cannot be claimed by the lessor for leased assets.

There is a special provision permitting a lessor to elect to pass investment tax credit on new property through to the lessee. That can be done even if the lessor is otherwise ineligible to claim the credit because of the noncorporate lessor rule. However, for the pass through to occur, the property must be *new* investment tax credit property in the hands of the lessor and a statement of election to allow the lessee to claim the investment tax credit must be filed within the prescribed time. And that time is the year of acquisition of the property. It is generally not possible to pass through the investment tax credit after the year of acquisition unless failure to pass the credit was an oversight and nothing was done inconsistent with passing the credit to the lessee.

For many lessors, other than regularly taxed corporations, the noncorporate lessor rules constitute a significant barrier to claiming investment tax credit. For some, the pass through of investment tax credit on new property is an acceptable solution, particularly if the lessee is closely related to the lessor or is the lessor's controlled partnership, corporation or other business entity. Note, however, that there is no pass through of investment tax credit for *used* property. Therefore, on purchase of farmland, it would not be possible for a lessor to pass through investment tax credit on the items accquired in the purchase.

Although the line is not drawn with clarity, it would appear that at some point a lease with active involvement by the lessor should no longer be characterized as a lease for this purpose. There is authority in other settings that a crop share or livestock share lease with active involvement by the lessor is no longer considered a lease but becomes a business relationship. Whether that will be true in the noncorporate lessor area remains to be seen. A good argument can be made, however, that a crop share lease with material participation by the lessor (or possibly by an agent of the lessor such as a farm manager) would elevate the relationship into a business status and thus avoid the noncorporate lessor limitations.

Also, it should be noted that for assets such as grain storage facilities to be used for the storage of the lessor's part of the crop under a crop share lease, arguably the facility is not leased to the lessee but rather is retained by the lessor for storage of the lessor's crop. Thus, investment tax credit on storage facilities under those conditions should be claimable.

To sum up, the noncorporate lessor rule warrants careful consideration any time investment tax credit property is acquired and leased to a lessee. The rule is particularly notable on the acquisition of real property because any investment tax credit items would be considered used property with no pass through of investment tax credit from lessor to lessee.

"At Risk" Limitation—For property placed in service after February 18, 1981, investment tax credit is not allowed for investments in new or used investment tax credit property to the extent that the amounts invested are not "at risk." Amounts are not considered "at risk" if—(1) The taxpayer is protected against the loss of the investment amount, (2) the amount was borrowed and the taxpayer is not personally liable for repayment, (3) the lender has an interest other than as a creditor or (4) the lender and the borrower are related parties.

When the credit is claimable—Investment tax credit is claimed in the year the eligible property is "placed in service." Generally, that is the year of acquisition. However, property is considered "placed in service" in the earliest of the following—

- In the taxable year in which the period for depreciation begins under the taxpayer's depreciation practice, or
- In the taxable year in which the property is placed in a condition of readiness and availability for a specifically assigned function.

Example: A contract is signed for purchase of a farm on November 1, 1981 with $500 in earnest money paid at that time. The contract calls for 25% of the purchase price to be paid the following March 1 with possession given on that date. The year of possession would generally be considered to be the year of acquisition and the year the property was placed in service for the purchaser as a taxpayer. That would be 1982 in this example.

Example: Elmer Jones ordered a new grain dryer on December 1, 1981 and paid the full purchase price. As of December 31, the grain dryer had not yet been built by the manufacturer. The asset could not be considered placed in service for investment tax credit purposes for that year. Likewise, if the grain dryer had been built and was sitting on a loading dock at the factory at the close of business on December 31, the asset would not be considered to have been placed in service that year by the taxpayer.

If the grain dryer was sitting on the dealer's lot on December 31 in crates, the investment tax credit would still not be claimable by the purchaser for that year. However, if the grain dryer had been set up and was delivered late on the afternoon of December 31, investment tax credit should be claimable for the taxable year ending that December 31 because the dryer was in a condition of readiness and availability for use.

Maximum credit claimable—For most taxpayers, investment tax credit can offset the first $25,000 of calculated income tax plus 80 percent of the tax liability above $25,000. This is the rule for 1981. The percentage figure rises to a permanent level of 90 percent of tax liability above $25,000 in 1982.

Recapture of investment tax credit—If property on which investment tax credit has been claimed is disposed of before the end of the estimated useful life or recovery period for investment tax credit purposes, the credit is recomputed using the actual number of years owned. If the allowable credit thus computed is less than the amount of credit originally claimed, the difference is added to the current year's tax bill.

Example: J purchased a 160 acre farm with possession obtained March 1, 1980. Investment tax credit was claimed on $80,000 of qualified investment including a silo, a grain drying and storage complex, feeding floors, tile lines, fences and outside power and light systems. All assets were entered on the depreciation schedule for investment tax credit purposes with a useful life of fifteen years. Thus, the maximum investment tax credit of $8,000 (10% of $80,000) was claimed by J. On May 1, 1982, J received an offer for the property at double the price paid in 1980. J accepted the offer. All investment tax credit would be recaptured and would have to be repaid as additional income tax.

For recovery property placed in service after 1980, the recapture percentage is determined from the following table.

When recovery property ceases to be eligible	The recapture percentage is— for 5, 10 or 15 year property	for 3-year property
One full year after placed in service	100	100
During second full year	80	66
During third full year	60	33
During fourth full year	40	0
During fifth full year	20	0
After five years	0	0

For recovery property, the increase in income tax because of recapture is limited to investment tax credit that was used to reduce tax liability. Investment tax credit carrybacks and carryovers are to be adjusted for investment tax credits not used to reduce tax liability.

For purposes of recapture of investment tax credit, a disposition includes a sale, a taxable exchange and a tax-free exchange such as a trade. Thus, most property dispositions trigger recapture of investment tax credit.

For property transferred to a new partnership or corporation in a tax-free exchange, there is no recapture of investment tax credit so long as the transfer is a "mere change in the form of doing business." Care is needed for such transfers to assure that any restructuring of the business (where some assets are left out of a newly created entity or assets are added to a new entity) does not trigger investment tax credit recapture.

In a 1976 revenue ruling, investment tax credit was recaptured on transfer of all investment tax credit assets at the time of incorporation of a dental practice where the building, representing 30 percent of the value of all assets in the dental practice, was not transferred to the new corporation. In three 1981 cases, investment tax credit was not recaptured on formation of farm corporations where the land was retained in individual ownership and not transferred to the newly formed corporation. Of course, there is no recapture of investment tax credit if the change of organizational form involves no alteration in the assets comprising the business activity.

Investment tax credit may also be recaptured if the property ceases to be eligible for the credit. For example, if an automobile acquired for business use is converted to use as a personal asset, the credit must be recomputed with the excess amount repaid as additional tax.

There is no recapture on death of the property owner. In fact, investment tax credit may be taken on property acquired in the year of death. Property owned by a deceased individual or held in a partnership or tax option corporation is treated as if held for its full estimated useful life on death of a partner or shareholder, as the case may be.

A change in the interest of a shareholder of a tax option (Subchapter S) corporation or a partner in a partnership, for example by gift, may result in a partial or total recapture of investment tax credit. Recapture occurs if corporate stock (or ownership of a partner's interest) is brought below two-thirds of the owner's interest on the date the property (with respect to which the credit was claimed) was acquired by the corporation (or partnership). The same basic rule applies to the beneficiary of an estate or trust.

Example: On October 1, 1980, at a time when shareholder A owned 90 percent of the stock in a tax option corporation owning farmland, the corporation acquired two new grain storage bins with dryer and augers. The estimated useful life of the equipment was 15 years with a 10 percent investment tax credit claimed on the total cost of $35,000. If at any time before October 1, 1987, the ownership of shareholder A drops below 60 percent of the stock in the corporation, part or all of the $3,500 of investment tax credit on the investment would be recaptured.

If a regularly taxed corporation elects to be taxed under Subchapter

S of the Internal Revenue Code (as a tax option corporation), investment tax credit is recaptured on the corporate assets unless agreements are executed by the corporation and the shareholders in which the signers agree to pay any recaptured investment tax credit. It should also be noted that, for investment tax credit purposes, a Subchapter S corporation apportions the qualified investment in new and used investment tax credit property pro rata among its shareholders as of the last day of the corporation's taxable year. Any recaptured investment tax credit is payable by the shareholders receiving the original benefit.

Depreciation—For any depreciable asset with a useful life of more than one year, a deduction can be claimed each year of the useful life of the asset representing recovery of the investment in the asset. Some items of property are not viewed as depreciable. Only property used in a trade or business or held for the production of income is depreciable. And, unless the life of the property is determinable, no depreciation may be claimed.

Normally, land is not depreciable. In a limited number of situations, land resources may be depletable. For natural resources such as sand, gravel, coal, oil, natural gas or other minerals, a depletion allowance can be claimed. Depletion may be calculated as a fraction of the cost investment in the depletable asset or as a percentage of gross income from the resource. Guideline percentages have been established for such percentage depletion allowances.

In one part of the country, in the area of the Ogallala water formation in the Southwest, an allowance has been permitted for the reduction in level of groundwater because of irrigation and other water uses. In general, however, soil is not otherwise depletable even if erosion occurs at a rate faster than soil formation takes place.

Amount depreciable—The amount of depreciation that may be claimed is limited to the taxpayer's income tax *basis* in the property. As noted earlier in this chapter, the income tax basis of the property depends upon how it is acquired. For purchased property the income tax basis is ordinarily the cost. For assets acquired by gift, the income tax basis is generally the donor's basis although there may be an adjustment for part or all of the federal gift tax paid. For assets acquired by inheritance, the income tax basis for the property is derived from the value placed on the property in the estate settlement process for federal estate tax purposes. In all instances, the original income tax basis is adjusted upward for improvements made and downward for depreciation taken.

"Expense Method" Depreciation—For depreciable tangible personal property placed in service after 1980, all or part of the income tax basis can be deducted currently. The aggregate basis amount eligible for the deduction is as follows:

	Maximum Expense Amount

1981	-0-
1982	$ 5,000
1983	5,000
1984	7,500
1985	7,500
1986 and thereafter	10,000

Several limitations and conditions are imposed on the property to be eligible for expense method depreciation—(1) property not subject to investment tax credit is generally not eligible, (2) property acquired by gift or inheritance is not eligible, (3) property acquuired by estates or trusts is not eligible, (4) noncorporate lessors may utilize the provision only if one of the two tests is met for noncorporate lessors to claim investment tax credit, (5) for property traded in, only the cash boot paid is eligible for expense method depreciation, and (6) the property was not acquired from ineligible party (spouse, ancestor or lineal descendant) or a controlled entity.

If expense method depreciation is disposed of, the rules on recapture of depreciation apply to the gain. Gain on the property is treated as ordinary income to the extent of the amount expensed and depreciation taken with respect to the property.

The maximum allowable deduction per year is limited to one-half the above amounts for married individuals filing a separate return. For a partnership the dollar limitations apply to both the partnership and to each of the partners. Members of a group of controlled corporations divide the expensed amount.

In general, expense method property is not eligible for investment tax credit. The portion of cost that is expensed is not eligible for the credit.

Accelerated Cost Recovery System (ACRS)—The Economic Recovery Tax Act of 1981 repealed The Asset Depreciation Range (ADR) system of depreciation that was rarely used except by large businesses and replaced the ADR system with The Accelerated Cost Recovery System (ACRS). Under ACRS, the cost of an asset may be depreciated or recovered over a period shorter than the assets' useful life.

A taxpayer can elect to exclude property from ACRS if for the first taxable year for which a deduction would be allowable the property is properly depreciated "under the unit-of-production method or any method of depreciation not expressed in years . . ."

The cost of most eligible personal property can be recovered over 3, 5, 10 or 15 years. Property is classified by recovery period as follows:

3-years Automobiles, light duty trucks, R&D equipment, breeding hogs and personal property (or other "Section 1245" property which means property treated as personal property for depreciation recapture purposes) with a class life of four years or less.

5-years Personal property with a class life of five or more years including all breeding stock (other than swine) and most other equipment except long-lived public utility property. Five-year property also includes farm storage facilities (corn cribs, grain bins and silos), fences, tile lines, water systems, outside power and light systems, depreciable parts of dams and ponds, paved drives and feeding floors. Also includes single purpose agricultural and horticultural structures (and petroleum storage facilities) which are designated as section 1245 property in the 1981 legislation. This is where most farm personal property falls. The term does not include buildings or their structured components.

10-years Public utility property with an ADR midpoint life greater than 18 but not greater than 25 years, railroad tank cars, residential manufactured homes and real property with an ADR midpoint life of 12.5 years or less.

15-years Depreciable real property that does not have a class life of 12.5 years or less; public utility property with a class life exceeding 25 years.

Taxpayers have the option of using straight line cost recovery over the regular or optional longer recovery periods or a prescribed accelerated method over the regular recovery period. The same recovery period must be used for all property (other than 15-year real property) acquired in a taxable year in the same class for which an optional period election has been made.

A taxpayer may elect to claim deductions over the regular recovery period or optional longer recovery periods. The rules vary, however, depending upon whether the property is personal property or real property. For personal property, the optional recovery periods are:

	Optional Periods
3-year property	5 or 12 years
5-year property	12 or 25 years
10-year property	25 or 35 years
15-year property	35 or 45 years

The amount of depreciation claimable depends upon the recovery period for the property and the number of years since the property was placed in service.

Any election under ACRS is made on the taxpayer's income tax return for the tax year the property is placed in service.

Property placed in service during 1981–84

Recovery year	Applicable percentage for the class of property		
	3-year	5-year	10 year
1	25%	15%	8%
2	38%	22%	14%
3	37%	21%	12%
4	—	21%	10%
5	—	21%	10%
6	—	—	10%
7	—	—	9%
8	—	—	9%
9	—	—	9%
10	—	—	9%
	100%	100%	100%

Property placed in service in 1985

Recovery year	Applicable percentage for the class of property		
	3-year	5-year	10-year
1	29%	18%	9%
2	47%	33%	19%
3	24%	25%	16%
4	—	16%	14%
5	—	8%	12%
6	—	—	10%
7	—	—	8%
8	—	—	6%
9	—	—	4%
10	—	—	2%
	100%	100%	100%

Property placed in service after 1985

Recovery year	Applicable percentage for the class of property		
	3-year	5-year	10-year
1	33%	20%	10%
2	45%	32%	18%
3	22%	24%	16%
4	—	16%	14%
5	—	8%	12%
6	—	—	10%
7	—	—	8%
8	—	—	6%
9	—	—	4%
10	—	—	2%
	100%	100%	100%

FARMLAND

Limits on depreciation for real property—Since 1969, limitations have been imposed on the rate at which most items of depreciable real property could be depreciated. In general, depreciable real property such as buildings, fences, and tile lines could not be depreciated under the double-declining balance or sum-of-the-years-digits methods. The only exception was for new residential rental housing. For other new depreciable real property, acquired after July 24, 1969, depreciation could be claimed no faster than the 150% declining balance method.

For used depreciable real property (except for used residential rental housing), acquired after July 24, 1969 and before 1981, depreciation was limited to the straight line method. For used residential rental housing, having a useful life of 20 years or more, depreciation could be claimed under the 125% declining balance method.

In general, the 1981 legislation assigns an ACRS 15-year recovery period for depreciable real property but a taxpayer may elect a 35-year or 45-year recovery period. Real property other than low-income housing, is depreciated using the 175% declining balance method changing to the straight line method to obtain the full depreciation benefit if accelerated cost recovery is utilized.

The following cost recovery table applies to all 15-year real estate except for low income housing and assumes accelerated cost recovery.

Table 16.1 ACRS Cost Recovery Tables for Real Estate (Except Low-Income Housing)

If the Recovery Year is:	The applicable percentage is: (Use the Column for the Month in the First Year the Property is Placed in Service)										
	1	2	3	4	5	6	7	8	9	10	11	12
1	12	11	10	9	8	7	6	5	4	3	2	1
2	10	10	11	11	11	11	11	11	11	11	11	12
3	9	9	9	9	10	10	10	10	10	10	10	10
4	8	8	8	8	8	8	9	9	9	9	9	9
5	7	7	7	7	7	7	8	8	8	8	8	8
6	6	6	6	6	7	7	7	7	7	7	7	7
7	6	6	6	6	6	6	6	6	6	6	6	6
8	6	6	6	6	6	6	5	6	6	6	6	6
9	6	6	6	6	5	6	5	5	5	6	6	6
10	5	6	5	6	5	5	5	5	5	5	6	5
11	5	5	5	5	5	5	5	5	5	5	5	5
12	5	5	5	5	5	5	5	5	5	5	5	5
13	5	5	5	5	5	5	5	5	5	5	5	5
14	5	5	5	5	5	5	5	5	5	5	5	5
15	5	5	5	5	5	5	5	5	5	5	5	5
16		1	1	2	2	3	3	4	4	4	5

(Note: This table does not apply for short taxable years of less than 12 months.)

As noted above, straight line cost recovery could be used over the regular or optional longer recovery periods.

If non residential real property is depreciated under an accelerated method of depreciation, all gain is treated as ordinary income to the extent of all recovery period deductions previously taken. However, if straight line cost recovery is used, *none* of the gain is recaptured as ordinary income on later sale.

Extra first year 20% depreciation—For some property, through 1980, a taxpayer could deduct in the first year up to 20% of the cost of new or used depreciable tangible personal property with a useful life of at least six years. This extra first year 20% depreciation deduction was repealed in 1981 effective at the end of 1980.

Salvage value and useful life—In establishing the depreciation schedule on acquisition of farmland, it has been necessary to set a useful life for each depreciable asset. Under the ACRS cost recovery system, salvage value is not taken into account, however.

Other Income Tax Deductions

Once the purchase of land has been completed, several income tax rules come into play that influence the amount of deductions claimable by the purchaser.

Deductibility of interest—In general, interest paid is income tax deductible by a purchaser of land who finances the acquisition by mortgage or under an installment land contract. Several limitations, however, exist on the interest deduction.

- If the farmland is acquired as an investment asset, and not for the purpose of carrying on the business of farming, the rules limiting investment interest would be applicable. Those rules provide that the amount of interest deductible to carry investment assets is limited to $10,000 plus the net investment income and excess expenses from "net lease" property. There is an exception where a family is attempting to acquire a 50% or greater interest in a partnership or corporation. Amounts of interest above the limitation can be carried forward to future years and deducted, again within the limitation, for those years.
- Limitations exist when interest can be claimed as an income tax deduction. For a number of years, a taxpayer could prepay up to one year's interest without the deduction being disallowed as distorting income. However, under legislation enacted in 1976, taxpayers are placed essentially on an accrual basis for purposes of deductibility of interest. A taxpayer may not prepay interest. Any interest prepaid is accrued and deducted in the year the interest represents the payment for the use of funds. The only exception to that rule is for the payment of "points" on a home mortgage where the payment of "points" is a common practice in the community.

FARMLAND

Soil and water conservation expense—A deduction may be claimed for expenditures for soil and water conservation with respect to land used in farming. The maximum deduction claimable in any year is 25% of gross income from farming. Excess amounts may be carried over to the next taxable year, again to offset 25% of gross income from farming in that year. There is no limit on the number of years of carryover.

The decision to deduct soil and water conservation expenditures is a one-time election. The way the expenses are handled the first time that a taxpayer has soil and water conservation expenses constitutes an election. Therefore, if a deduction is not made in the first year the taxpayer has soil and water conservation expenses, the expenditures automatically become part of the income tax basis for the land with no benefits received until the land is later sold and then the expenses merely offset capital gain. Thus, in the first year that a taxpayer has soil and water conservation expenses, attention should be given to the matter of how the expenses should be handled.

In general, most taxpayers benefit to a greater degree from handling the expense as a deduction. One instance in which a taxpayer would not benefit as greatly from a deduction would be where a large soil and water conservation investment is anticipated and the taxpayer contemplates not having gross farm income in the carryover years. In that event, the carryover could be lost to the extent the individual did not have gross income from farming in a future year. Adding the expenditures to the income tax basis might be a better choice in that situation.

If soil and water conservation expenditures are deducted, and the land is held by the taxpayer for less than ten years, part or all of the expenditures are recaptured as ordinary income on disposition of the land. Note that the minimum holding period for avoiding recapture of deductions is for a holding period of more than ten years *for the land*. It is not a question of the number of years since the deduction was claimed. If the land was held for less than five years. 100% of the deduction amount is recaptured as ordinary income on sale. If the land is held between five and ten years, the amount recaptured on sale reduces from 100% to 0%.

Land clearing expenditures—A taxpayer may deduct expenditures for clearing land to make it suitable for use in farming. The deduction is limited to $5,000 or 25% of *taxable* income from farming. Although the land clearing deduction is similar in many respects to the soil and water conservation expense deduction, the two are different in several aspects. One difference is the maximum amount claimable as noted.

The land clearing expense deduction can be claimed in any year or can be added to income tax basis for the land. The election is not binding on all future years as it is with soil and water conservation expense deductibility. Another difference is that there is no carryover of excess amounts above the allowable deduction for land clearing expenses. Amounts above the allowable level each year are added to the income tax basis of the land.

There is a similar recapture rule for land clearing expenses as for soil and water conservation expenses. If the land is held for less than five years, 100% of the land clearing expense is recaptured as ordinary income on sale. If the land is held for more than ten years, none of the land clearing expenses is recaptured as ordinary income. For holding periods between five and ten years, the recapture amount is reduced from 100% to 0%.

As with soil and water conservation expense, land clearing expense deductibility is usually preferable to adding the expense amounts to the income tax basis of the land.

CHAPTER 17
DISPOSITION OF FARMLAND BY SALE OR EXCHANGE

Disposition of farmland deserves careful attention. The transaction typically represents a substantial amount of capital and a large amount of gain for income tax purposes. The range of alternatives for disposal of land is impressive with the choices for farmland or other real estate including sale for cash, installment sale, disposition by private annuity or exchange as well as transfer by gift and disposition at death. The choices may be differentiated by income tax liability, involvement of the seller in the financing of the transaction and the risks assumed by the seller.

In this chapter, emphasis is placed upon disposition of farmland by sale during life; in Chapter 18, discussion focuses on disposition of farmland at death. Disposition by gift is covered in Chapter 19.

Preparing for Sale
Several steps should be taken by a farmland owner anticipating disposition by sale or exchange before the land is listed for sale or otherwise placed on the market. Such planning may influence the method of disposition selected.

Calculating gain—The amount of gain on land that is under consideration for transfer should be calculated early in the process of planning for its disposition. As noted in Chapter 16, the amount of potential taxable gain depends upon (1) the adjusted income tax basis of the property and (2) the selling price. The selling price is adjusted downward for costs of sale including commissions, fixing up expenses and other costs involved in the transaction.

Calculation of the adjusted income tax basis is discussed in the preceding chapter. In brief, the income tax basis depends upon how the property was acquired, the amount of improvements made since acquisition and the allowable depreciation on those assets for which depreciation may be claimed.

- For *purchased* property, the adjusted income tax basis is equal to the purchase price plus improvements made and minus depreciation allowable. It is noted that depreciation is assumed *even though depreciation may not have been actually claimed*.

Example: M had purchased 320 acres of farmland in 1940 for $32,000. Over the years, $60,000 of improvements in livestock facilities had been made and $45,000 of depreciation had been claimed. In addition, a $15,000 tenant house built in 1946 with a 30-year useful life was inadvertently omitted from the depreciation schedule. The current adjusted income tax basis would be:
$32,000 + ($60,000 − $45,000) + ($15,000 − $15,000) = $47,000

Thus, any gain received by M above $47,000 would be taxable gain. The tenant house is treated as though it had been depreciated.

- For property acquired by gift, in general the income tax basis is that acquired from the donor of the property. Again, improvements since the date of the gift are added to the donor's basis and depreciation is subtracted.

Example: T acquired 80 acres of land by gift from her mother in 1963. T's mother had acquired the tract, comprised only of bare land, by purchase in 1942 for $3,000. There was no gift tax paid on the gift transaction; fair market value as of the date of the gift was $33,000. The mother's income tax basis of $3,000 would carry over and become T's income tax basis for the property.

- For property acquired by inheritance, the income tax basis goes back to the federal estate tax value placed on the property in the estate plus improvements made since the date of death and minus depreciation taken.

Example: D received 40 acres of land by inheritance at the death of her grandfather in 1950. The value placed on the property for federal estate tax purposes was $4,000. In 1952, D built a grain bin on the 40 acre tract at a cost of $2,000 which had been fully depreciated by 1970. D's adjusted income tax basis presently would be $4,000.

Recapture of investment tax credit—Disposition of farmland by sale or exchange generally triggers recapture of investment tax credit. Recapture occurs on all transfers, including a tax-free trade, except for transfers at death or transfers to another entity such as a partnership or corporation in a tax-free exchange where it is a "mere change in the form of doing business." Thus, if investment had been made in property eligible for investment tax credit within the recapture period prior to sale, the amount of investment tax credit that would be recaptured should be calculated. As noted in Chapter 16, the investment tax credit is refigured using the number of years the property was actually held. The investment tax credit appropriate for that holding period is subtracted from the investment tax credit actually claimed with the difference being recaptured in the year of disposition. Most items eligible for investment tax credit that would be sold in conjunction with farmland would likely have a useful life of five years or longer. Therefore, the maximum investment tax credit has generally been claimed on the items.

Property typically eligible for investment tax credit includes storage facilities, fences, tile lines, water wells for livestock, paved feeding floors and drives, outside power and light systems, silos, "single-purpose agricultural structures" (such as confinement livestock units) and certified pollution control facilities.

Recapture of depreciation—Part (or all in some instances) of the gain on disposition of farmland may be recaptured as ordinary income rather than being taxed as capital gain. For purposes of recapture of depreciation, there are two rules that could be applicable.

- For any depreciable personal property (machinery, depreciable livestock and equipment) all gain on sale that represents depreciation claimed since 1961 (since 1969 for livestock) and cost recovery since 1980 is recaptured as ordinary income on sale. The amount of the gain to be reported as ordinary income because of the recapture rules is calculated on Form 4797, Part III, with the total transferred to Schedule D and then to the individual income tax return, Form 1040, or the appropriate partnership or corporation return.

The same recapture rules apply to storage facilities such as corn cribs, grain bins, and silos.

Example: R is contemplating the sale of a 320-acre farm which has been operated under a livestock share lease with a resident tenant. As part of the sale, R anticipates selling 100 head of stock cows, all which were purchased in 1974. The cows were purchased for $24,000 and $18,000 of depreciation and cost recovery has been claimed on the cows. If the portion of the selling price applicable to the cows is $52,000, the total gain on the cows would be $46,000 of which $18,000 would be recaptured as ordinary income. That amount represents the depreciation and cost recovery claimed since 1969. The rest of the gain, or $28,000, would be reportable as long-term capital gain. Again, the calculations would be handled in Part III of Form 4797.

The same procedure would be followed for machinery and equipment disposed of in connection with the sale.

- For most items of depreciable real property (except storage facilities and items included under the rule discussed above) a different set of recapture provisions applies. In general, recapture of depreciation is limited to the amount of "excess" depreciation claimed since 1963. Excess depreciation is defined as depreciation claimed in excess of the straight-line rate. However, the amount of depreciation actually recaptured as ordinary income is determined by four rules, depending upon when the depreciation was claimed:

    (1) For depreciation claimed after 1963 and before 1970, all

FARMLAND

excess depreciation is recaptured as ordinary income except that the amount of recapture phases out on the basis of one percent per month for each month over 20 months the property was held. Thus, if property was held for 19 months, the *entire* amount of excess depreciation would be recaptured as ordinary income. If the property was held 120 months, there would be no recapture. Between 20 and 120 months, the amount of excess depreciation recaptured reduces by one percent per month.

(2) For depreciation claimed after 1969 and before 1976, the full amount of excess depreciation is recaptured as ordinary income with no phase out except for depreciable residential property. For depreciable residential property, a special phase-out applies commencing 100 months after acquisition and ending 200 months after acquisition. Thus, for real property held 99 months, all excess depreciation would be fully recaptured as ordinary income. Property held 200 months would face no recapture of depreciation on depreciable real property. Between 100 and 200 months, there is a reduction of one percent per month for each month above 100 the property was held.

(3) For depreciation claimed after 1975, all excess depreciation is recaptured as ordinary income with no phase out applicable except for low income housing.

(4) After 1980, if non-residential real property is depreciated under an accelerated method of depreciation, all gain is treated as ordinary income to the extent of all recovery deductions previously taken. If 15-year real property is depreciated under straight line cost recovery for 15, 35 or 45 years, no part of the gain or sale is recaptured as ordinary income. Thus the selection of a method of depreciation has important income tax implications if the property is likely to be sold.

Recapture of soil and water conservation expense and land clearing expense—Disposition of land within 10 years after acquisition may lead to recapture of all expenditures made for soil and water conservation or land clearing purposes. Although the provisions for soil and water conservation expense and land clearing expense *deductibility* are different in several respects, the *recapture* provisions are identical. If the land is disposed of within five years after acquisition, 100% of the amount of soil and water conservation expense or land clearing expense claimed as an income tax deduction is recaptured as ordinary income. That assumes, of course, there is that much gain on the disposition. If the gain on disposition is less than the amount of such expenses, only the amount of the gain would be recaptured as ordinary income. If the land had been held for 10 years or more, no part of the expense amounts would be recaptured as ordinary income. Between 5 and 10 years, a portion of the expense amount (a 20% reduction for each year beyond five) would be recaptured as ordinary income.

Note that the recapture provision for soil and water conservation expense or land clearing expense deductibility is based upon *the number of years the land had been held*. It is immaterial how many years have elapsed since the expenditures for soil and water conservation expense or land clearing expense had been incurred.

Excess Deductions Account—For taxpayers who had incurred net farm losses and the net farm losses had created a balance in their Excess Deductions Account, gain on sale of "farm recapture property" may be recaptured as ordinary income rather than being taxed as capital gain. This was an "anti-tax shelter" provision enacted in 1969. Gain on disposition *of land* is not recaptured under the Excess Deductions Account provision, except to the extent of soil and water conservation expense or land clearing expense deductions claimed. Gain attributable to machinery, equipment, breeding stock or buildings could be recaptured to the extent that the taxpayer has a balance in the taxpayer's Excess Deductions Account. For most farmers, there is no balance in their EDA account because only net farm losses in excess of $25,000 each year were added to the EDA and then only if the taxpayer had nonfarm income in excess of $50,000. Those exemptions, however, do not apply to regularly taxed corporations which could have an Excess Deductions Account balance and hence could encounter recapture on sale of "farm recapture property."

Government cost sharing payments—On sale or other disposition of land, amounts representing government cost sharing benefits that were previously excluded from income may be recaptured as ordinary income. Payments from several different federal and state programs dealing with soil and water conservation, forestry and the environment may be excluded from income. Whether to exclude the payment amounts from income or to report the benefits as income with allowable deductions and credits claimed as an offset is a matter for each taxpayer to decide.

If excluded from income, on sale or other disposition within 20 years after receipt of the payments, part or all of the gain is recaptured as ordinary income representing the benefit amounts previously excluded. For disposition within 10 years, all of the benefits are subject to recapture. Beyond 10 years, the recapture amount is reduced by 10% for each year the property was held after receipt of payments.

Expenses of crop production—As a general rule, if land used in farming or ranching and held for more than one year is sold with an unharvested crop, the entire gain, both that applicable to the land and that applicable to the growing crop, is long-term capital gain. For that result, the crop and the land must be sold at the same time to the same person in one transaction.

However, the costs of growing the unharvested crop must be capitalized (added to the income tax basis of the property) and not deducted. The same holds true for depreciation attributable to production of the crop. If the production expenses are for another taxable year, an amended income tax return is to be filed reflecting the disallowed deductions.

Reporting Income Tax Gain

Upon sale or other disposition of land, the income tax liability must be paid in the year of the transaction or the gain deferred under installment sale or private annuity rules. Those arrangements for reporting gain are discussed in the sections following. In this section, the reporting of gain as capital gain or ordinary income is discussed.

For land held as an investment, the asset is a capital asset and gain is entitled to capital gain treatment. For land held for one year or less, any gain would be short-term capital gain. For land held more than one year, any gain would be long-term capital gain except to the extent of recapture of depreciation, investment tax credit, soil and water conservation expense or land clearing expense, government cost sharing payments or, on sale of "farm recapture property," to the extent of the balance in the taxpayer's Excess Deductions Account.

For land used in a trade or business, a special set of rules applies for the taxation of gain. Net gains for such property held for the requisite period for long-term capital gain treatment (more than one year for land) are entitled to be taxed as long-term capital gains. Net losses on such property, however, are treated as ordinary losses. This distinction can be quite important if a taxpayer has a net loss on property used in the trade or business.

The way in which long-term capital gains are taxed depends upon whether the taxpayer is an individual or a corporation.

Individual taxpayers—For individuals, long-term capital gains are eligible for a 60% deduction with 40% of the long-term capital gain taxable as ordinary income. The capital gain is reported on schedule D with the deduction made on that form. For individuals, the top income tax rate is reduced from 70% to 50% in 1982 with repeal of the maximum tax also in 1982. The maximum tax rate on capital gains is, therefore, reduced from 28% to 20% (50% rate times 40% of capital gain taxable as ordinary income). A special alternative tax for 1981 provides for a maximum 20% rate on long-term capital gains for sales or exchanges after June 9, 1981.

For individuals, capital gains are no longer subject to the "add on" minimum tax on preference income. Since the beginning of 1979, capital gains (and excess itemized deductions) have been subject to an alternative minimum tax. It should be pointed out that the alternative minimum tax is not limited merely to instances in which taxpayers have capital gains income. The alternative minimum tax can be applied in any situation in which the alternative minimum tax produces a greater tax than the taxpayer's regular tax for the year.

The alternative minimum tax is calculated by adding to the taxpayer's other income the excluded part of the capital gains and excess

itemized deductions. From that is subtracted a $20,000 exemption. The tax is applied at a rate of 10% on the next $40,000 of income and 20% on all above $60,000. If the alternative minimum tax is greater than the regular tax, the taxpayer pays the greater amount. Most credits, such as investment tax credit, now reduce the alternative minimum tax.
as investment tax credit, now reduce the alternative minimum tax.

Example: Larry and Barbara Warren file a joint income tax return, using the zero bracket amount (no itemized deductions) showing wages of $20,000, business income of $14,000 and a long-term capital gain of $100,000. The regular tax and the alternative minimum tax would be computed as follows:

	Regular tax	Alternative minimum tax
Wages	$20,000	
Business income	14,000	
Capital gain (60% deductible)	40,000	
Adjusted gross income	$74,000	74,000
Regular tax	$24,102	
Plus capital gains exclusion		60,000
Alternative minimum taxable income		134,000
Exemption		−20,000
Balance		114,000
Calculation of alternative minimum tax:		
	$ 40,000 × 10% =	4,000
	74,000 × 20% =	14,800
Total	$114,000	$18,800

Because the regular tax of $24,102 exceeds the alternative minimum tax of $18,800, the regular tax would be paid.

Corporations—For corporations, the 60% deduction for long-term capital gains income does not apply. That deduction is only applicable to individual taxpayers. For regularly taxed corporations, long-term capital gains income is taxed as ordinary income if the corporation is in a tax bracket of 20% or below. In the event the corporation is in a tax bracket higher than 20%, the flat rate of 28% applies on corporate long-term capital gains. Thus, a corporation should never pay a greater rate of federal income tax than 28% on its long-term capital gains income.

For a corporation, the "add on" minimum tax continues to apply to capital gains. The alternative minimum tax, effective in 1979 for individuals, is not applicable to corporate taxpayers. Thus, for corporations, a minimum tax of 15% is imposed on the total of tax preference items (including capital gains) reduced by the greater of $10,000 or the full amount of the regular income tax.

    Example: ABC Farm, Inc. has preference income of $90,000. The regular income tax for the year is $32,000. The minimum tax would be computed as follows—

Tax preference income	$90,000
Exemptions	32,000
Amount subject to tax	$58,000
Minimum tax (15% rate)	$8,700

The minimum tax does not apply to tax-option corporations (those taxed under Subchapter S of the Internal Revenue Code) or to personal holding companies.

Installment sale

Many sellers of farmland find the installment sale alternative to be an attractive way to dispose of farmland. In addition to the advantage of a low down payment for the buyer, and a fast-moving remedy if the buyer defaults, the seller can spread the income tax liability over the term of the installment obligation. For transactions involving a large amount of gain, the income tax spreading feature can be an especially attractive aspect of disposition of land by installment sale.

    Income tax aspects—To be eligible for the installment reporting of gain, it is not necessary for the transaction to involve a *contract*. If the requirements are met, a transaction involving a deed given by the seller with the seller taking back a mortgage for the unpaid purchase price can be qualified for installment reporting, also.

    Until enactment of legislation in late 1980, it was necessary for several requirements to be met for installment reporting of gain: (1) the seller could receive no more than 30% of the selling price in the year of sale and (2) there must have been at least two payments spread over at least two taxable years. Both requirements were repealed in the 1980 legislation effective for all of 1980.

    Transactions are presumed to be subject to installment reporting rules regardless of the amount received by the seller in the year of sale. A special election is necessary if installment reporting of gain is *not* desired.

    The year of sale is still a matter of substantial importance. It determines when initial payments are taxed. In general, the "year of sale" is the year of transfer of the benefits and burdens of ownership, usually the year of transfer of possession, unless title transfers in an earlier year.

    EXAMPLE: V agrees on November 1, 1981, to sell a tract of land for $100,000 with $500 in earnest money paid on November 1. The agreement is that the buyer would make an additional payment of $10,000 on March 1, 1982, and get possession. The year of 1982 would be the year of sale. The $500 in earnest money would be carried over and reported in 1982 along with other payments received that year.

Example: Assume that V, in addition to receiving $500 on November 1, 1981, received $30,000 on March 1, 1982. Would the transaction be eligible for installment reporting of gain? The answer is yes under the 1980 amendment. The limit of 30% of the selling price in the year of sale is no longer applicable. Formerly, the seller had to be careful to assure that the total of payments received in the year of sale, including any payments as earnest money received in a prior year, did not exceed 30% of the selling price.

In general, mortgages on the land are not considered as payments in the year of sale. However, if the indebtedness on the property, in the form of a mortgage or otherwise, exceeds the seller's income tax basis in the land, the excess of indebtedness over income tax basis would be considered a payment in the year of sale.

Example: Assume, in the above illustration, that V had a mortgage of $60,000 on the land in question. The selling price, as noted above, was $100,000 with $500 paid as earnest money on November 1, 1981. Assume that V's income tax basis in the property is $20,000. In that case, the excess of indebtedness over basis ($60,000 − $20,000 = $40,000) would be considered a payment in the year of sale. The transaction would be eligible for installment reporting but the amount of $40,500 would be taxable in the year of sale.

If indebtedness exceeds the income tax basis, and it is desired to avoid the extra income tax liability, one solution has been for the buyer to delay taking over the mortgage or other indebtedness until some year after the year of sale. Thus, if the buyer were to delay assuming a mortgage until the third year, for example, the excess of indebtedness over basis would be income in that year. However, if that route is taken, the buyer should have no association with the holder of the mortgage (the mortgagee) until the time arrives for the buyer to take over the mortgage indebtedness.

For transactions that are qualified for installment reporting of gain, part of each payment received is reported as return of income tax basis, which is not subject to income tax, part is capital gain, usually taxed as long-term capital gain, and part is interest which is taxed as ordinary income. Any recaptured depreciation or other amounts are reported first before any gain is reported for capital gain purposes.

The calculation of taxable gain on behalf of the seller requires that four amounts be determined: (1) the selling price, (2) total contract price, (3) the income tax basis for the property, and (4) the amount of any mortgage or other indebtedness on the property to be taken over by the buyer.

- The *selling price,* of course, is the total amount to be paid for the property and includes the amount of any mortgage or other indebtedness to be paid off by the buyer.

Example: M agrees to sell a tract of land for a total of $100,000 with the buyer paying $30,000 in cash and taking over a $70,000 mortgage. The "selling price" would be $100,000.

- The *total contract price* is the amount the buyer is going to pay the seller directly. It does not include the amount of any mortgage or other indebtedness to be paid by the buyer to someone else.

Example: Returning to the above illustration involving M who agreed to sell a tract of land for $100,000 with a $70,000 mortgage, the total contract price would be—
$100,000 − $70,000
= $30,000.
Thus, the total contract price would be $30,000.

- The *income tax basis* is, as has been noted previously, the amount of property value that is not subject to income tax on a taxable disposition. The income tax basis for a tract of land depends upon whether the property had been acquired by purchase, by inheritance, or by gift and whether improvements had been made to the property and depreciation had been claimed. All of that information enters into the calculation of income tax basis as discussed in Chapter 16.
- The amount of any *mortgage* or other indebtedness on the property is also a necessary part of the calculation of the seller's income tax liability under installment reporting of gain.

With those four elements of information, it is possible to make the necessary calculations for determining the seller's income tax liability.

The first step is to determine the *gross profit* in the transaction. That term is defined as the selling price less the income tax basis.

Example: Assume that P, the seller, agreed to sell a tract of land for $500,000 with an income tax basis of $200,000. The gross profit would be—
$500,000 − $200,000
= $300,000.

The second step is to determine the *gross profit percentage*. The term "gross profit percentage" is defined as the gross profit divided by the total contract price.

Example: Assume in the above example that P agrees to sell the $500,000 tract of land with a mortgage of $100,000 which is to be taken over by the buyer. P's income tax basis for the property is $200,000. Thus, the gross profit percentage would be—

$$\frac{\text{gross profit}}{\text{total contract price}} = \frac{\$500,000 - \$200,000}{\$500,000 - \$100,000} = \frac{\$300,000}{\$400,000} = \frac{3}{4}$$

Thus, three-fourths of every principal payment would be reported as gain and one-fourth of every principal payment would be return of basis which would not be taxable. Thus, if the seller received a payment of $50,000 in the year of sale, three-fourths of that or $37,500 would be reported as capital gain (long-term capital gain if the land had been held for more than one year) with $12,500 being return of basis which is not subject to income tax. There would, of course, be income tax liability for any interest paid as ordinary income.

Under the "unstated interest rule," contracts for more than $3,000 that have a term of more than one year must bear interest of 9% or more. If not, interest is recalculated at 10% compounded semi-annually. These figures are adjusted periodically by the Internal Revenue Service. Before July 1, 1981 the minimum rate was 6%. For transactions between controlled entities as taxpayers, the interest rate is to fall between 11% and 13%.

Under legislation adopted in 1981, an installment sale of land qualifies for a maximum interest rate of 7% if the sale is between members of the same family and if the sales price of land sold or exchanged between the same family members during the calendar year does not exceed $500,000. In the event the $500,000 limit is exceeded, the 7% maximum rate is available on sales or exchanges up to that limit. This provision is effective for payments made after June 30, 1981, pursuant to sales or exchanges after that date. IRS can publish regulations permitting a rate lower than 7% by 1% or more but has not done so to date. That would suggest a minimum rate of 6% for such family transactions. There is some indication that IRS intends to interpret "land" as only the soil and not as including depreciable improvements. Unless resolved in favor of the usual broad definition of land as including depreciable improvements, related sellers and buyers of farm land may wish to include a provision within the installment contract that if "land" is interpreted to only include soil, the portion of the sale representing depreciable improvements would bear interest at 9% minimum.

Occasionally, the purchaser of land wishes to receive title to the property and yet the seller does not wish to receive all the payment in the year of sale. This sometimes leads to the creation of an escrow arrangement with the buyer paying the full purchase price to the escrow agent and with the escrow agent transmitting the deed to the buyer once that payment has been made. The hope, of course, on the part of the seller is that the seller can continue to defer the reporting of gain for income tax purposes until the payments are actually received by the seller. However, in many instances, escrow arrangements have led to "constructive receipt" of the amounts involved by the seller. That means that the full amount would be reported as income in the year of payment by the buyer to the escrow holder. Most escrows that have succeeded have involved an independent trustee. Therefore, escrow arrangements should be planned very carefully if it is hoped that the seller would be able to defer the reporting of gain until payments are actually received.

Under the 1980 amendments, if land (or other property) is sold to a related party and within two years the property is sold by the purchaser, the result is taxable gain for the original seller. This change was designed to discourage sales within the family as a type of escrow agreement.

Example: Dan Jones sold 160 acres to his daughter for $480,000 with installment reporting of gain and gave the daughter a deed to the property. Two weeks later the daughter sold the land to a developer and gave the developer a deed to the tract of land. Before the 1980 amendments became effective, sale of the land by the daughter would not have disturbed the deferral of gain for her father. Under the 1980 change, however, sale by the daughter within two years creates taxable gain for the father.

Disposition of contract—In general, the right to defer the reporting of gain for income tax purposes under the installment sale procedure is personal to the seller. If the seller transfers the contract to another, in general that triggers immediate taxability of all the gain in the contract. That is the outcome if the seller makes a gift of the contract, sells the contract, pledges the contract for an amount approximately equal to its value or exchanges the installment obligation for a private annuity. There is no acceleration of income tax liability if the installment obligation is transferred to a newly-formed partnership or corporation in a tax-free exchange, however. Also, there is no triggering of gain if the seller dies and the contract passes into the hands of an estate or heirs—so long as the heirs are not the obligor under the contract.

The installment obligation as an asset of the decedent's estate does not receive a new income tax basis, which most other assets would receive, but instead takes on the status of "income in respect of decedent," and the gain continues to be taxable as income after death. Thus, those who inherit an installment obligation would continue to report the payments in the same way the seller would have done had he or she lived. Thus, a portion of each payment continues to be reported as capital gain figured in the same way that the decedent figured such amounts prior to death. A part is return of basis which is not taxable and a portion of interest would continue to be reported as ordinary income.

This feature of installment obligation taxation constitutes a substantial disadvantage for older taxpayers holding property with a great deal of gain. A sale of land prior to death thus "locks" the gain into the installment obligation and those who inherit the obligation must eventually report the gain as taxable gain. The installment obligation does not receive a new income tax basis at death.

Under the 1980 amendments to the installment sale rules, disposition of an installment sale obligation to the obligor at death is a taxable disposition to the estate. Any previously unreported gain from an installment sale is taxable to the deceased seller's estate if the obligation is transferred by will or by inheritance under state law to the obligor or is cancelled by the executor or administrator of the estate.

There is some authority that mere renegotiation of terms under an installment obligation (for example, a stretch-out in the time for making principal payments and an increase in the interest rate) would not be considered a taxable disposition such as would trigger immediate taxability of the gain. Likewise, there is authority that substituting a deed and mortgage having the same terms and conditions as the original contract would not trigger taxability of gain, either. Because the triggering of income tax potentially due under an installment obligation could create a substantial income tax liability, it is important for sellers under an installment obligation to use great care in planning for the disposition of the obligation, including a disposition at death.

The election—To be eligible for installment reporting of gain, the seller need no longer make an election on the income tax return for the year of sale. Rather, the installment sale rules automatically apply to a qualified sale unless the seller elects not to have the provision apply with respect to a deferred payment sale.

Private Annuity

Land may also be disposed of under a private annuity. Private annuity arrangements are typically family transactions, with the annuitant (usually a parent or parents) conveying an asset such as land to an obligor (often a child or children) in exchange for the obligor's unsecured but firm promise to make payments to the annuitant as long as the annuitant lives.

A private annuity is similar to a commercial annuity where a specified premium is paid annually or in a single premium amount to an insurance company in exchange for the insurance company's promise to make payments to the annuitant so long as the annuitant lives. With a private annuity, however, the obligor is typically a family member rather than an insurance company and the amount paid for the annuity is usually not cash but a tangible asset such as land. The common feature of commercial and private annuities, however, is that the obligor continues to make payments so long as the annuitant lives. In the case of a two-person annuity, the payments may continue for so long as either annuitant survives.

Private annuities may encounter two types of problems: (1) an array of non-tax problems and (2) income tax, gift tax and estate tax problems.

Non-tax considerations—Looking first at the non-tax considerations, problems of fairness may arise if one child in a family of several children is the obligor.

Example: The mother, as the surviving parent, deeds 160 acres of land to her son in exchange for the son's promise to pay the mother $30,000 per year for the rest of the mother's life. At the time of the transaction, the mother's life expectancy was 15 years. If the mother dies prematurely, the obligor has a windfall and the other children may be quite unhappy. If the mother lives a longer than normal life, the obligor may pay substantially more for the property than it is worth. Thus, the only situation in which unfairness does not arise is where the parent dies right on schedule or all children are obligors.

Another non-tax disadvantage involves the risk taken by the annuitant. If the obligor encounters financial difficulty, the property funding the annuity might be lost to the obligor's creditors, with the result that the annuitant's retirement plan would be effectively eliminated. An annuitant may not retain any control or security interest in the property forming the basis for the private annuity. If such interest or control is retained, the transaction would be treated as a sale. Moreover, the full value of the property might be included in the annuitant's estate for federal estate tax purposes at death.

Another non-tax feature involves the premature death of the obligor. In that event, the obligor's estate could suffer substantially, particularly if the annuitant lives a longer than normal life.

As a practical matter, payment amounts needed to support a private annuity (and avoid a gift) may call for annual payments that may be quite large. This may be an unacceptable burden to the obligor.

Tax considerations—From the standpoint of tax consequences, the obligor is acquiring the property with payments of principal only; there is no interest paid and no income tax deduction for interest is allowed the obligor as the purchaser of the asset involved. Over the period of the annuity, this can become a sizable item from the standpoint of the purchaser.

If the payments made by the obligor are too large, taking into account the value of the asset and the life expectancy of the annuitant, a gift occurs from obligor to annuitant and gift tax could be due. If the payments are too small, the gift runs the other way—from annuitant to obligor.

The principal tax reason for a private annuity is that there is no federal estate tax liability at the death of the annuitant. Annuity payments terminate at the annuitant's death and there is nothing to be taxed in the estate of the annuitant. However, if the value of the property transferred exceeds substantially the value needed to support annuity payments, the annuitant retains control over the property transferred or the use of the property by the obligor is otherwise limited, the private annuity could be recharacterized as a transfer of property with a retained life estate. The result of such a recharacterization would be that the value of the property would be included in the annuitant's estate at death for federal estate tax purposes.

The annuitant's income tax treatment is parallel—but not identical—to the income tax treatment of an installment sale. Thus, the income tax liability from a private annuity transaction is computed by spreading the gain over a period of years measured by the annuitant's life expectancy. The gain is capital gain if the transferred property was a capital asset.

The amount of gain is computed by subtracting the annuitant's income tax basis for the property from the present value of the annuity.

Example: Assume a transfer of a $120,000 farm with a $40,000 income tax basis by a 74 year old male annuitant in exchange for an annuity payment of $14,400 per year. The total expected return under the private annuity would be $145,440 ($14,400 per year times a life expectancy of 10.1 years). The present value of the annuity, figured from the annuity tables, would be $87,823.

The "investment in the contract" or the income tax basis for the property ($40,000 from the example) is divided by the expected return of $145,440 to produce the exclusion ratio of 27.5%. Out of each annual payment of $14,400, the annuitant would report 27.5% or $3,960 as return of basis which is not subject to income tax. This portion of each payment continues to be excludible even if the annuitant survives beyond the life expectancy. The capital gains income of $47,823 ($87,823 − $40,000) would be spread over the life expectancy of the annuitant (10.1 years in this example) with $4,735 reported as capital gain each year. The remaining part of each annuity payment would be reportable as ordinary income. For the first 10.1 years, that would total $5,705 of ordinary income per year.

After the gain of $47,823 has been fully reported, the amount of payments received (less the excluded portion of $3,960 each year) would be taxable as ordinary income.

For purposes of starting a cost recovery (depreciation) schedule, the obligor's income basis for the property involved is the value of the expected annuity payments to be made under the private annuity agreement. As excess payments are made, they are added to the income tax basis. After the death of the annuitant, depreciation is calculated using the total payments actually made. Similarly, the income tax basis for computing gain or loss on sale after death is the total of all payments actually made. Thus, a premature death of the annuitant means not only a bargain to the obligor in terms of payments made, but also a low income tax basis for the property in the hands of the obligor.

For sale of the property before the annuitant's death, the income tax basis for calculating gain is more involved. The obligor uses as an income tax basis the total of payments actually made plus the present value of payments remaining based on the annuitant's life expectancy at the date of disposition of the property.

Tax-Free Exchange
A sale of property, other than sale under threat or imminence of

condemnation, **generally results in taxable gain to the seller even though the proceeds are reinvested in similar property.** There are additional exceptions for sale of the personal residence and reinvestment of the proceeds in another personal residence (within a period commencing two years before or after sale of the old residence, and for the sale of a personal residence after age 55). In the latter situation, up to $125,000 of gain in a personal residence may be excluded from income once in the taxpayer's lifetime.

With an exchange of property, transfers can take place without income tax liability if the requirements are met.

> **Example:** Farmer A with 160 acres of land located adjacent to a highway interchange, wishes to trade the low basis, high value land for a larger tract elsewhere. The 160-acre tract has a fair market value of $800,000 and an income basis of $25,000. If A locates a larger tract of land with a fair market value of $800,000 and makes a trade with B, the owner of the larger tract, there will be no income tax liability except to the extent of boot received in the transaction. If A paid boot, that adds on to A's basis for the land. If A received boot, that would be taxable gain.
>
> With a tax-free exchange, the income tax basis in each property carries over and becomes the income tax basis for the new tract of land. Thus, in the above example, A's income tax basis of $25,000 on the tract of land given up would become the income tax basis for the larger tract of land received in exchange. Thus, the $25,000 would be allocated over the depreciable and nondepreciable components of the new tract. There would, therefore, be only limited opportunity for depreciation to be claimed on the depreciable components of the new tract.

To be eligible for a tax-free exchange, the property must have been held for productive use in a trade or business or for investment and must have been traded solely for property of a like kind to be held either for productive use in a business or for investment. The tax-free exchange rule does not cover stock in trade or other property held primarily for sale or to stocks, bonds, or other securities. An exchange of real property for personal property is not a "like kind" exchange. However, the exchange of a farm or ranch for urban real estate involves property of "like kind." Similarly, the exchange of an outright interest (a fee simple) in real estate for a leasehold interest for 30 years or more to run would be a like kind exchange.

If mortgaged real estate is exchanged for other real estate in an exchange that would otherwise be a tax-free transfer, the amount of the mortgage of which a property owner is relieved is treated as "other property or money."

> **Example:** Assume that D owns 80 acres of land with an income tax basis of $85,000 and a $70,000 mortgage. The tract of land has a fair market value of $130,000. That tract is exchanged for another tract of land with a fair market value of $80,000 which

is subject to a $32,000 mortgage. D also receives a second mortgage of $12,000 on the first tract. The calculation would be handled as follows:

Fair market value of property received	$ 80,000
Less mortgage on that property	32,000
	$ 48,000
Fair market value of second mortgage received	$ 12,000
Mortgage on property transferred by taxpayer	70,000
	$130,000
Less basis of property transferred	85,000
Gain realized	$ 45,000

The entire amount of gain would be recognized because it is less than the sum of the $38,000 net mortgage reduction and the $12,000 second mortgage received.

If property is taken by threat or imminence of condemnation or by involuntary conversion (such as from destruction, theft or other casualty) the proceeds may be reinvested in property similar or related in service or use to the property converted without paying income tax on the gain. For land, the proceeds must be reinvested within three years after the year in which any part of the proceeds was received. For other property, the reinvestment period is two years.

Income Tax Treatment of Entity Owning Land

The time of acquisition of land is a good time to review the choices for organizational structure of the land owning entity. The income tax treatment of the entities is an important factor to consider in making the decision.

Individual ownership—For individual ownership of land, the ordinary income from the farm or ranch operation and the rental amounts received under a lease are taxed at the state and federal income tax rates applicable to individual taxpayers. The federal rates range from 13 percent to 70 percent (for 1981) with a 50 percent maximum tax on personal service income. For this purpose, "personal service income" includes wages, salaries, professional fees and reasonable compensation for services rendered in a business. Passive investment income such as from cash rents is not, however, considered to be personal service income. The individual income tax rate for married individuals filing jointly is scheduled to drop to a range of 12% to 50% for 1982.

Long-term capital gains income is 60% deductible with the remaining 40% taxable as ordinary income. A maximum 20% rate applies to net long term capital gains in sales after June 9, 1981. Capital gains income is no longer a preference item for the minimum tax on preference income for individuals but is subject to the alternative minimum tax as discussed earlier in this chapter.

Partnerships, estates and trusts—General and limited partnerships, estates and trusts operate under unique rules for taxing ordinary income and capital gains. However, the rules for individuals are generally applicable with the partners and beneficiaries, as the case may be, reporting ordinary income and capital gains as individuals. Limited opportunities exist for estates and trusts to be taxed as separate taxpayers.

Subchapter S corporations—For corporations taxed under Subchapter S of the Internal Revenue Code (as tax-option corporations), ordinary income, long-term capital gains, operating losses and investment tax credits pass through to the shareholders to be reported on their own, individual income tax returns. Only under very limited circumstances is a Subchapter S corporation a taxpayer.

Regularly taxed corporations—For corporations taxed under the rules for regularly taxed corporations, corporate taxable income for federal income tax purposes is taxed under a special set of graduated rate brackets as follows:

Corporate taxable income	Income Tax Rate		
	1981	1982	1983 and later
0–25,000	17%	16%	15%
25,000–50,000	20%	19%	18%
50,000–75,000	30%	30%	30%
75,000–100,000	40%	40%	40%
Over 100,000	46%	46%	46%

In general, "controlled corporations" with substantial identity of ownership may share one set of graduated rate brackets.

Regularly taxed corporations may be subject to two penalty taxes. The accumulated earnings tax is imposed for accumulations of earnings and profits beyond the reasonable needs of the business (over $150,000 in amount through 1981, rising to $250,000 thereafter except for service corporations [such as law firms where the figure remains at $150,000]). The personal holding company tax is levied (at a 70% rate) on closely held corporations with substantial amounts of passive investment income. After 1981, the personal holding company tax drops to 50%.

For regularly taxed corporations, long-term capital gains are taxed at a rate of 28% (or the lower rate applicable to ordinary income). The 60% deduction available to individual taxpayers on long-term capital gains may not be claimed by regularly taxed corporations.

Corporate capital gains are subject to the 15% minimum tax on preference income (after the preference income is reduced by the greater of $10,000 or the corporation's income tax bill).

Many states have state income taxes, as well.

CHAPTER 18
TRANSFERRING LAND AT DEATH

Viewed by many investors as a long-term investment, land is often held until death and becomes part of the owner's estate. As a matter of estate planning prior to death, a great deal of interest is typically shown in how land is valued at death, the likely federal estate and state inheritance tax liability, the effects of land ownership at death on estate settlement costs and the ultimate disposition of land to the desired beneficiaries.

For some purposes, land is treated as any other asset of comparable value that is held until death. However, for other purposes, including the special methods for valuing land at death, land as an asset is treated uniquely.

Use Valuation of Land

Traditionally, land held until death has been valued at fair market value. That is defined as the price at which the land would change hands between a willing buyer and a willing seller, neither being under any compulsion to buy or to sell. Because of the sharp rise in federal estate tax liability for farms, which was attributable in no small measure to increases in land value, Congress in the Tax Reform Act of 1976 enacted a special method for valuing land. That method is referred to as "use" valuation. Actually, the new method of valuation of farmland, bases land value on capitalization of gross cash rents on comparable land in the locality. The other method, new also in 1976, utilizes a five-factor formula for determining the value of land.

The original justification for Congressional consideration of valuation of land at death as part of the overall effort to ease the death tax burden on farms and small businesses was the influence of large urban places on land value. The provision as enacted, however, was not limited to instances where the land value was influenced by the presence of metropolitan or developed resort areas. The valuation techniques are available for any situation in which the requirements can be met.

It should be kept in mind that the two new methods for valuing land at death are only available for federal estate tax purposes. The valuation techniques are not applicable to valuation of land for federal gift tax purposes or for income tax purposes. Because use valuation of land often produces a value at 40 to 60% of comparable sale price, substantial savings in federal estate tax are possible if the pre-death requirements can be met and if recapture can be avoided in the 15 years following death (10 years for deaths after 1981). Therefore, a substantial incentive may exist for land to be retained at death, especially if it is believed that the land could be qualified for use valuation in the estate of the owner. The maximum limit on reduction of gross estate from applying use valuation was increased from $500,000 to $600,000 for deaths in 1981, to $700,000 for 1982 and to $750,000 for 1983 and thereafter.

Capitalization of cash rents—The formula most likely to be used for farmland involves the capitalization of gross cash rents (minus property tax) on comparable land in the locality, capitalized at the average annual effective Federal Land Bank interest rate. All calculations are to use the last five full calendar years before death.

Example: X died on October 15, 1981. The conditions for use valuation of land can be met. For purposes of calculating the use value, cash rents and property taxes would be used for the years 1976, 1977, 1978, 1979, and 1980.

The formula for determining use valuation for land, using capitalized cash rent figures, utilizes the following formula—

$$V = \frac{\text{gross cash rent—property tax}}{\text{effective Federal Land Bank interest rate}}$$

Example: Assume the average annual gross cash rent on comparable land in the locality is $74 per acre with an average of $4 per acre of property tax during the same five year period. Assume further that the effective Federal Land Bank interest rate is 8¾%. The formula for determining use valuation of land would be—

$$V = \frac{\$74.00 - \$4.00}{.0875}$$
$$= \$800.00 \text{ per acre.}$$

As can be seen, the value of land is very sensitive to changes in the interest rate. With a doubling of the interest rate, the value of land is cut in half. If the interest rate is reduced by half, the land value is doubled. Thus, land values are very sensitive to changes in the interest rate under that formula.

To be eligible to use the cash rent capitalization formula, it is necessary to have available—(1) cash rent information on comparable land in the locality, (2) property tax data on comparable land in the locality and (3) "effective" Federal Land Bank interest rates. Obtaining cash rent information on comparable land is typically the greatest hurdle to applying use valuation of land based on the cash rent capitalization formula. Several factors have been identified by the Internal Revenue Service as guidelines for determining what is comparable land.

- For those states that have developed an index number system for ranking land in terms of productivity, that index can be used to help determine what is comparable land. Several states have developed such systems, motivated in most instances by a need for a method for equalizing land values for property tax purposes. Typically, the index system ranks land on a scale between 0 and 100 with 100 being the most productive land. The index system takes into account the type of soil, amount of erosion that has occurred, average rainfall, and other factors bearing upon productivity for production of the major crop or crops in the area. The index approach provides an objective means for determining what is comparable land.
- It is permissible to consider whether soil depleting crops have been grown equally on the two tracts.
- The guidelines suggest that attention be given to whether soil and water conservation practices have been used comparably on the two tracts.
- It is also permissible to consider whether flooding possibilities are comparable on the two tracts.
- The slope of land may be taken into account as a factor influencing productivity.
- For livestock operations, the carrying capacity may be considered as a factor bearing upon comparability.
- For tracts that are partially or totally timbered, comparability of the timber may be taken into account.
- The number, type and condition of buildings are viewed as factors to be considered when determining what is comparable land to the extent that it affects "efficient management and use of property and value per se."
- Finally, the proximity of the land tracts to transportation facilities is a factor. That is because tracts of land located near major shipping terminals have a higher value. Grain produced near shipping terminals is worth more since it does not have to be shipped long distances. Part of that extra value of production is capitalized into the value of land.

The effective Federal Land Bank interest rate, which is used in the cash rent capitalization formula, is published annually by the Internal Revenue Service. Table 18.1 contains the effective Federal Land Bank interest figures, by Federal Land Bank district, to be used in the capitalization formula. The rate to be used is the district rate where the land is located. Essentially, the *effective* Federal Land Bank interest rate is derived from the stated rate of interest charged borrowers adjusted for the cost to a borrower of owning Federal Land Bank stock.

Crop Share Rents—For deaths after 1981, if there is no comparable land from which average cash rentals may be obtained, "av-

Table 18.1 Effective Federal Land Bank Interest Rates

Federal Land Bank District	Average annual effective interest rates for "use" valuation				
	Death in 1977	Death in 1978	Death in 1979	Death in 1980	Death in 1981
Baltimore	8.65%	8.86%	9.04%	9.24%	9.66%
Columbia	8.58%	8.79%	8.96%	9.17%	9.40%
Houston	8.29%	8.48%	8.60%	8.76%	9.09%
Louisville	8.64%	8.80%	8.88%	9.21%	9.53%
New Orleans	8.26%	8.48%	8.72%	8.96%	9.33%
Omaha	8.70%	8.92%	9.05%	9.25%	9.59%
Sacramento	8.67%	8.82%	9.04%	9.35%	9.63%
St. Louis	8.50%	8.71%	8.93%	9.20%	9.77%
St. Paul	8.21%	8.47%	8.69%	8.95%	9.30%
Spokane	8.63%	8.88%	9.10%	9.31%	9.60%
Springfield	8.42%	8.55%	8.65%	8.81%	9.10%
Wichita	8.52%	8.72%	8.88%	9.08%	9.36%

erage net share rentals" from crop share leases may be used. The term "net share rental" means that landowner's portion of the crop share return from the land minus the "cash operating expenses which, under the lease, are paid by the lessor."

Five factor formula—For those who cannot locate cash rented tracts of comparable farmland, and for nonfarm tracts which are ineligible for the cash rent capitalization approach, it is possible to elect the five factor formula for determining use valuation of land. That formula includes (1) capitalization of income that the property could be expected to yield over a reasonable period under prudent management; (2) the capitalization of a fair rental value for the property; (3) the assessed valuation of the land for property tax purposes if the state bases property tax assessments on current use for the property; (4) comparable sales of property in the same geographical area but without significant influence from metropolitan or resort areas and (5) any other factor that would fairly value the real property in question. For farmland, it is doubted that the five factor formula will be used often. The rent capitalization approach is generally viewed as more advantageous to the estate in terms of reducing federal estate tax liability.

Pre-death requirements for eligibility—To be eligible for use valuation, several requirements must be met in the period prior to death.
- At least 50% of the estate (using fair market value figures) must be comprised of business real and personal property and that amount must pass to the qualified heirs by inheritance or by purchase.

Example (1): Grandfather Smith died on December 5, 1980 owning 320 acres of land and $50,000 in cash. The cash was used to pay death taxes and estate settlement costs. Of the three children, two lived off the farm and preferred cash. The on-farm heir preferred to receive the land. The administrator of the estate, during estate settlement, sold the farm to the on-farm son. A 1981 amendment has made the two-thirds of the farm representing the portion purchased from the estate eligible for use valuation. The 1981 amendment was retroactive to January 1, 1977. Before enactment of the 1981 amendment property passing by purchase to qualified heirs was not eligible for use valuation.

Example (2): At the time of death, Grandfather Smith's land was owned by a corporation. At his death, the corporate stock passed to his estate for distribution to the three sons. If the on-farm son purchases the two-thirds of the stock that would otherwise pass eventually to the off-farm children, at least the two-thirds of the land represented by the two-thirds of the stock acquired by purchase would be eligible because of the 1981 amendments. Of course, the land represented by the one-third of the stock held by the purchasing son should also be eligible for use valuation.

Example (3): Grandmother Elkins died owning 160 acres of farmland (making up 45% of her estate), breeding stock (representing 10% of her estate) and money market certificates (representing 45% of the estate). Unless at least one-half of the livestock passes to the qualified heirs, the requirement would not be met that at least 50% of the estate must be comprised of farm real or personal property and must pass to the qualified heirs.

For deaths through 1981, qualified heirs are defined as members of the family which include all lineal ascendants of the decedent-to-be, all lineal descendants, the descendants of the grandparents of the decedent-to-be, the spouse of the individual and the spouse of any descendants. Thus, "member of family" is a broad term, indeed. After 1981, the definition is narrowed to include only the ancestors of the decedent-to-be, the person's spouse, a lineal descendant of the individual, a lineal descendant of the individual's spouse, a lineal descendant of the parents of the individual and the spouse of any lineal descendant.

- The second major pre-death requirement is that the land must comprise at least 25% of the estate, again using fair market value figures. For both this calculation and the preceding one, indebtedness attributable to the property is subtracted.
- The third pre-death requirement is that the land must have been owned by the decedent or member of the family for 5 or more of the last 8 years before death and used in the business of farming (or another eligible use) for 5 or more of the last 8 years before death. This requirement means that land acquired shortly before death would not be eligible for use valuation. A minimum of five years' ownership by the decedent or member of the family is necessary to make the land eligible for use valuation. For deaths after 1981, period of ownership of land acquired in a tax-free exchange or involuntary conversion can be added to that for the property given up.

- The deceased or member of the family must have "participated materially" in the production of income for at least 5 of the last 8 years before death. This requirement can be met if the deceased or member of the deceased's family had been operating the land prior to death, the land had been leased to a member of the family, or the land had been leased to a non-family member under a "material participation" lease. In general, a "material participation" lease includes crop share or livestock share leases with a strong record of involvement in decision making under the lease. Because of an amendment made to federal law in 1974, material participation cannot be achieved through an agent for purposes of use valuation of land.

  For decedents dying after 1981, landowners who are retired and receiving social security benefits or disabled need meet the material participation test only for five or more of the last eight years before the date of disability or retirement. The amendment also specifies that a surviving spouse (who acquired qualified real property from a decedent) who is involved in "active management" of the farm or other business meets the material participation test. The term "active management" is defined as "the making of the management decisions of a business (other than the daily operating decisions)." Congressional committee reports (but not the law itself) state that the active management test can be met without reporting income under a lease as self-employment income.

- In regulations published in 1980, the Internal Revenue Service identified an additional test that must be met for use valuation eligibility. It is the "qualified use" test which requires that the decedent (in the period before death) and each qualified heir (in the recapture period after death) have an "equity interest" in the farm operation. The test must be met: (1) at the time of death, (2) for five or more of the last eight years before death and (3) by each qualified heir for the recapture period after death.

  In effect, a cash rent lease by the decedent (or qualified heir after death) failed to meet the test under the IRS interpretation. On April 27, 1981 the IRS announced that regulations would be published permitting the qualified use test to be met *in the predeath period* by the decedent *or a member of the decedent's family*. The practical result of that announcement was to permit cash rent leases to family members in the predeath period. An amendment in 1981 gives that interpretation the force of law, retroactive to January 1, 1977.

Example (1): Elizabeth Moore died on December 1, 1981, owning a 320 acre farm. The farm had been cash rented to her son since March 1, 1976. The land would meet the qualified use test.

Example (2): The same facts as in Example (1) except that Eliz-

abeth Moore had cash rented the farm to an unrelated farm tenant, from March 1, 1970, through March 1, 1979, and had then shifted to a crop share lease with the same tenant which continued until the time of her death. The land would meet the qualified use test as of the date of death but not for at least five of the last eight years before death.
- A qualified heir must receive a "present interest" in the land. If the land is left in trust, for example, that has meant the qualified heir holding the life estate must be assured of receiving the income interest from the property. In the event a trustee has discretion in paying income or principal to various family members, IRS had taken the position that the land would not qualify for use valuation. However, on April 27, 1981, IRS announced a change in interpretation permitting trustee discretion in paying income and principal if all actual and potential beneficiaries were members of the family. A 1981 amendment gives that interpretation the force of law retroactive to January 1, 1977.

Avoiding recapture of the use value benefit—As with the pre-death requirements, several post death requirements are imposed to prevent mere investors from taking advantage of the tax break created for land held until death that is used in the farm or other closely held business and valued under the use valuation rules. In the recapture period following death, several requirements must be met in order to maintain use valuation of land. The 15-year recapture period applies for deaths through 1981 (full recapture for the first 10 years with a phaseout between 10 and 15 years. For deaths after 1981, a 10-year recapture period applies except where the 10-year period is extended for up to two years for use of the two-year "grace period" for the qualified use test after death.

To avoid recapture, the land under use valuation must not be disposed of outside the family. It is permissible to transfer land by gift or sale to members of the family but transfer of land outside the family terminates use valuation. The only exception is for land taken by involuntary conversion (such as by condemnation) and land exchanged in a tax-free exchange. For involuntary conversions, recapture of use value benefit does not occur if the proceeds from involuntary conversion are reinvested in property used for the same purpose within the statutory period for reinvestment (three years).

Absence of material participation for more than three years in any eight year period ending after death leads to recapture of the use value benefit.

Example: F retired in 1979 and went under a nonmaterial participation crop share lease so that income received under the lease would not reduce social security benefits. Death occurred in 1981, with two years of nonmaterial participation under the lease. If the nonmaterial participation lease continues for more than one more year after death, the condition is met for recapture of the use valuation benefit. Thus, heirs do not have eight years after death to amass five years of material participation. The conditions for recapture of use valuation benefits could be met during the first year after death if there were more than two years of nonmaterial participation immediately before death. Therefore, it is important to maintain surveillance over the material participation status of the land in the period before death as well as the period following death.

In a provision added in 1981 but retroactive to 1977, for a qualified heir who is the surviving spouse of the decedent, person who has not reached age 21, disabled individual or student, the material participation test may be met by "active management" by the qualified heir (or a fiduciary if the qualified heir is a person under age 21 or a disabled individual).

Failure to meet the qualified use test in the recapture period after death results in recapture of the federal estate tax benefit, also. However, a 1981 amendment created a two year grace period after the decedents' death for meeting the qualified use test.

Example: Ezra Thomas died on April 1, 1980. The estate was closed on February 25, 1981, with the land, which had been rented to son Elmer under a cash rent lease, distributed to the two children, Ellen and Elmer. Commencing March 1, 1981, Elmer farmed the tract with cash rental paid to Ellen for her one-half interest. Ellen's undivided interest would be subject to recapture of the federal estate tax benefit from use valuation for failure to meet the qualified use test after the death of her father, if the cash rent lease continued after April 1, 1982. The announcement by IRS on April 27, 1981, permitting the qualified use test to be met by the decedent or a member of the decedent's family in the predeath period was not extended to the post-death period. The 1981 amendments likewise did not deal with the qualified use test in the post-death period except to institute the two-year grace period.

Property Interests Subject to Federal Estate Tax

Typically, where land is owned by a decedent, it is an asset of substantial value in the estate. The part of the total value of land that is included in the gross estate for federal estate tax purposes depends upon how the land was owned at the time of death.

Method of ownership or co-ownership—For property owned in the name of the decedent alone, the full value of the property, valued either at its fair market value or its "use" value, is included in the gross estate. For assets held in tenancy in common, the decedent's fractional interest is included in the estate for federal estate tax purposes.

Example: A 160-acre farm valued at $480,000 is owned in tenancy in common by A and B with equal undivided interests. At the death of A, $240,000 of the total value would be included in A's estate.

For assets held in joint tenancy (or tenancy by the entirety) the portion of value of an asset subject to tax is determined by one of three rules.

- Under the "consideration furnished" rule, which is the general rule for taxation of joint tenancy property at death, the entire value of the jointly owned property is subject to federal estate tax at the death of the first joint tenant to die except to the extent that the surviving joint tenant can prove contribution toward property acquisition. The burden of proving contribution is on the surviving joint tenant. Traditionally, this rule has placed a heavy burden on the surviving joint tenant with the survivor often unable to prove contribution as needed for a portion of the value to be excluded from the deceased's estate.
- Dissatisfaction with the consideration furnished rule led in 1976 to enactment of the "fractional share" rule. Under that provision, one half the value of joint tenancy or tenancy by the entirety property held by a husband and wife is subject to federal estate tax at the death of the first to die. It is an arbitrary rule with one-half being taxed regardless of which joint owner dies first. To be able to invoke the fractional share rule, the joint ownership arrangement must have been created after 1976 and the transaction must have been subject to federal gift tax. Thus, the fractional interest rule did not apply to joint tenancy arrangements created before 1977. Moreover, the requirement that the joint tenancy arrangement must have been subject to federal gift tax on acquisition meant that joint tenancies in land created after 1976 involving a husband and wife were not subject to the rule unless the transaction was specifically reported as a gift on a gift tax return timely filed. Since 1954, joint tenancies in land by husbands and wives have not involved a gift on acquisition even if contributions were unequal unless reported as a gift. Apparently, few joint tenancies in land involving husbands and wives have been so reported. Therefore, few tracts of land have been subject to the fractional share rule.

Under a special provision that expired at the end of 1979, pre-1977 joint tenancies could be subjected to federal gift tax in order for the property involved to be brought under the fractional share rule. That was permitted with gifts reported for any calendar quarter through 1979.

For deaths after 1981, the fractional share rule has been amended to provide that one-half the value of joint tenancy or tenancy by the entirety property held by a husband and wife

is taxed at the death of the first spouse to die. The amended rule applies to all types of property regardless of when acquired.
- Continuing Congressional concern about the taxation of joint tenancy property led in 1978 to a third rule, "the credit for services" rule which has been repealed effective for deaths after 1981. Under the provision, a surviving spouse as a surviving joint tenant is credited with services to a business in which the jointly owned property was used. For each year the spouse participated materially in the production of income in the business and the property was committed to the business the surviving spouse is credited with two percent of the value of the property in excess of the original consideration furnished by husband and wife increased by six percent simple interest for each year.

Under the credit for services rule, at least 50% of the value of the property must be included in the deceased's estate. And the provision cannot reduce the deceased's estate by more than $500,000.

Example: John and Shirley Thomas purchased 160 acres of farmland in 1970, shortly after they were married, with title taken in joint tenancy. Shirley provided $12,000 from years as a bookkeeper at a local bank. John furnished $24,000 in accumulated earnings from military service and farming with his father before marriage. Both John and Shirley participated materially in the operation. John died in 1980, survived by Shirley, with the land valued at $400,000. Application of the two percent per year credit for services rule would be handled as follows, assuming John's estate elects to apply the "credit for services" rule (see Figure 18.1):

1. Together, John and Shirley are credited with $57,600 in the property—
   a. John's original contribution of $24,000 increases to $38,400 as a result of figuring growth for ten years at six percent per year.
   b. Shirley's original contribution of $12,000 grows to $19,200 at six percent per year for 10 years.
2. That leaves $342,400 in property value above their joint contribution.
3. For 10 years of material participation at two percent per year, the credit-for-services amount is 20% of $342,400 or $68,480.
4. The total amount credited to Shirley is—
   $= 12,000 + 7,200 + 68,480$
   $= 87,680$

   That amount is excluded from John's estate.
5. The amount included in John's estate would be—
   $= 400,000 - 87,680 = 312,320$

Figure 18.1 "Credit for services" rule for taxing joint tenancy property at death.

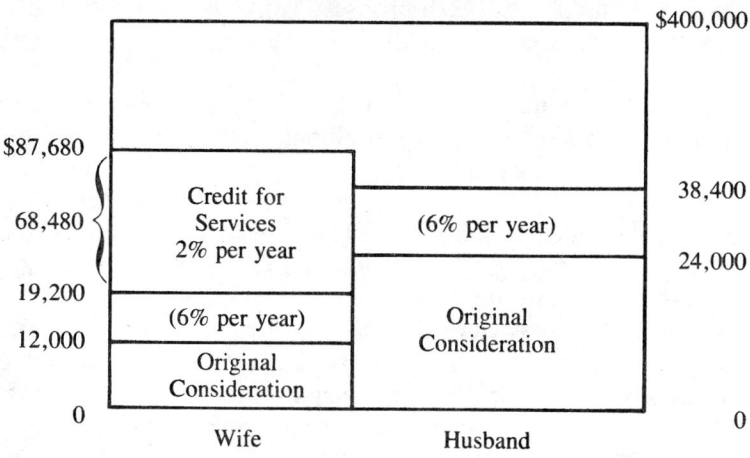

As a practical matter, the credit for services rule has been of only limited usefulness.
- For the estate to be eligible to elect the credit for services rule, the surviving spouse must have participated materially in the business for each year counted at the two percent rate. Whether the surviving spouse has participated materially is to be determined in a manner similar to the way the concept of material participation is interpreted for social security purposes—whether income is self-employment income and hence subject to self-employment tax.

  In general, it is believed that relatively few wives of sole proprietors could meet the material participation test, at least if payment by the wife of social security tax on part of the business earnings would be necessary to meet the test. Paying self-employment tax at 9.3% of the first $29,700 of self-employment income (1981 rate) could be a substantial barrier to material participation.
- If some business earnings are channeled to the wife in order to assure qualification for the credit for services rule, and the husband's social security earnings are reduced as a result, the outcome could be lower social security benefits for the husband. Moreover, because it is often advantageous for a wife to claim social security benefits on the basis of 50% of the husband's benefit, rather than claiming benefits on the basis of her own earnings record, reducing the husband's earnings record could reduce the wife's social security benefits, also.

FARMLAND

The credit for services rule only applies for federal estate tax purposes. It cannot be relied upon in the event of joint tenancy severance for federal gift tax purposes. Moreover, as noted above the credit for services rule has been repealed effective at the end of 1981.

Perhaps the key point is that any couple with sufficient property to face significant federal estate tax problems at one or both deaths probably should move away from joint tenancy ownership. Joint tenancy or tenancy by the entirety ownership leaves the property to the surviving co-owner outright. That result is inconsistent with any of the techniques for minimizing death taxes at the death of the surviving spouse. And that is where the major death tax burden arises.

Gifts within three years of death—For deaths before 1982, the gross estate also includes, for federal estate tax purposes, the value of all property given away within three years of death except for gifts for which a federal gift tax return was not required to be filed (other than for life insurance policies). In general, that means that any gift within three years of death in excess of $3000 to an individual each year is included in the estate for federal estate tax purposes at death.

Example: A owns 80 acres of land which was deeded to her daughter, X, two years before A's death in 1981. The value of the land at the date of the gift was $80,000. The full value would be included in A's federal estate tax gross estate.

Example: Grandmother T, one year before her death in 1981, made a gift of $3000 in cash to each of her 50 grandchildren. No part of the $150,000 involved would be included in Grandmother T's estate for federal estate tax purposes.

Example: L, holding a remainder interest in a 160-acre farm (the life estate was held by the individual's mother), made a gift of the remainder interest to his children 30 months before death in 1981. The value of the remainder interest was $2500. Because the gift of the remainder interest would be a gift of a *future* interest, a federal estate tax return would be required to be filed and the full amount of the gift would be included in the gross estate for federal estate tax purposes at death.

For deaths after 1981, the three year rule has been amended to include in the donor's gross estate only property given away within three years of death over which powers or interests were retained, life insurance policies and powers of appointment. The three year rule also applies for purposes of calculations to determine eligibility for stock redemption (at capital gains rates) after death to pay death taxes and estate settlement costs, "use" valuation for land and property representing an interest in a closely held business for installment payment of federal estate tax. The new rule means that, in general, appreciation after the date of a gift is not subject to federal estate tax.

For gifts before 1977, it was a "gift in contemplation of death" rule with the result that inclusion in the estate was dependent upon whether the gift transfer was motivated by thoughts of death (which would make the value of the gift included in the gross estate) or motivated by thoughts of life (the amount of the gift could avoid inclusion in the federal estate tax gross estate).

Transfers with retained powers—Property transferred during life is also included in the gross estate at death if the transfer involved retained powers, rights or interests by the transferor over the property.

>**Example:** L and M, husband and wife, deeded 320 acres of land to their only son but retained the right to receive the income from the property. The retention of the right to receive the income would assure that the full value of the property would be included in their estates for federal estate tax purposes.

Transfer of ownership of a residence from one spouse to another with retention of the right by the transferor to occupy the residence has generally not resulted in taxation of the value in the estate of the transferor if that individual ran the risk of being evicted from the premises at any time. Therefore, most transfers of ownership of the residence from one spouse to another have not resulted in inclusion of the value of the residence in that individual's estate. Transfers of a residence from parent to child with retention of the right to occupy the residence have encountered a greater likelihood that the value of the property would be included in the estate of the transferor. In such situations, the arrangement should be handled as any other lease arrangement, with reasonable rental paid, if the arrangement is to have any chance of succeeding in terms of the residence not being included in the transferor's estate.

The Adjusted Gross Estate

Once the assets are valued, attention shifts to the deductions that can be claimed from the gross estate to produce the adjusted gross estate. Amounts may be subtracted for debts of the decedent, property losses during estate settlement, and the various administration expenses in settling the estate. Thus, the attorney's fee, executor's fee, court costs, costs of last illness, death and burial and other ordinary and reasonable expenses of administering the estate are deductible for federal estate tax purposes. The result is the adjusted gross estate. For deaths after 1981, the concept of "adjusted gross estate" has been repealed as a matter of federal law for purposes of determining the size of the marital deduction and other deductions. The concept will continue to be useful for planning purposes but should be defined *wherever* used in a will or trust.

Several deductions may be claimed from the adjusted gross estate to produce the taxable estate.

Marital deduction—If a spouse survives, a marital deduction may be subtracted from the adjusted gross estate. For deaths before 1982, the marital deduction is the greater of $250,000 or 50% of the adjusted gross estate if that much or more of the property passed to the surviving spouse. In general, property passing outright to the surviving spouse is eligible for the marital deduction. Property left to the surviving spouse to the extent of a life estate only would not be eligible for the marital deduction on the grounds that a life estate is a "terminable interest" and would terminate at the death of the surviving spouse. For deaths after 1981, 100% of the value of the property passing to the surviving spouse is deductible under the federal estate tax marital deduction. After 1981, even life estates may be made eligible for the federal estate tax marital deduction if desired.

Orphan's deduction—If a minor child or children under the age 21 are left surviving, an orphan's deduction may be claimed for deaths through 1981 equal to $5000 times the number of years each minor child is under the age of 21. To be eligible for the deduction, the child must be left without parent or spouse of parent surviving.

> **Example:** R, a 25 year old widow, was killed in an automobile accident leaving a child, age five. An orphan's deduction could be claimed equal to 16 times $5000 or $80,000 if that amount or more passed to the child.

The orphans deduction was repealed in 1981 effective for deaths after December 31, 1981.

Charitable deduction—A deduction, unlimited in amount, may be claimed for amounts passing to a qualified charity at death. In general, the recipient must be a church, subdivision of government, or an organization recognized as tax exempt by the Internal Revenue Service. Outright bequests to a charity are eligible for the deduction. In addition, a part of the value of a farm or personal residence is eligible for the deduction even though the property was not left outright to the charity. If a farm or personal residence is left to a family member for life, remainder interest to the charity, the value of the remainder interest of the farm or personal residence would be eligible for the federal estate tax charitable deduction. Thus, remainder interests in farms or personal residences can be left to a charitable organization in a relatively simple arrangement with the remainder interest eligible for a federal tax deduction. If done during life, the value of the charitable remainder may be deducted for federal income tax purposes and for federal gift tax purposes. If transferred at death, the value of the charitable interest in the farm or personal residence is eligible for a federal estate tax deduction.

For other property left to a charity, the value of a remainder interest is deductible for federal income, gift and estate tax purposes only if the property is left in a charitable remainder annuity trust, unitrust or pooled income fund. The latter techniques involve greater formality than a life estate-remainder arrangement that is possible for a farm or personal residence.

For the special exception for the farm or personal residence to apply, the asset must pass to the charitable organization. The exception does not apply if the property is to be sold and the proceeds passed to the charitable organization.

The term "personal residence" may include a vacation residence as well as the taxpayer's principal residence.

Calculating the Taxable Estate and Tax Due

With all allowable deductions claimed, the resulting "taxable" estate is subject to federal estate tax. Taxable gifts made during life (those gifts not covered by the federal gift tax annual exclusion, the marital deduction or the gift tax charitable deduction), are included in the federal estate tax taxable estate.

Example: In 1978, C made a gift of 320 acres of land valued at $640,000 to her daughter. After claiming the federal gift tax annual exclusion of $3000, the remaining amount of $637,000 was subject to federal gift tax. At the death of C in 1982, the amount of $637,000 would be included in C's estate for purposes of calculating the federal estate tax. The amount of the gift tax paid by C in 1978 would be deducted from the amount of federal estate tax due. In effect, taxable gifts made since 1976 are taken into account at death and move the individual up the tax rate schedule.

With the amount of the tax calculated, the unified federal estate and gift tax credit may be subtracted from the calculated amount. The unified credit can be used during life to offset federal gift tax due or can be used at death to reduce federal estate tax. The amount of the unified credit phases in over a six-year period as discussed in Chapter 19. By 1987, the unified credit of $192,800 will be "worth" the same as a deduction of $600,000.

Credits may also be claimed for part of the estate inheritance or state estate tax paid, a credit for federal estate tax on prior transfers of the same property within the prior 10 years and a credit for death taxes paid to a foreign government.

Federal Estate Tax Return

The federal estate tax return is to be filed within nine months after death. The tax due is paid at that time unless provision has been made for deferral of the tax due under the 10 or 15 year installment provision (for interests in a closely-held business) or one-year extensions of time have been obtained (for reasonable cause). Ten year installment payment has been repealed effective for deaths after 1981.

CHAPTER 19
TRANSFERRING LAND BY GIFT

The decision to make gifts of land during life, utilize one or more of the provisions for sale of the property or retain the land until death is a complex issue influenced by several factors. A desire to provide financial assistance to family members (or others), a desire to avoid income tax liability for income generated by income-producing property, and a desire to reduce management responsibilities may all influence the decision of whether to make a gift. Federal gift tax considerations also typically enter into the decision whether to transfer property by gift.

Federal Gift Tax

Although gifts were made substantially less attractive by federal tax legislation enacted in 1976, gifts still constitute an alternative for disposing of land and other property. For many property owners, the major concern about gifts is a threshold question: Is a gift program preferable to sale of land or retention until death? A part gift/part sale, a commonly used approach for family transfers, is another possibility for land.

In addition to programs of intended gift making, gift tax concerns can also arise with shifts in property ownership or co-ownership such as on conversion of joint tenancy to tenancy in common.

How Valued

For federal gift tax purposes, transfers of assets during life, including land, are valued at their "fair market value." That is defined as the price at which the asset would change hands between a willing buyer and a willing seller, neither being under any compulsion to buy or to sell. For land, valuation is heavily influenced by comparable sales.

It is notable that gifts of land are not eligible for "use" valuation which has been available to value eligible land for federal estate tax purposes since 1976. Use valuation is discussed in detail in Chapter 18.

With land transferred by gift not eligible for use valuation, a substantial incentive exists to retain land until death. That is because land transferred during life by gift (or by sale) is valued at 100% of fair market value whereas land retained until death may be valued at a fraction of fair market value, perhaps as low as 40 to 60%, if the requirements for use valuation can be met. Especially now that federal gift tax and federal estate tax use the same rate schedule and the same unified credit, the difference in tax treatments because of different valuation rules is indeed striking.

Another disadvantage of transferring by gift is the income tax treatment of the transfer. For property given away during life, the former income tax basis carries over to the recipient and becomes the donee's income tax basis. Thus, any gain in the property is simply transferred to the recipient of the gift. For assets, such as land, retained until death, a new income tax basis is obtained equal to the value used for federal estate tax purposes. That's the case except for gifts of appreciated property within one year of the recipient's death where the property (or the proceeds) are received back after death by the donor or donor's spouse. With that exception, a substantial amount of income tax gain is typically eliminated at the death of the property owner. Again, this factor can represent a substantial disadvantage for gifts compared to retention of property until death. Of course, these factors should be considered along with all other reasons favoring or disfavoring gift making.

Major legislative changes, effective in 1977, altered substantially the strategy for gift making for many individuals concerned with federal gift tax liability. However, some federal tax provisions, notably the federal gift tax annual exclusion, were left unchanged by the 1976 legislation. Amendments in 1981 expanded the annual exclusion as noted below.

Annual exclusion—For years, it has been possible for property owners to make gifts of up to $3,000 per recipient per year without paying federal gift tax. That provision largely survived the 1976 legislation. Moreover, for gifts of $3,000 or less of present interests in property to each recipient, the property is not included in the gross estate for federal estate tax purposes, even if given within three years of death. That rule is applicable to all property except for transfers of life insurance policies. For life insurance policies transferred within three years of death, the proceeds are included in the gross estate regardless of the gift value of the policy at the time of the gift transaction.

For husbands and wives as donors, the gift amount can be $6,000 per recipient per year without federal gift tax liability even though made out of one spouse's property.

Example: A husband and wife with four adult children, each married, can give a total of $48,000 to the four children and their spouses each year. The husband and wife as donors could give each child $6,000 and each child's spouse $6,000 for a total of $12,000 per couple per year.

Effective in 1982, the federal gift tax annual exclusion is $10,000 per recipient per year rather than $3,000. After 1981, a husband and wife may give each recipient $20,000 per year under the gift tax annual exclusion. Moreover, commencing in 1982, payment of medical expenses and tuition aren't considered gifts.

The major problem with effective use of the federal gift tax annual exclusion is in making gifts of $3,000 (or $10,000) or less in value if much of the wealth is tied up in assets such as land. For such assets, the choice is usually either a gift of an undivided interest (such as a 1/64th interest in a 40-acre tract), a conveyance of a few acres of land each year, or formation of a corporation or partnership with a gift of corporate stock or partnership interests.

It should be noted that the federal gift annual exclusion is only available for gifts of present interests. Gifts of *future* interests such as gifts of a remainder interest, are not eligible for the federal gift tax annual exclusion.

Example: Grandmother, wishing to secure her own economic future and still make a transfer to children, conveys a remainder interest in 160 acres to the four children but retains a life estate for herself for so long as she lives. Retention of the life estate would mean that the interest transferred to the children would be a future interest and would not be eligible for the federal gift tax annual exclusion. Moreover, because of the retained life estate, the full value of the entire property would be included in grandmother's estate at her death.

Example: Grandmother Smith left 160 acres to her granddaughter, Dorothy, but with a life estate to her daughter, Ann. The life estate to daughter Ann would be eligible for the federal gift tax annual exclusion because it is a gift of a present interest. The income from the property would pass to Ann after the transaction was completed. The gift of the remainder interest to grandaughter Dorothy, however, would be a gift of a future interest and would not be eligible for the federal gift tax annual exclusion. Assuming Grandmother Smith lived more than three years after the transfer, no part of the value would be included in her gross estate for federal estate tax purposes.

Marital deduction—For gifts to a spouse, a federal gift tax marital deduction is available through 1981 to cover the first $100,000 of such gifts and 50% of gifts to a spouse above $200,000. There is no federal gift tax marital deduction for gifts between $100,000 and $200,000 to a spouse.

To the extent that the federal gift tax marital deduction exceeds 50% of the amount of the gift to the spouse that must be reported on a federal gift tax return, the excess is subtracted from the maximum allowable federal estate tax marital deduction at death. This "cutdown" provision was enacted in 1976 to enable couples to more nearly equalize estates with up to $50,000 advanced out of the federal estate tax marital deduction to be used during life.

Example: John Allen in 1981 made a gift of 80 acres of land valued at $103,000 to his wife, Susan. The federal gift tax annual exclusion of $3,000 would reduce the amount of the gift to $100,000. The federal gift tax marital deduction would reduce the taxable gift to zero. Thus, no federal gift tax would be due and no reduction of the federal estate and gift tax unified credit would occur. However, 50% of the value of the gift to be reported on a federal gift tax return (50% to $103,000) would be subtracted from the federal gift tax marital deduction claimed (which would be $100,000 in the example) with the excess ($48,500) reducing the federal estate tax marital deduction at death. Therefore, $48,500 would be subtracted from the maximum allowable federal estate tax marital deduction. Had the federal estate tax marital deduction at death for John been $450,000, the making of a gift of $103,000 to his wife would reduce the federal estate tax marital deduction to $401,500. No "cut-down" of the federal estate tax marital deduction occurs for gifts to the spouse in excess of $200,000.

Of course, the reduction of federal estate tax marital deduction at death is of no concern if the donor survives. It is only when the donor dies first that the reduction in federal estate tax marital deduction would increase federal estate tax due at death. Moreover, if the estates of husband and wife are relatively well-balanced during life it may be that a full marital deduction at death would not be optimal in any event. Therefore, the reduction in federal estate tax marital deduction because of gifts during life should not always be viewed as disadvantageous.

Commencing with gifts in 1982, there is no limit on the amount of property that can be given to a spouse. Even life estates given to a spouse can be made eligible for the gift tax marital deduction. And use of the 100% gift tax marital deduction does not reduce the federal estate tax marital deduction available at death.

Charitable deductions—A federal gift tax charitable deduction is available for gifts to a qualified charity. In general, that means a deduction is available for gifts to a church, a subdivision of government, or an organization approved by the Internal Revenue Service as exempt and eligible for a federal gift tax charitable deduction. Not all organizations that are organized as nonprofit under state law are recognized as tax exempt by the Internal Revenue Service. Moreover, not all organizations that are approved as tax exempt can assure a tax deduction for contributions. For contributions to be deductible for income tax, gift tax, and estate tax purposes, it must be an organization approved under Section 501(c)(3) of the Internal Revenue Code.

In many instances, gifts to charity are of an outright bequest or transfer during life of the entire amount of property. However, in some instances, property owners may prefer to make a transfer to the charity during life but with a retained life estate for a spouse or other member of the family. Since 1969, for most types of property, it has been necessary to establish a special type of trust in order to obtain a charitable deduction for the value of the remainder interest where the life estate or other present interest was held by a noncharitable beneficiary. In general, it has been necessary to establish a charitable remainder annuity trust, unitrust, or a pooled income fund in order to receive an income, gift or estate tax deduction for the value of the remainder interest.

However, for transfers involving a farm or personal residence, it is possible to obtain a deduction for a remainder interest passing to a charitable organization when the property is left in a legal life estate (life estate not in trust) to a noncharitable beneficiary (such as a family member) with the remainder to a charitable organization. That option, which is relatively simple to create, is only open for gifts of farmland and gifts of a personal residence (including a vacation residence).

Directing that the property be sold at the conclusion of the life interest held for the family member with the cash proceeds payable to the charity would make the property ineligible for the simplified provision, however.

Unified credit—Taxable gifts after 1976 are added back in the taxable estate at death for purposes of calculating the federal estate tax due. This is the mechanism by which account is taken of the use of the unified credit to cover gifts during life. Any unused unified credit can be used at death to cover federal estate tax due. Thus, gifts during life not covered by the federal gift tax annual exclusion, marital deduction or charitable deduction become "taxable gifts" with the federal gift tax reduced by the unified credit. However, at death all taxable gifts after 1976 become part of the taxable estate and the unified credit is restored and may be claimed fully at death. Account is also taken of any gift tax paid during life as a credit against federal estate tax due at death.

As enacted in 1976, the unified credit phases in over a five-year period as follows:

- For gifts made between January 1, 1977 and June 30, 1977, $6,000;
- For gifts made after December 31, 1976 and before January 1, 1978 and for death during 1977, $30,000;
- For gifts made or deaths occurring in 1978, $34,000;
- For gifts made or deaths occurring in 1979, $38,000;
- For gifts made or deaths occurring in 1980, $42,500;
- For gifts made or deaths occurring in 1981, $47,000.

Legislation in 1981 raised the level of the unified gift and estate tax credit for each year through 1987 as follows for gifts made or deaths occurring in:

	Credit Amount	Deduction Equivalent
1982	$ 62,800	$225,000
1983	79,300	275,000
1984	96,300	325,000
1985	121,800	400,000
1986	155,800	500,000
1987	192,800	600,000

The unified credit is equivalent to a deduction of $175,625 for gifts made or deaths occurring in 1981. By 1987, the unified credit will be "worth" $600,000. The unified credit offsets that much in property value whether made as a gift during life or as property passing at death. However, use of the unified credit during life to cover gifts is not "costless." If the unified credit is used fully to cover gifts during life, a donor's estate at death is at the 32% federal estate tax bracket (for purposes of paying federal estate tax in 1981), 37% by 1987. Thus, a "phantom tax" is exacted when taxable gifts are made during life that are covered by the unified credit.

The effect of the use of the unified credit during life is to move the taxpayer up the federal estate tax schedule at death. This "phantom tax" is seen most clearly in situations where gifts are made during life to an individual, such as a brother or sister, with a life expectancy approximately the same as the donor. A phantom tax would be exacted from the donor at the time of the gift with an actual federal estate tax imposed on the property at the death of the donee. Thus, two taxes would be due on the property within a period of perhaps a few months or a few years. For that reason, in general it is preferred from a tax savings standpoint for any use of the unified credit during life to be to cover gifts to an individual in a subsequent generation such as a child or grandchild. In that event, the second death (and the second tax) should be several years away.

Changing Co-Ownership Patterns

Thus far in this chapter, attention has been focused on intentional gift making. Gift tax liability can also be incurred for inadvertent gifts such as can occur on shifts in ownership or co-ownership patterns.

For property owners with assets held in joint tenancy (or tenancy by the entirety) and with estates of sufficient size to be concerned about federal estate tax liability, it is generally advisable to terminate the joint ownership pattern. The co-ownership patterns, including joint tenancy and tenancy in common, are discussed in Chapter 11.

The paragraphs following discuss first the federal gift tax treatment of severances of joint ownership before 1982. The consequences of severances after 1981 are discussed thereafter.

Land purchased before 1955. For land bought—and paid for—before 1955, the rule is clear. Whatever gift was involved on acquisition occurred then, whether or not reported as a gift. For joint tenancy purposes, it would have been a gift of one-half the amount involved if the husband, for example, had provided all of the funds for the purchase.

In the few states recognizing tenancy by the entirety (the special brand of joint tenancy involving husbands and wives), it is a bit more complicated. The amount of any gift depends upon the ages of the spouses.

But in either situation, the gift, if any, occurred on acquisition. So it is possible to sever the joint tenancy into tenancy in common without a gift. For joint tenancy, there is no gift if it is severed into a tenancy in common with equal co-ownership as shown in Figure 19.1. Transfer of title to the husband's name alone would trigger a gift from wife to husband of *one-half the current value*. Likewise, a transfer to the wife's name alone means a gift of one-half the value from husband to wife. Again, where tenancy by the entirety is at stake, the "safe harbor" of no gift depends on the age of the spouses.

For land bought before 1955, improvements to the land made before 1955 involved gifts unless the funds for the improvements were furnished the same way the land was owned. The same rule applies to payments on mortgages or land contracts to reduce the indebtedness. Thus, if the husband provided all the funds for a new $10,000 silo built in 1954 and placed on land owned in joint tenancy, it would have involved a gift of $5,000 to the wife as co-owner of the land.

Land acquired after 1954. The rule on gifts involving joint tenancy property was changed for land acquired after 1954 and before 1982. For land acquired by a husband and wife in joint tenancy, it has not been a gift—even though contributions were all by one spouse, for example—unless reported as a gift on a federal gift tax return filed on time. And apparently very few have done that as noted above.

So, if the husband had provided all of the funds on acquisition, and it was not reported then as a gift, *for gift tax purposes* it is still viewed as the husband's property. Therefore, the only "safe harbor" would be to sever the joint tenancy with title transferred to the husband's name alone. A shift to tenancy in common would mean a gift of 50% of the current value of the property from husband to wife as

Figure 19.1 Gift on severance of joint tenancy where husband provided money for purchase.

	Property in husband's name	H and W Property in tenancy in common	Property in wife's name
Pre-1955 acquisition	50%	—0—	50%
Acquisition after 1954 and before 1982	—0—	50%	100%

shown in Figure 19.1. Severance to the wife's name alone would trigger a gift of 100% of the current value.

But what about improvements or reduction in indebtedness after 1954?

- If improvements were made (or indebtedness reduced) before 1955 on land acquired before 1955, it was a gift at that time if contributions were unequal. For improvements made (or indebtedness reduced) after 1954 on land acquired before 1955, there is apparently no gift at the time—unless reported as a gift on a timely filed federal gift tax return. So a $50,000 grain storage facility built in 1958 from the husband's funds on land owned in joint tenancy and acquired before 1955 is still viewed as the husband's property for federal gift tax purposes. Where that has occurred, there is likely to be a gift if the pre-1955 joint tenancy is severed into tenancy in common.
- For land acquired in joint tenancy *after 1954* and reported as a gift, improvements made (or indebtedness reduced) are not treated as gifts (if contributions were unequal) through 1976 *unless reported as gifts*.

  But the law was changed in 1976. Commencing in 1977, an election to treat the creation of joint ownership in land as a gift applies also to all later additions to the property as improvements or reduction in indebtedness. One election handles both—the gift on creation of the joint tenancy and later increases in value from paying off a mortgage or land contract or from land improvements.
- Finally, for land bought after 1954 and not reported as a gift, improvements or reductions in indebtedness are not treated as gifts even though contributions were unequal.

Inadvertent severance of joint ownership. Deliberate severance of joint tenancies creates enough problems. But inadvertent break ups may be even more troublesome.

- There is no severance if joint tenancy land is sold or exchanged for other real property *held identically*. And there is no taxable termination if the proceeds of sale go into improvements to land held in identical joint ownership by the spouses.

  So there is no federal gift tax problem if 40 acres of land bought in 1956 are sold and the proceeds go into a new confinement feeding unit on the remaining part of the farm owned the same way as the tract that was sold.
- An increase in indebtedness on joint tenancy land is treated as a severance. That is the rule unless offset by additions or improvements "within a reasonable time."
- It appears that transfer of joint tenancy land to a new partnership causes a severance.

Example: A farm was purchased by a husband and wife in joint tenancy in 1958 with the husband providing all the funds. It was

not reported as a gift. If the land is transferred in 1981 to a newly formed partnership with partnership interests owned equally by husband and wife, there would be a gift from husband to wife of one-half the current value of the land.
- The same rule apparently applies to land transferred to a newly formed corporation. It is a severance of the joint tenancy.
- If joint tenancy land is sold under installment land contract, that also appears to sever the joint tenancy.

Example: A husband and wife acquired 320 acres of land in 1960 for $160,000 in cash, all provided by the wife from an inheritance. Title was taken in joint tenancy. It was not reported as a gift on a federal gift tax return.

If the land is sold in early 1981 for $640,000 on an installment contract, even though the husband and wife own the *contract* in joint tenancy, it is still a gift of one-half the value from wife to husband. And that could trigger a sizeable federal gift tax.

- There is no severance, according to a recent case, if joint tenancy land is transferred to a revocable living trust. There was a right retained to revoke the trust and receive back the land. IRS had insisted even that was a taxable termination but lost the case.

Severance After 1981—Commencing in 1982, several changes have been made in severance of joint tenancy and tenancy by the entirety property. All acquisitions of property are treated as a gift at the time of acquisition if contributions are unequal. For husband-wife transfers (from one spouse to another) the 100% federal gift tax marital deduction is available to cover gifts to the other spouse including gifts arising by virtue of severances of co-ownership patterns.

Part Gift/Part Sale

Gifts can also occur in what are sometimes termed "part-gift/part-sale" transactions. Such transactions can arise—(1) when assets, such as land, are undervalued in a sale, (2) when part of an asset, such as land, is conveyed by gift and the rest of the asset is transferred by sale, or (3) an asset is sold under an installment sale arrangement with part or all of the payments forgiven as they come due.

In any event, the purchaser's income tax basis is the greater of the giver's basis or the purchase price for the property. For that reason it is usually preferable from an income tax point of view for a "part-gift/part-sale" transaction to be handled with a specific part of the asset conveyed by gift and the rest of the asset transferred by sale. In that event, the purchaser often ends up with a higher total income tax basis for the property.

Example: George and Sylvia Fox are considering a transfer of 160 acres of land valued at $300,000 to their daughter. The tract has an income tax basis of $100,000. They are willing to sell the 160 acres to their daughter for $150,000. Their daughter's income tax basis would be $150,000—the greater of their basis ($100,000) or the purchase price ($150,000). On the other hand, if the parents were to sell 80 acres to their daughter for $150,000 and make a gift of the other $150,000 to their daughter, her income tax basis would be $150,000 (from the purchase) plus $50,000 (the basis from the gift amount) or $200,000.

In recent years, the major tax concern with part-gift/part-sale transactions has been the forgiveness of installment payments. In 1977, the Internal Revenue Service withdrew approval of the outcome of two earlier Tax Court cases. The position taken by IRS at that time was that consistent forgiveness of installment payments due, such as on a land transaction, could be viewed as a present intent to make a series of future gifts with the entire transaction treated as a gift in the year of the initial "sale" transaction.

That outcome led to federal gift tax liability for the "seller" and, presumably, an income tax basis for the recipient based upon the donor's income tax basis for the property rather than a new basis derived from the purchase price. For that reason, it was generally advisable for anyone forgiving installments not to forgive all payments as they came due. Certainly, it seemed prudent for the transaction to be characterized carefully in the year of sale as a sale with the election made for installment payment purposes. After the first year, it seemed advisable for payments not to be forgiven for very many years before another payment was collected and reported for income tax purposes if it was desired to maintain initial characterization of the transaction as a sale.

In 1980, as part of the general revision of installment sale rules, federal tax law was changed to require forgiveness or cancellation of payments to be treated as a taxable disposition. Thus, amounts forgiven or cancelled are nonetheless income to the seller. Thus to avoid recharacterization of installment sales as gifts, it seems advisable for sellers under installment obligations to collect all payments, report the payments for income tax purposes and make a gift of part or all of the proceeds in a separate transaction as desired.

APPENDICES

APPENDIX A

INTEREST FACTORS FOR COMPOUNDING AND DISCOUNTING

Table 1. Interest Factors for *Compounding* a Single Amount: $(1 + i)^n$

N	0.5%	0.75%	1%	1.5%	2%	2.5%	3%	4%	5%	6%
1	1.005	1.007	1.010	1.015	1.020	1.025	1.030	1.040	1.050	1.060
2	1.010	1.015	1.020	1.030	1.040	1.051	1.061	1.082	1.102	1.124
3	1.015	1.023	1.030	1.046	1.061	1.077	1.093	1.125	1.158	1.191
4	1.020	1.030	1.041	1.061	1.082	1.104	1.126	1.170	1.216	1.262
5	1.025	1.038	1.051	1.077	1.104	1.131	1.159	1.217	1.276	1.338
6	1.030	1.046	1.062	1.093	1.126	1.160	1.194	1.265	1.340	1.419
7	1.036	1.054	1.072	1.110	1.149	1.189	1.230	1.316	1.407	1.504
8	1.041	1.062	1.083	1.126	1.172	1.218	1.267	1.369	1.477	1.594
9	1.046	1.070	1.094	1.143	1.195	1.249	1.305	1.423	1.551	1.689
10	1.051	1.078	1.105	1.161	1.219	1.280	1.344	1.480	1.629	1.791
11	1.056	1.086	1.116	1.178	1.243	1.312	1.384	1.539	1.710	1.898
12	1.062	1.094	1.127	1.196	1.268	1.345	1.426	1.601	1.796	2.012
13	1.067	1.102	1.138	1.214	1.294	1.379	1.469	1.665	1.886	2.133
14	1.072	1.110	1.149	1.232	1.319	1.413	1.513	1.732	1.980	2.261
15	1.078	1.119	1.161	1.250	1.346	1.448	1.558	1.801	2.079	2.397
16	1.083	1.127	1.173	1.269	1.373	1.484	1.605	1.873	2.183	2.540
17	1.088	1.135	1.184	1.288	1.400	1.522	1.653	1.948	2.292	2.693
18	1.094	1.144	1.196	1.307	1.428	1.560	1.702	2.026	2.407	2.854
19	1.099	1.153	1.208	1.327	1.457	1.599	1.753	2.107	2.527	3.026
20	1.105	1.161	1.220	1.347	1.486	1.639	1.806	2.191	2.653	3.207
24	1.127	1.196	1.270	1.429	1.608	1.809	2.033	2.563	3.225	4.049
28	1.150	1.233	1.321	1.517	1.741	1.996	2.288	2.999	3.920	5.112
30	1.161	1.251	1.348	1.563	1.811	2.098	2.427	3.243	4.322	5.743
32	1.173	1.270	1.375	1.610	1.885	2.204	2.575	3.508	4.765	6.453
36	1.197	1.309	1.431	1.709	2.040	2.433	2.898	4.104	5.792	8.147
48	1.270	1.431	1.612	2.043	2.587	3.271	4.132	6.571	10.401	16.393
60	1.349	1.566	1.817	2.443	3.281	4.400	5.892	10.520	18.678	32.987

Table 1. (continued)

N	7%	8%	9%	10%	12%	14%	16%	20%	25%
1	1.070	1.080	1.090	1.100	1.120	1.140	1.160	1.200	1.250
2	1.145	1.166	1.188	1.210	1.254	1.300	1.346	1.440	1.563
3	1.225	1.260	1.295	1.331	1.405	1.482	1.561	1.728	1.953
4	1.311	1.360	1.412	1.464	1.574	1.689	1.811	2.074	2.441
5	1.403	1.469	1.539	1.611	1.762	1.925	2.100	2.488	3.052
6	1.501	1.587	1.677	1.772	1.974	2.195	2.436	2.986	3.815
7	1.606	1.714	1.828	1.949	2.211	2.502	2.826	3.583	4.768
8	1.718	1.851	1.993	2.144	2.476	2.853	3.278	4.300	5.960
9	1.838	1.999	2.172	2.358	2.773	3.252	3.803	5.160	7.451
10	1.967	2.159	2.367	2.594	3.106	3.707	4.411	6.192	9.313
11	2.105	2.332	2.580	2.853	3.479	4.226	5.117	7.430	11.642
12	2.252	2.518	2.813	3.138	3.896	4.818	5.936	8.916	14.552
13	2.410	2.720	3.066	3.452	4.363	5.492	6.886	10.699	18.190
14	2.579	2.937	3.342	3.797	4.887	6.261	7.988	12.839	22.737
15	2.759	3.172	3.642	4.177	5.474	7.138	9.266	15.407	28.422
16	2.952	3.426	3.970	4.595	6.130	8.137	10.748	18.488	35.527
17	3.159	3.700	4.328	5.054	6.866	9.276	12.468	22.186	44.409
18	3.380	3.996	4.717	5.560	7.690	10.575	14.462	26.623	55.511
19	3.617	4.316	5.142	6.116	8.613	12.056	16.776	31.948	69.389
20	3.870	4.661	5.604	6.727	9.646	13.743	19.461	38.337	86.736
24	5.072	6.341	7.911	9.850	15.179	23.212	35.236	79.497	211.758
28	6.649	8.627	11.167	14.421	23.884	39.204	63.800	164.844	516.988
30	7.612	10.063	13.267	17.449	29.960	50.949	85.850	237.375	807.793
32	8.715	11.737	15.763	21.113	37.582	66.214	115.519	341.820	1262.177
36	11.424	15.968	22.251	30.912	59.135	111.832	209.163	708.798	3081.488
48	25.729	40.210	62.583	97.015	230.290	538.792	1241.597	6319.699	44841.550
60	57.945	101.257	176.024	304.472	897.591	2595.835	7370.141	56346.970	652530.400

APPENDIX A (continued)

Table 2. Interest Factors for *Compounding* a Series of Equal Series of Payments: [EFIF$_{i,n}$]

N	0.5%	0.75%	1%	1.5%	2%	2.5%	3%	4%	5%	6%
1	1.000	1.000	1.000	1.000	1.000	1.000	1.000	1.000	1.000	1.000
2	2.005	2.007	2.010	2.015	2.020	2.025	2.030	2.040	2.050	2.060
3	3.014	3.022	3.030	3.045	3.060	3.076	3.091	3.122	3.152	3.184
4	4.029	4.045	4.060	4.091	4.121	4.152	4.184	4.246	4.310	4.375
5	5.049	5.075	5.101	5.152	5.204	5.256	5.309	5.416	5.526	5.637
6	6.074	6.113	6.151	6.229	6.308	6.388	6.468	6.633	6.802	6.975
7	7.104	7.159	7.213	7.323	7.434	7.547	7.662	7.898	8.142	8.394
8	8.140	8.213	8.285	8.432	8.583	8.736	8.892	9.214	9.549	9.897
9	9.180	9.274	9.368	9.559	9.754	9.954	10.159	10.583	11.026	11.491
10	10.226	10.344	10.461	10.702	10.949	11.203	11.464	12.006	12.578	13.181
11	11.277	11.421	11.566	11.863	12.168	12.483	12.808	13.486	14.207	14.971
12	12.333	12.507	12.681	13.041	13.412	13.795	14.192	15.026	15.917	16.870
13	13.395	13.601	13.808	14.236	14.680	15.140	15.618	16.627	17.713	18.882
14	14.462	14.703	14.946	15.450	15.973	16.519	17.086	18.292	19.598	21.015
15	15.534	15.813	16.096	16.681	17.293	17.932	18.599	20.024	21.578	23.276
16	16.611	16.932	17.256	17.931	18.639	19.380	20.157	21.824	23.657	25.672
17	17.694	18.058	18.429	19.200	20.011	30.864	21.761	23.697	25.840	28.212
18	18.782	19.194	19.613	20.488	21.412	22.386	23.414	25.645	28.132	30.905
19	19.876	20.338	20.809	21.796	22.840	23.946	25.117	27.671	30.538	33.759
20	20.975	21.490	22.017	23.123	24.297	25.544	26.870	29.778	33.065	36.785
24	25.427	26.187	26.971	28.632	30.421	32.348	34.426	39.083	44.501	50.815
28	29.969	31.027	32.126	34.480	37.050	39.859	42.930	49.967	58.401	68.527
30	32.274	33.501	34.782	37.537	40.567	43.902	47.575	56.085	66.437	79.057
32	34.602	36.013	37.491	40.686	44.226	48.149	52.502	62.701	75.297	90.888
36	39.329	41.151	43.073	47.273	51.993	57.300	63.275	77.598	95.833	119.118
48	54.088	57.518	61.217	69.561	79.350	90.857	104.407	139.263	188.018	256.558
60	69.757	75.420	81.662	96.209	114.047	135.988	163.050	237.990	353.567	533.111

Table 2. (continued)

N	7%	8%	9%	10	12%	14%	16%	20%	25%
1	1.000	1.000	1.000	1.000	1.000	1.000	1.000	1.000	1.000
2	2.070	2.080	2.090	2.100	2.120	2.140	2.160	2.200	2.250
3	3.215	3.246	3.278	3.310	3.374	3.440	3.506	3.640	3.813
4	4.440	4.506	4.573	4.641	4.779	4.921	5.066	5.368	5.766
5	5.751	5.867	5.985	6.105	6.353	6.610	6.877	7.442	8.207
6	7.153	7.336	7.523	7.716	8.115	8.535	8.977	9.930	11.259
7	8.654	8.923	9.200	9.487	10.089	10.730	11.414	12.916	15.073
8	10.260	10.637	11.028	11.436	12.300	13.233	14.240	16.499	19.842
9	11.978	12.488	13.021	13.579	14.776	16.085	17.518	20.799	25.802
10	13.816	14.487	15.193	15.937	17.549	19.337	21.321	25.959	33.253
11	15.784	16.645	17.560	18.531	20.655	23.044	25.733	32.150	42.566
12	17.888	18.977	20.140	21.384	24.133	27.271	30.850	39.580	54.208
13	20.141	21.495	22.953	24.522	28.029	32.088	36.786	48.496	68.760
14	22.550	24.215	26.019	27.975	32.393	37.581	43.672	59.196	86.949
15	25.129	27.152	29.360	31.772	37.280	43.842	51.659	72.035	109.687
16	27.888	30.324	33.003	35.949	42.753	50.980	60.925	87.442	138.109
17	30.840	33.750	36.973	40.544	48.884	59.117	71.673	105.930	173.636
18	33.999	37.450	41.301	45.599	55.750	68.393	84.140	128.116	218.045
19	37.379	41.446	46.018	51.158	63.440	78.968	98.603	154.739	273.556
20	40.995	45.762	51.159	57.274	72.052	91.024	115.379	186.687	342.945
24	58.176	66.765	76.788	88.496	118.155	158.656	213.977	392.483	843.033
28	80.697	95.339	112.966	134.208	190.698	272.885	392.501	819.219	2063.951
30	94.460	113.283	135.304	164.491	241.332	356.781	530.309	1181.876	3227.174
32	110.217	134.213	164.033	201.134	304.846	465.812	715.744	1704.100	5044.707
36	148.912	187.102	236.118	299.121	484.461	791.657	1301.021	3538.988	12321.940
48	353.265	490.130	684.256	960.148	1911.580	3841.375	7753.730	31593.490	179362.100
60	813.506	1253.208	1944.706	3034.720	7471.594	18534.530	46057.120	281729.800	2610117.000

APPENDIX A (continued)

Table 3. Interest Factors for *Discounting* a Single Amount: $\dfrac{1}{(1+i)^n}$

N	0.5%	0.75%	1%	1.5%	2%	2.5%	3%	4%	5%	6%
1	0.995	0.993	0.990	0.985	0.980	0.976	0.971	0.962	0.952	0.943
2	0.990	0.985	0.980	0.971	0.961	0.952	0.943	0.925	0.907	0.890
3	0.985	0.978	0.971	0.956	0.942	0.929	0.915	0.889	0.864	0.840
4	0.980	0.971	0.961	0.942	0.924	0.906	0.808	0.855	0.823	0.792
5	0.975	0.963	0.951	0.928	0.906	0.884	0.863	0.822	0.784	0.747
6	0.971	0.956	0.942	0.915	0.888	0.862	0.837	0.790	0.746	0.705
7	0.966	0.949	0.933	0.901	0.871	0.841	0.813	0.760	0.711	0.665
8	0.961	0.942	0.923	0.888	0.853	0.821	0.789	0.731	0.677	0.627
9	0.956	0.935	0.914	0.875	0.837	0.801	0.766	0.703	0.645	0.592
10	0.951	0.928	0.905	0.862	0.820	0.781	0.744	0.676	0.614	0.558
11	0.947	0.921	0.896	0.849	0.804	0.762	0.722	0.650	0.585	0.527
12	0.942	0.914	0.887	0.836	0.788	0.744	0.701	0.625	0.557	0.497
13	0.937	0.907	0.879	0.824	0.773	0.725	0.681	0.601	0.530	0.469
14	0.933	0.901	0.870	0.812	0.758	0.708	0.661	0.577	0.505	0.442
15	0.928	0.894	0.861	0.800	0.743	0.690	0.642	0.555	0.481	0.417
16	0.923	0.887	0.853	0.788	0.728	0.674	0.623	0.534	0.458	0.394
17	0.919	0.881	0.844	0.776	0.714	0.657	0.605	0.513	0.436	0.371
18	0.914	0.874	0.836	0.765	0.700	0.641	0.587	0.494	0.416	0.350
19	0.910	0.868	0.828	0.754	0.686	0.626	0.570	0.475	0.396	0.331
20	0.905	0.861	0.820	0.742	0.673	0.610	0.554	0.456	0.377	0.312
24	0.887	0.836	0.788	0.700	0.622	0.553	0.492	0.390	0.310	0.247
28	0.870	0.811	0.757	0.659	0.574	0.501	0.437	0.333	0.255	0.196
30	0.861	0.799	0.742	0.640	0.552	0.477	0.412	0.308	0.231	0.174
32	0.853	0.787	0.727	0.621	0.531	0.454	0.388	0.285	0.210	0.155
36	0.836	0.764	0.699	0.585	0.490	0.411	0.345	0.244	0.173	0.123
48	0.787	0.699	0.620	0.489	0.387	0.306	0.242	0.152	0.096	0.061
60	0.741	0.639	0.550	0.409	0.305	0.227	0.170	0.095	0.054	0.030

Table 3. (continued)

N	7%	8%	9%	10%	12%	14%	16%	20%	25%
1	0.935	0.926	0.917	0.909	0.893	0.877	0.862	0.833	0.800
2	0.873	0.857	0.842	0.826	0.797	0.769	0.743	0.694	0.640
3	0.816	0.794	0.772	0.751	0.712	0.675	0.641	0.579	0.512
4	0.763	0.735	0.708	0.683	0.636	0.592	0.552	0.482	0.410
5	0.713	0.681	0.650	0.621	0.567	0.519	0.476	0.402	0.328
6	0.666	0.630	0.596	0.564	0.507	0.456	0.410	0.335	0.262
7	0.623	0.583	0.547	0.513	0.452	0.400	0.354	0.279	0.210
8	0.582	0.540	0.502	0.467	0.404	0.351	0.305	0.233	0.168
9	0.544	0.500	0.460	0.424	0.361	0.308	0.263	0.194	0.134
10	0.508	0.463	0.422	0.386	0.322	0.270	0.227	0.162	0.107
11	0.475	0.429	0.388	0.350	0.287	0.237	0.195	0.135	0.086
12	0.444	0.397	0.356	0.319	0.257	0.208	0.168	0.112	0.069
13	0.415	0.368	0.326	0.290	0.229	0.182	0.145	0.093	0.055
14	0.388	0.340	0.299	0.263	0.205	0.160	0.125	0.078	0.044
15	0.362	0.315	0.275	0.239	0.183	0.140	0.108	0.065	0.035
16	0.339	0.292	0.252	0.218	0.163	0.123	0.093	0.054	0.028
17	0.317	0.270	0.231	0.198	0.146	0.108	0.080	0.045	0.023
18	0.296	0.250	0.212	0.180	0.130	0.095	0.069	0.038	0.018
19	0.277	0.232	0.194	0.164	0.116	0.083	0.060	0.031	0.014
20	0.258	0.215	0.178	0.149	0.104	0.073	0.051	0.026	0.012
24	0.197	0.158	0.126	0.102	0.066	0.043	0.028	0.013	0.005
28	0.150	0.116	0.090	0.069	0.042	0.026	0.016	0.006	0.002
30	0.131	0.099	0.075	0.057	0.033	0.020	0.012	0.004	0.001
32	0.115	0.085	0.063	0.047	0.027	0.015	0.009	0.003	0.001
36	0.088	0.063	0.045	0.032	0.017	0.009	0.005	0.001	0.000
48	0.039	0.025	0.016	0.010	0.004	0.002	0.001	0.000	0.000
60	0.017	0.010	0.006	0.003	0.001	0.000	0.000	0.000	0.000

APPENDIX A (continued)

Table 4. Interest Factors for *Discounting* a Series of Equal Payments: [EPIF$_{1,n}$]

N	0.5%	0.75%	1%	1.5%	2%	2.5%	3%	4%	5%	6%
1	0.995	0.993	0.990	0.985	0.980	0.976	0.971	0.962	0.952	0.943
2	1.985	1.978	1.970	1.956	1.942	1.927	1.913	1.886	1.859	1.833
3	2.970	2.955	2.941	2.912	2.884	2.856	2.829	2.775	2.723	2.673
4	3.950	3.926	3.902	3.854	3.808	3.762	3.717	3.630	3.546	3.465
5	4.925	4.889	4.853	4.782	4.713	4.646	4.580	4.452	4.329	4.212
6	5.895	5.845	5.795	5.697	5.601	5.508	5.417	5.242	5.076	4.917
7	6.861	6.794	6.728	6.598	6.472	6.349	6.230	6.002	5.786	5.582
8	7.822	7.736	7.651	7.486	7.325	7.170	7.020	6.733	6.463	6.210
9	8.778	8.671	8.565	8.360	8.162	7.971	7.786	7.435	7.108	6.802
10	9.729	9.599	9.471	9.222	8.982	8.752	8.530	8.111	7.722	7.360
11	10.675	10.520	10.367	10.071	9.787	9.514	9.253	8.760	8.306	7.887
12	11.617	11.434	11.254	10.907	10.575	10.258	9.954	9.385	8.863	8.384
13	12.554	12.342	12.133	11.731	11.348	10.983	10.635	9.986	9.393	8.853
14	13.487	13.243	13.003	12.543	12.106	11.691	11.296	10.563	9.899	9.295
15	14.414	14.136	13.864	13.343	12.849	12.381	11.938	11.118	10.380	9.712
16	15.337	15.024	14.717	14.131	13.577	13.055	12.561	11.652	10.838	10.106
17	16.256	15.904	15.561	14.907	14.292	13.712	13.166	12.166	11.274	10.477
18	17.170	16.779	16.397	15.672	14.992	14.353	13.753	12.659	11.689	10.828
19	18.079	17.646	17.225	16.426	15.678	14.979	14.324	13.134	12.085	11.158
20	18.984	18.507	18.044	17.168	16.351	15.589	14.877	13.590	12.462	11.470
24	22.559	21.888	21.242	20.030	18.914	17.885	16.935	15.247	13.799	12.550
28	26.064	25.170	24.315	22.726	21.281	19.965	18.764	16.663	14.898	13.406
30	27.790	26.774	25.806	24.015	22.396	20.930	19.600	17.292	15.372	13.765
32	29.499	28.355	27.268	25.266	23.468	21.849	20.389	17.874	15.803	14.084
36	32.866	31.446	30.106	27.660	25.488	23.556	21.832	18.908	16.547	14.621
48	42.574	40.183	37.972	34.042	30.673	27.773	25.267	21.195	18.077	15.650
60	51.718	18.172	44.953	39.379	34.760	30.908	27.675	22.623	18.929	16.161

Table 4. (continued)

N	7%	8%	9%	10%	12%	14%	16%	20%	25%
1	0.935	0.926	0.917	0.909	0.893	0.877	0.862	0.833	0.800
2	1.808	1.783	1.759	1.736	1.690	1.647	1.605	1.528	1.440
3	2.624	2.577	2.531	2.487	2.402	2.322	2.246	2.106	1.952
4	3.387	3.312	3.240	3.170	3.037	2.914	2.798	2.589	2.362
5	4.100	3.993	3.890	3.791	3.605	3.433	3.274	2.991	2.689
6	4.767	4.623	4.486	4.355	4.111	3.889	3.685	3.326	2.951
7	5.389	5.206	5.033	4.868	4.564	4.288	4.039	3.605	3.161
8	5.971	5.747	5.535	5.335	4.968	4.639	4.344	3.837	3.329
9	6.515	6.247	5.995	5.759	5.328	4.946	4.607	4.031	3.463
10	7.024	6.710	6.418	6.145	5.650	5.216	4.833	4.192	3.571
11	7.499	7.139	6.805	6.495	5.938	5.453	5.029	4.327	3.656
12	7.943	7.536	7.161	6.814	6.194	5.660	5.197	4.439	3.725
13	8.358	7.904	7.487	7.103	6.424	5.842	5.342	4.533	3.780
14	8.745	8.244	7.786	7.367	6.628	6.002	5.468	4.611	3.824
15	9.108	8.559	8.061	7.606	6.811	6.142	5.575	4.675	3.859
16	9.447	8.851	8.313	7.824	6.974	6.265	5.668	4.730	3.887
17	9.763	9.122	8.544	8.022	7.120	6.373	5.749	4.775	3.910
18	10.059	9.372	8.756	8.201	7.250	6.467	5.818	4.812	3.928
19	10.336	9.604	8.950	8.365	7.366	6.550	5.877	4.843	3.942
20	10.594	9.818	9.129	8.514	7.469	6.623	5.929	4.870	3.954
24	11.469	10.529	9.707	8.985	7.784	6.835	6.073	4.937	3.981
28	12.137	11.051	10.116	9.307	7.984	6.961	6.152	4.970	3.992
30	12.409	11.258	10.274	9.427	8.055	7.003	6.177	4.979	3.995
32	12.647	11.435	10.406	9.526	8.112	7.035	6.196	4.985	3.997
36	13.035	11.717	10.612	9.677	8.192	7.079	6.220	4.993	3.999
48	13.730	12.189	10.934	9.897	8.297	7.130	6.245	4.999	4.000
60	14.039	12.377	11.048	9.967	8.324	7.140	6.249	5.000	4.000

APPENDIX B

SAMPLE OFFER TO BUY CONTRACT

IOWA STATE BAR ASSOCIATION
Official Form No. 26 (Trade-Mark Registered, State of Iowa, 1957)

FOR THE LEGAL EFFECT OF THE USE
OF THIS FORM, CONSULT YOUR LAWYER

OFFER TO BUY REAL ESTATE AND ACCEPTANCE

_____, Iowa, _____ 19____

TO _____(herein designated as Sellers):
(Insert names of Seller and Spouse)

THE UNDERSIGNED (herein designated as Buyers) hereby offer to buy the real estate situated in _____ County, Iowa, described as follows:

together with any easements and servient estates appurtenant thereto, but with reservations and exceptions only as follows:
(Strike out inapplicable parts, if any, of (a), (b) or (c) below.)
    (a) Title shall be taken subject to applicable zoning restrictions, except as in 1, below;
    (b) And subject to any reasonable, customary and appropriate restrictive covenants as may be shown of record, except as in 1, below;
    (c) And subject to easements of record for public utilities, public roads and public highways;
    (d) And subject to _____
                  (Liens?)   (Mineral reservation of record?)   (Covenants of record running with the land?)

(Easements not recorded?) (Driveway or other easement of record?) (Interests of other parties?) (Lessees?) (See paragraph No. 19)

for the total sum of $_____ payable at _____
County, Iowa, as follows:

Select (A) or (B) or (C) below:

A. INSTALLMENT PLAN: By payment of $_____ herewith, to be held by _____
Sellers' Agent, pending delivery of final papers, and $_____ at the rate of $_____ or more each month, including interest to date of each payment, until the entire purchase price, with interest at ____% per annum is paid; the first such payment to be made on the _____ day of _____ 19____, and thereafter on the _____ day of each month _____ until all sums due are paid in full; provided that in any event final payment of full balance under this contract shall be made on or before the _____ day of _____ 19____.

B. DOWN PAYMENT AND SETTLEMENT PAYMENT ONLY: By payment of $_____ herewith to be held by _____ Sellers' Agent, pending delivery of final papers, and the balance of $_____ upon performance by Sellers, all on the _____ day of _____ 19____.

C. OTHER PLAN:

1. SPECIAL USE. This offer is void unless Buyers are permitted, under any existing zoning and building restrictions, immediately to make the following conforming use of said real estate _____
2. TAXES. Sellers shall pay _____ (% or fraction) of real estate taxes payable in the year 19____, and all unpaid taxes for prior years. Any balance of taxes and/or subsequent taxes shall be paid by Buyers.
3. SPECIAL ASSESSMENTS. (a) Sellers shall pay all special assessments which are a lien on the date of acceptance of this offer.
  (b) If (a) hereof is stricken, then Sellers shall pay all installments of special assessments which, if not paid, would become delinquent the year this offer is accepted, and all prior installments thereof.
  All other special assessments shall be paid by Buyers.
4. INSURANCE. Sellers shall maintain $_____ of fire, windstorm and extended coverage insurance until possession is given and shall forthwith secure endorsements to the policies in such amount making loss payable to the parties as their interests may appear. Risk of loss from such hazards is on Buyers only when and as soon as (1) this offer is signed by both Sellers and Buyers and (2) performance of this paragraph by Sellers, and (3) after a copy hereof is delivered to Buyers. (See also paragraphs 10 and 20.) Buyers, if they desire, may obtain additional insurance to cover such risk.
5. POSSESSION. If Buyers timely perform all obligations on or before the _____ day of _____ 19____, possession shall on said date be delivered to Buyers, with adjustments of rent, insurance and interest as of date of transfer of possession. If Buyers are taking subject to right of Lessees, so indicate by "Yes" in the space following: _____; in which event, Sellers shall forthwith produce any written lease or leases on said premises for examination, and assignment.
6. FIXTURES. (a) All personal property that integrally belongs to or is part of said real estate, whether attached or detached, such as light fixtures (including fluorescent tubes but not mazda bulbs), shades, rods, blinds, venetian blinds, awnings, storm doors, storm sashes, screens, attached linoleum, plumbing fixtures, water heaters, water softeners, automatic heating equipment, air conditioning equipment other than window type, door chimes, built-in items and electrical service cable, fencing, gates and other attached fixtures, trees, bushes, shrubs and plants, shall be considered a part of real estate and included in this sale except _____
                        (Rented items?)
  (b) Wall to wall carpeting fastened to floor or walls shall _____ be a part of and included in this sale.
  (c) Outside television towers and antenna shall _____ be a part of and included in this sale.
7. ADDITIONAL PROVISIONS. This offer is made subject to the additional terms and provisions of Paragraphs 10 to 22, inclusive, printed on the reverse side hereof, without requirement of additional signatures, but Paragraph 23, or any additional provisions, or any changes of said Paragraphs 10 to 22, inclusive, other than the insertion of the amount of insurance in Paragraph 20, shall require the additional signatures of the parties on the reverse side hereof.
8. PURCHASE PRICE. It is agreed that at time of settlement, funds of the purchase price may be used to pay taxes, other liens and to acquire outstanding interests, if any, of other parties.
9. If this offer is not accepted by Sellers on or before _____ 19____, it shall become null and void and all payments shall be repaid to the Buyers.

Buyer _____   Buyer's Wife or Husband _____

Address _____   Phone _____

The foregoing offer is accepted this _____ day of _____, 19____

Seller _____   Seller's Wife or Husband _____

Address _____   Phone _____

B-6355 Copyright 1960 by The Iowa State Bar Association.

26. OFFER TO BUY
This Printing: August 18, 1964

ADDITIONAL PROVISIONS

The foregoing offer is subject to the following further conditions and provisions:

10. STATUS QUO MAINTAINED. Said real estate (and any personal property contracted for) as of date of this offer, and in its present condition will be preserved and delivered intact at the time possession is given. Except, however, in case of loss or destruction of part or all of said premises from causes covered by the insurance thereon, Buyers agree to accept such insurance recovery (proceeds to be applied as the interests of the parties appear) in lieu of that part of the damaged or destroyed improvements and Sellers shall not be required to repair or replace same. Buyers shall thereupon complete the contract and accept the property. (See paragraphs 4 and 20.)

11. ABSTRACT AND TITLE. Sellers shall promptly continue and pay for the abstract of title to and including date of acceptance of this offer, and deliver to Buyers for examination. The abstract shall become the property of the Buyers when the purchase price is listed in full, and shall show merchantable title in conformity with this agreement, the land title law of the State of Iowa and Iowa Title Standards of the Iowa State Bar Association. Sellers shall pay costs of additional abstracting and/or title work due to act or omission of Sellers, including transfers of death of Sellers or assigns.

12. DEED. Upon payment of purchase price, Sellers shall convey title by_____warranty deed, with terms and provisions as per form approved by the Iowa State Bar Association, free and clear of liens and incumbrances, reservations, exceptions or modifications except as in this instrument otherwise expressly provided. All warranties shall extend to time of acceptance of this offer, with special warranties as to acts of Seller up to time of delivery of deed.

13. FOR THE SELLERS: JOINT TENANCY IN PROCEEDS AND IN SECURITY RIGHTS IN REAL ESTATE. If, and only if, the Sellers, immediately preceding this offer, hold the title to the above described property in joint tenancy, and such joint tenancy is not later destroyed by operation of law or by acts of the Sellers (1) then the proceeds of this sale, and any continuing and/or recaptured rights of Sellers in said real estate shall be and continue in Sellers as joint tenants with rights of survivorship and not as tenants in common; and (2) Buyers, in the event of the death of either Seller agree to pay any balance of the proceeds of this sale to the surviving Seller and to accept deed from such surviving Seller consistent with paragraph 11, above; unless and except this paragraph 13 is stricken from this agreement.

13½. "SELLERS." Spouse, if not a titleholder immediately preceding this agreement, shall be presumed to have executed this instrument only for the purpose of relinquishing all rights of dower, homestead and distributive share and/or in compliance with section 561.13 I.C.A.; and the use of the word "Sellers" in the printed portion of this contract, without more, shall not rebut such presumption, nor in any way enlarge or extend the previous interest of such spouse in said property or in the sale proceeds thereof, nor bind such spouse except as aforesaid, to the terms and provisions of this contract.

14. TIME IS OF THE ESSENCE. Time is of the essence in this Agreement.

15. REMEDIES OF THE PARTIES — FORFEITURE — FORECLOSURE — REAL ESTATE COMMISSIONS:

(a) If Buyers fail to fulfill this agreement, the Sellers may forfeit the same as provided in the Code of Iowa, and all payments made hereunder shall be forfeited. To the extent in amount of any real estate commission owing by Sellers on account of this transaction all payments made hereunder shall be paid by the Seller to the person entitled, in full discharge of Sellers' obligation for such commission.

(b) If Sellers fail to fulfill this agreement, they shall nevertheless pay the regular real estate commission, if any be due, to the person entitled, but the Buyers shall have the right to have all their payments made hereunder returned to them.

(c) In addition to the foregoing remedies, Buyers and Sellers each shall be entitled to any and all other remedies, or action at law or in equity, including foreclosure, and the party at fault shall pay costs and attorney fees, and a receiver may be appointed.

16. EQUITY. If Buyers assume or take subject to a lien on this property, or are purchasing an interest of an equity holder, the Sellers, or their Broker, or Realtor, shall furnish Buyers with a statement, or statements, in writing from the holder of such lien or interest, showing the correct and agreed balance or balances.

17. If this instrument is to be followed by or to be replaced by an installment real estate contract, same shall be as per terms and provisions of the Official Form of the Iowa State Bar Association now in effect, but conformable to this instrument.

18. ALLOCATION OF VALUE OF ASSETS. Buyers and Sellers shall cooperate to make a reasonable allocation of values for the assets herein purchased; but failure to reach an agreement shall not in any manner delay or invalidate this contract or its performance.

19. APPROVAL OF COURT. If this property is an asset of any estate, trust or guardianship, this contract shall be subject to Court approval, unless declared unnecessary by the Buyers' attorney. If necessary, the appropriate fiduciary shall proceed promptly and diligently to bring the matter on for hearing for Court approval. (In that event the Court Officer's Deed shall be used.)

20. INSURANCE POLICIES. If Buyers purchase on installment contract, they shall, at their own expense, after possession, keep in effect fire, windstorm and tornado insurance, with extended coverage, for the benefit of the parties hereto, in an amount not less than the unpaid balance of the purchase price, or $_____ whichever may be less. The policies shall be delivered to the Sellers. (See also paragraphs 4 and 10.)

21. CONTRACT BINDING ON SUCCESSORS IN INTEREST. This contract shall apply to and bind the heirs, executors, administrators, assigns and successors in interest of the respective parties.

22. Words and phrases herein, including any acknowledgment hereof, shall be construed as in the singular or plural number, and as masculine, feminine or neuter gender, according to the context.

23. OTHER PROVISIONS.● (Personal Property?)

(If paragraph 23 is used, and/or if any changes are made in printed paragraphs 10 to 22 inclusive, other than the insertion of the amount of insurance in paragraph 20, sign below, as required in paragraph 7 above.)

_____ _____
Buyer   **Buyer's Wife or Husband**

_____ _____
Seller   **Seller's Wife or Husband**

Please type or print names under signatures as per Sec. 235.2 I.O.A. as amended.

STATE OF IOWA, _____ COUNTY, ss:
On this _____ day of _____ A. D. 19____, before me, the undersigned, a Notary Public in and for said County, in said State, personally appeared _____

_____
to me known to be the identical persons named in and who executed the within and foregoing instrument "Offer to Buy Real Estate and Acceptance" in its entirety and acknowledged that they executed the same as their voluntary act and deed.

_____, Notary Public in and for said County.

●Optional provisions: (a) Buyers understand that there is a mortgage of record with present balance of approximately $_____ payable to _____, which mortgage is to be timely paid by Sellers, (b) If Buyers, before paying _____% on the principal or total price of this sale, shall sell or assign their interest in this instrument, or in the real estate therein described, without the written consent of Sellers, which consent shall not be unreasonably withheld, the whole amount due herein, at the option of Sellers, shall immediately become due and payable. (Caveat: If such an accelerating clause is used, consider whether you have elected to proceed by foreclosure rather than by forfeiture.) (c) Buyers will _____ purchase Sellers existing insurance mentioned in numbered paragraph 4, above, and pay pro rata for the unexpired portion of said policies, as of and after date of possession.

FARMLAND

APPENDIX C
SAMPLE PAGE OF ABSTRACT TITLE

Cyclone Abstract and Title Co.

Excerpt from Abstract of Title

No. 1	The United States of America To William R. Cox.	ORIGINAL ENTRY Dated: August 25, 1854 Recorded Book OE-Page 164 Entry of E½ of NE¼ Section 23 Township 83 N Range 24W. Containing 80 Acres.

No. 2	The United States of America To William R. Cox	CERTIFIED COPY OF PATENT Dated: May 15, 1855 Filed: July 8, 1966 Recorded Book 4-Page 237 Conveys the E½ of the NE¼ of Section 23 in Township 83

North of Range 24 West, in the District of Lands, Subject to Sale at Fort Des Plaines, Iowa.

No. 3	Ames Sand and Gravel Co. By: Ben J. Cole, President By: E. J. Engeldinger, Secretary-Treasurer (No Seal) To O. C. Cramblit, Eva J. Nowlin, Mamie Bell and Della Bell	QUIT CLAIM DEED $1.00 Dated: January 31, 1930 Filed: February 8, 1930 Recorded Book 71-Page 196 Quit Claims the SE¼ of the NE¼ of Section 23 and SW¼ of NW¼ of Section 24, all in Township 83 North of Range 24 West of the 5th P.M.

Grantor hereby relinquishes all her right of downer.

Ben J. Cole appears as B.J. Cole in Acknowledgment.
Acknowledgement states "The Ames Sand and Gravel Co., A corporation without a seal."

FARMLAND

No. 8   Sern Sorenson and   MORTGAGE $9,600.00
        Viola M. Sorenson,
        Husband and Wife,   Dated: July 10, 1947
                                         Filed: July 15, 1947
                To
                                         Recorded Book 98-Page 471
        Connecticut General Life
        Insurance Company.   Conveys the E½ of the NE¼
                                         of Section 23 and W½ of NW¼
                                         of Section 24, all in Township
                                         83 North, Range 24 West of the
                                         5th P.M.
                Due $9,600.00 with interest according to the tenor
        and effect of the Promissory Note.
                This Mortgage and the note secured hereby are given
        in extension and renewal, and not in payment or discharge, of
        a note and mortgage dated March 4, 1946, executed by Sern
        Sorenson and Viola M. Sorenson, husband and wife, and evidencing
        and securing a loan in the principal amount of $7,700.00
        now reduced to $7,550.00 on which the interest has been paid
        only to March 1, 1946 said Mortgage was recorded on March 7,
        1946, in Book 98, Page 149.
                Indexed in Chattel Mortgage Index 12, Page SO-1.

No. 9   Connecticut General Life   RELEASE OF MORTGAGE
        Insurance Company

        By: P.H. Finley, Asst.   Dated: August 11, 1964
        Secretary   Filed: August 17, 1964
        (Corporate Seal)
                                         Filed Book 97-Page 269
                To
                                         Release of Mortgage recorded
        Sern Sorenson and   in Book 98 of Land Mortgages
        Viola M. Sorenson,   Page 471 and also indexed
        Husband and Wife.   in Chattel Mortgage Index 12,
                                         Page So-1.

Cyclone Abstract and Title Co.

APPENDIX D

SAMPLE DEED

IOWA STATE BAR ASSOCIATION
Official Form No. 1.1 (Trade-Mark Registered, State of Iowa, 1967)

FOR THE LEGAL EFFECT OF THE USE
OF THIS FORM, CONSULT YOUR LAWYER

WARRANTY DEED

Know All Men by These Presents: That _____

_____

_____, in consideration

of the sum of _____

in hand paid do hereby **Convey** unto _____

_____

the following described real estate, situated in _____ County, Iowa, to-wit:

And the grantors do **Hereby Covenant** with the said grantees, and successors in interest, that said grantors hold said real estate by title in fee simple; that they have good and lawful authority to sell and convey the same; that said premises are **Free and Clear of all Liens and Encumbrances Whatsoever** except as may be above stated; and said grantors Covenant to **Warrant and Defend** the said premises against the lawful claims of all persons whomsoever, except as may be above stated.

Each of the undersigned hereby relinquishes all rights of dower, homestead and distributive share in and to the described premises.

Words and phrases herein, including acknowledgment hereof shall be construed as in the singular or plural number, and as masculine or feminine gender, according to the context.

Signed this _____ day of _____, 19____.

_____

_____

Please type
or print
names
under
signatures
as per
Sec.
335.2
Code of
Iowa

STATE OF IOWA, } ss.
COUNTY OF _____ }

On this _____ day of _____, 19___ before me, the undersigned, a Notary Public in and for said County, in said State, personally appeared _____

_____

_____

to me known to be the identical persons named in and who executed the foregoing instrument, and acknowledged that they executed the same as their voluntary act and deed.

_____, Notary Public in and for said County

F-5450

1.1 **WARRANTY DEED**
This Printing: January 27, 1969

FARMLAND

STATE OF_____, _____COUNTY, ss:

On this_____day of_____, 19_____, before me, the undersigned, a Notary Public in and for said County, in said State, personally appeared _____

_____

_____ to me known to be the identical persons named in and who executed the foregoing instrument, and acknowledged that they executed the same as their voluntary act and deed.

_____, Notary Public in and for said County.

STATE OF_____, _____COUNTY, ss:

On this_____day of_____, 19_____, before me, the undersigned, a Notary Public in and for said County, in said State, personally appeared _____

_____

_____ to me known to be the identical persons named in and who executed the foregoing instrument, and acknowledged that they executed the same as their voluntary act and deed.

_____, Notary Public in and for said County.

Warranty Deed

TO

Entered upon transfer books and for taxation this_____day of_____19____ Auditor

By_____ Deputy

Filed for record, indexed and delivered to County Auditor this_____day of_____19____ at_____o'clock____M., and recorded in Book_____of_____on page_____County Records

Recorder's and Auditor's Fee $_____ PAID. Recorder

By_____ Deputy

WHEN RECORDED RETURN TO

FARMLAND 301

APPENDIX E

SAMPLE MORTGAGE

IOWA STATE BAR ASSOCIATION
Official Form No. 13.2

FOR THE LEGAL EFFECT OF THE USE
OF THIS FORM, CONSULT YOUR LAWYER

NOTE: Use this form only when a 12-month period of redemption is desired. Use Form 13.1 for the six-month period.

REAL ESTATE MORTGAGE—IOWA

This Indenture made this _____ day of _____, A. D. 19____
between _____
_____Mortgagors
of the County of _____, and State of Iowa, and _____
_____Mortgagee,
of the County of _____, and State of _____.
WITNESSETH: That the said Mortgagors in consideration of _____
_____DOLLARS
($_____) loaned by Mortgagee, received by Mortgagors, and evidenced by the promissory note hereinafter referred to, do, by these presents **SELL, CONVEY AND MORTGAGE**, unto the said Mortgagee, his successors and assigns forever, _____

the following described Real Estate situated in the County of _____, State of Iowa, to-wit:

together with all personal property that may integrally belong to, or be or hereafter become an integral part of said real estate, and whether attached or detached (that is, light fixtures, shades, rods, blinds, venetian blinds, awnings, storm windows, storm doors, screens, linoleum, water heater, water softener, automatic heating equipment and other attached fixtures), and hereby granting, conveying and mortgaging also all of the easements, servient estates appurtenant thereto, rents, issues, uses, profits and right to possession of said real estate, and all crops raised thereon from now until the debt secured thereby shall be paid in full.

Said Mortgagors hereby covenant with Mortgagee, or successor in interest, that said Mortgagors hold said real estate by title in fee simple; that they have good and lawful authority to sell, convey and mortgage the same; that said premises are **Free and Clear of all Liens and Encumbrances Whatsoever** except as may be above stated; and said Mortgagors Covenant to Warrant and Defend the said premises against the lawful claims of all persons whomsoever, except as may be above stated.

Each of the undersigned hereby relinquishes all rights of dower, homestead and distributive share in and to the above described premises, and waives any rights of exemption, as to any of said property.

CONDITIONED HOWEVER, That if said Mortgagors shall pay or cause to be paid to said Mortgagee, or his successor, or assigns, said sum of money which shall be legal tender in payment of all debts and dues, public and private, at time of payment, all at the time, place, and upon the terms provided by one[1] promissory note of Mortgagors to Mortgagee, of even date herewith, and shall perform the other provisions hereof, then these presents will be void, otherwise to remain in full force and effect.

1. **TAXES.** Mortgagors shall pay each installment of all taxes and special assessments of every kind, now or hereafter levied against said property, or any part thereof, before same become delinquent, without notice or demand; and shall procure and deliver to said Mortgagee, on or before the tenth day of October of each year, duplicate receipts of the proper officers for the payment of all such taxes and assessments then due.

2. **INSURANCE.** Mortgagors shall keep in force insurance on all buildings in companies to be approved by Mortgagee against loss by fire, tornado and other hazards, casualties and contingencies as Mortgagee may require, in an amount not less than $_____. Mortgagors shall deposit such policies with proper riders with the Mortgagee and shall pay all premiums therefor when due without notice or demand.

3. **REPAIRS TO PROPERTY.** Mortgagors shall keep the buildings and other improvements on said premises in as good repair and condition, as same may now be, or are hereafter placed; ordinary wear and tear only excepted, and shall not suffer or commit waste on or to said security.

4. **ATTORNEYS' FEES.** In case of any action, or in any proceedings in any court, to collect any sums payable or secured by this mortgage, or to protect the lien or title herein of the Mortgagee, or in any other case permitted by law in which attorney fees may be collected from Mortgagors, or charged upon the above described property, they agree to pay reasonable attorney fees.

5. **CONTINUATION OF ABSTRACT.** In event of any default herein by Mortgagors, Mortgagee may, at the expense of Mortgagors, procure an abstract of title, or continuation thereof, for said premises, and charge and add to the mortgage debt the cost of such abstract or continuation with interest upon such expense at the highest legal rate.

Copyright 1961 by The Iowa State Bar Association
B-5868

13.2 REAL ESTATE MORTGAGE
This Printing: July 6, 1964

FARMLAND

6. ADVANCES OPTIONAL WITH MORTGAGEE. It is expressly understood and agreed that if the insurance above provided for is not promptly effected, or if the taxes or special assessments assessed against said property shall become delinquent, Mortgagee (whether electing to declare the whole mortgage due and collectible or not), may (but need not) effect the insurance above provided for, and need not, but may and is hereby authorized to pay said taxes and special assessments (irregularities in the levy or assessment of said taxes being expressly waived), and all such payments with interest thereon at the highest legal rate from time of payment shall be a lien against said premises.

7. ACCELERATION OF MATURITY AND RECEIVERSHIP. And it is agreed that if default shall be made in the payment of said note, or any part of the interest thereon, or any other advance or obligation which may be secured hereby or any agreed protective disbursement, such as taxes, special assessments, insurance and repairs, or if Mortgagors shall suffer or commit waste on or to said security, or if there shall be a failure to comply with any and every condition of this mortgage, then, at the option of the Mortgagee, said note and the whole of the indebtedness secured by this mortgage, including all payments for taxes, assessments or insurance premiums, shall become due and shall become collectible at once by foreclosure or otherwise after such default or failure, and without notice of broken conditions; and at any time after the commencement of an action in foreclosure, or during the period of redemption, the court having jurisdiction of the case shall, at the request of the Mortgagee, appoint a receiver to take immediate possession of said property, and of the rents and profits accruing therefrom, and to rent or cultivate the same as he may deem best for the interest of all parties concerned, and shall be liable to account to said Mortgagors only for the net profits, from application of rents, issues and profits upon the costs and expenses of the receivership and foreclosure and the indebtedness, charges and expenses hereby secured and herein mentioned. And it is hereby agreed, that after any default in the payment of either principal or interest, such sums in default secured by this mortgage shall draw interest at the highest legal rate.

8. DEFINITION OF TERMS. Unless otherwise expressly stated, the word "Mortgagors", as used herein, includes heirs, executors, administrators, assigns and successors in interest of such "Mortgagors"; the word "Mortgagee", as used herein, unless otherwise expressly stated, includes the heirs, executors, administrators, assigns and successors in interest of such "Mortgagee". All words referring to "Mortgagors" or "Mortgagee" shall be construed to be of the appropriate gender and number, according to the context. This construction shall include the acknowledgment hereof.

9. The address of the Mortgagee is _____
(Street and Number)

_____ _____ _____ (See last sentence of Section 447.9 I.C.A.)
(City)   (Zone)   (State)

10. ADDITIONAL PROVISIONS. The following additional provisions are hereby incorporated herein: **(Insert due date or due dates if desired)**[2] The principal obligation herein, the one promissory note above referred to is payable $ _____ on _____ and $ _____ on _____

IN WITNESS WHEREOF, said Mortgagors have hereunto set their hands the day and year first above written.

_____

_____
Mortgagors

STATE OF IOWA, _____ COUNTY, ss:
On this _____ day of _____, A. D. 19____, before me, the undersigned, a Notary Public in and for said County, in said State, personally appeared _____

to me known to be the identical persons named in and who executed the foregoing instrument, and acknowledged that they executed the same as their voluntary act and deed.

_____, Notary Public in and for said County

[1]Only one original promissory note is contemplated with the use of this mortgage form.

[2]CONSIDER THE STATUTE OF LIMITATIONS. If this loan constitutes a long term transaction (over ten years), consider the advisability of making the maturity date or dates in the original note a matter of public record by insertion in this mortgage. See Iowa Land Title Examination Standards, Problems 10.4 and 10.5.

FARMLAND

APPENDIX F

SAMPLE INSTALLMENT CONTRACT

IOWA STATE BAR ASSOCIATION
Official Form No. 21½ (Trade-Mark Registered, State of Iowa, 1957)

FOR THE LEGAL EFFECT OF THE USE OF THIS FORM, CONSULT YOUR LAWYER

REAL ESTATE CONTRACT—INSTALLMENTS

IT IS AGREED this _____ day of _____, 19___, by and between _____

_____

of the County _____, State of Iowa, Sellers; and _____

_____

of the County of _____, State of Iowa, Buyers:

That the Sellers, as in this contract provided, agree to sell to the Buyers, and the Buyers in consideration of the premises, hereby agree with the Sellers to Purchase the following described real estate situated in the County of _____, State of Iowa, to-wit:

together with any easements and servient estates appurtenant thereto, but with such reservations and exceptions of title as may be below stated, and certain personal property if and as may be herein described or if and as an itemized list is attached hereto and marked "Exhibit A" all upon the terms and conditions following:

1. TOTAL PURCHASE PRICE. The buyer agrees to pay for said property the total of $ _____ due and payable at _____ County, Iowa, as follows:

(a) **DOWN PAYMENT** of $ _____ **RECEIPT OF WHICH IS HEREBY ACKNOWLEDGED;** and

(b) **BALANCE OF PURCHASE PRICE,** $ _____ as follows $ _____

2. POSSESSION. Buyers, concurrently with due performance on their part shall be entitled to possession of said premises on the _____ day of _____, 19___; and thereafter so long as they shall perform the obligations of this contract. If Buyers are taking subject to the rights of lessees and are entitled to rentals therefrom on and after date of possession, so indicate "yes" in the space following _____

3. TAXES. Sellers shall pay _____

_____

and any unpaid taxes thereon payable in prior years. Buyers shall pay any taxes not assumed by Sellers and all subsequent taxes before same become delinquent. Whoever may be responsible for the payment of said taxes, and the special assessments, if any, each year, shall furnish to the other parties evidence of payment of such items not later than July 15 of each year. **Any proration of taxes shall be based upon the taxes for the year currently payable unless the parties state otherwise.**

(Decide, for yourself, if that formula is fair if Buyers are purchasing a lot with newly built improvements.)

4. SPECIAL ASSESSMENTS. Sellers shall pay the special assessments against this property: (Strike out either (a) or (b) below.)

(a) Which, if not paid in the year 19_____ would become delinquent and all assessments payable prior thereto.

(b) Which are a lien thereon as of _____ (Date)

(c) Including all sewage disposal assessments for overage charge heretofore assessed by any municipality having jurisdiction as of date of possession.

Buyers, except as above stated shall pay all subsequent special assessments and charges, before they become delinquent.

5. MORTGAGE. Any mortgage or encumbrance of a similar nature against the said property being timely paid by Sellers so as not to prejudice the Buyers' equity herein. Should Sellers fail to pay Buyers may pay any such sums in default and shall receive credit on this contract for such sums so paid. MORTGAGE BY SELLERS. Sellers, their successors in interest or assigns may, and hereby reserve the right to at any time mortgage their right, title or interest in such premises or to renew or extend any existing mortgage for any amount not exceeding _____ %, of the then unpaid balance of the purchase price herein provided. The interest rate and amortization thereof shall be no more onerous than the installment requirements of this contract. Buyers hereby expressly consent to such a mortgage and agree to execute and deliver all necessary papers to aid Sellers in securing such a mortgage which shall be prior and paramount to any of Buyers' then rights in said property. DEED FOR BUYERS SUBJECT TO MORTGAGE. If Buyers have reduced the balance of this contract to the amount of any existing mortgage balance on said premises, they may at their option, assume and agree to pay said mortgage according to its terms, and subject to such mortgage shall receive a deed to said premises; or Sellers at their option, any time before Buyers have made such a mortgage commitment, may reduce or pay off such mortgage. ALLOCATED PAYMENTS. Buyers in the event of acquiring this property from an equity holder instead of a holder of the fee title, or in the event of a mortgage against said premises reserve the right, if reasonably necessary for their protection to divide or allocate the payments to the interested parties as their interests may appear. SELLERS AS TRUSTEES. Sellers agree that they will collect no money hereunder in excess of the amount of the unpaid balance under the terms of this contract less the total amount of the encumbrance on the interest of Sellers or their assigns in said real estate; and if Sellers shall hereafter collect or receive any moneys hereunder beyond such amount, they shall be considered and held as collecting and receiving said money as the agent and trustee of the Buyers for the use and benefit of the Buyers.

6. INSURANCE. Except as may be otherwise included in the last sentence of paragraph 1(b) above, Buyers as and from said date of possession, shall constantly keep in force, insurance, premiums therefor to be prepaid by Buyers (without notice or demand) against loss by fire, tornado and other hazards, casualties and contingencies as Seller may reasonably require on all buildings and improvements, now on or hereafter placed on said premises and any personal property which may be the subject of this contract in companies to be reasonably approved by Sellers in an amount not less than the full insurable value of such improvements and personal property or not less than the unpaid purchase price herein whichever amount is smaller with such insurance payable to Sellers and Buyers as their interests may appear. BUYERS SHALL PROMPTLY DEPOSIT SUCH POLICY WITH PROPER RIDERS WITH SELLERS for the further security for the payment of the sums herein mentioned. In the event of any such casualty loss, the insurance proceeds may be used under the supervision of the Sellers to replace or repair the loss if the proceeds be adequate; if not then some other reasonable application of such funds shall be made; but in any event such proceeds shall stand as security for the payment of the obligations herein.

7. CARE OF PROPERTY. Buyers shall take good care of this property; shall keep the buildings and other improvements now or hereafter placed on the said premises in good and reasonable repair and shall not injure, destroy or remove the same during the life of this contract. Buyers shall not make any material alteration in said premises without the written consent of the Sellers. Buyers shall not use or permit said premises to be used for any illegal purpose.

8. LIENS. No mechanics' lien shall be imposed upon or foreclosed against the real estate described herein.

9. ADVANCEMENT BY SELLERS. If Buyers fail to pay such taxes, special assessments and insurance and effect necessary repairs, as above agreed, Sellers may, but need not, pay such taxes, special assessments, insurance and make necessary repairs, and all sums so advanced shall be due and payable on demand or such sums so advanced may at the election of Sellers, be added to the principal amount due hereunder and so secured (For Buyers' rights to make advancements, see paragraph 5 above.)

Copyright 1958 by The Iowa State Bar Association
909010

21½ REAL ESTATE CONTRACT
This Printing: September, 1979

FARMLAND

10. JOINT TENANCY IN PROCEEDS AND SECURITY RIGHTS IN REAL ESTATE. If and only if, the Sellers immediately preceding this sale, hold the title to the above described property in joint tenancy, and such joint tenancy has not later been destroyed by operation of law or by acts of the Sellers, this sale shall not constitute such destruction and the proceeds of this contract, and any continuing and/or recaptured rights of Sellers in said real estate, shall be and continue in Sellers as joint tenants with rights of survivorship and not as tenants in common; and Buyers, in the event of the death of one of such joint tenants, agree to pay any balance of the proceeds of this contract to the surviving Seller (or Sellers) and to accept deed solely from him or them consistent with paragraph 13 below unless and except this paragraph is stricken from this agreement.

10½. "SELLERS." Spouse, if not titleholder immediately preceding this sale, shall be presumed to have executed this instrument only for the purpose of relinquishing all rights of dower, homestead and distributive share and/or in compliance with section 561.13 Code of Iowa; and the use of the word "Sellers" in the printed portion of this contract, without more, shall not rebut such presumption, nor in any way enlarge or extend the previous interest of such spouse in said property, or in the sale proceeds, nor bind such spouse except as aforesaid, to the terms and provisions of this contract.

11. TIME IS OF THE ESSENCE of this Agreement. Failure to promptly assert rights of Sellers herein shall not, however, be a waiver of such rights or a waiver of any existing or subsequent default.

12. EXCEPTIONS TO WARRANTIES OF TITLE. The warranties of title in any Deed made pursuant to this contract (See paragraph 13) shall be without reservation or qualification EXCEPT: (a) Zoning ordinances; (b) Such restrictive covenants as may be shown of record; (c) Easements of record, if any; (d) A limited by paragraphs 1, 2, 3 and 4 of this contract; (e) Sellers shall give Special Warranty as to the period after equitable title passes to Buyers; (f) Spouse if not a titleholder, need not join in any warranties of the deed unless otherwise stipulated: (g) _____

_____
(Mineral reservations of record?)

(h) _____
(Liens?)   (Easements not recorded?)   (Interests of other parties?)   (Lessees?)

13. DEED AND ABSTRACT, BILL OF SALE. If all said sums of money and interest are paid to Sellers during the life of this contract, and all other agreements for performance by Buyers have been complied with, Sellers will execute and deliver to Buyers a _____ Warranty Deed conveying said premises in fee simple pursuant to and in conformity with this contract; and Sellers shall at this time deliver to Buyers an abstract showing merchantable title, in conformity with this contract. Such abstract shall begin with the government patent (unless pursuant to the Iowa State Bar Association title standards there is a lesser requirement as to period of abstracting) to said premises and shall show title thereto in Sellers as of the date of this contract; or as of such earlier date if and as designated in the next sentence. This contract supersedes the previous written offer of Buyers to buy the above described property which was accepted by Sellers on the _____ day of _____, 19____. Sellers shall also pay the cost of any abstracting due to any act or change in the personal affairs of Sellers resulting in a change of title by operation of law or otherwise. If any personal property is a part of this agreement, then upon due performance by Buyers, Sellers shall execute and deliver a Bill of Sale consistent with the terms of this contract. Sellers shall pay all taxes on any such personal property payable in 19____ and all taxes thereon payable prior thereto.

14. APPROVAL OF ABSTRACT. Buyers have _____ examined the abstract of title to this property and such abstract is _____ accepted.

15.1. FORFEITURE. If Buyers (a) fail to make the payments aforesaid, or any part thereof, as same become due; or (b) fail to pay the taxes or special assessments or charges, or any part thereof, levied upon said property, or assessed against it, by any taxing body before any of such items become delinquent; or (c) fail to keep the property insured; or (d) fail to keep it in reasonable repair as herein required; or (e) fail to perform any of the agreements as herein made or required; then sellers, in addition to any and all other legal and equitable remedies which they may have, at their option, may proceed to forfeit and cancel this contract as provided by law (Chapter 656 Code of Iowa). Upon completion of such forfeiture Buyers shall have no right of reclamation or compensation for money paid, or improvements made; but such payments and/or improvements if any shall be retained and kept by Sellers as compensation for the use of said property, and/or as liquidated damages for breach of this contract; and upon completion of such forfeiture, if the Buyers, or any other person or persons shall be in possession of said real estate or any part thereof, such party or parties in possession shall at once peacefully remove therefrom, or failing to do so may be treated as tenants holding over, unlawfully after the expiration of a lease, and may accordingly be ousted and removed as such as provided by law.

15.2. FORECLOSURE. If Buyers fail, in any one or more of the specified ways to comply with this contract, as in (a), (b), (c), (d) or (e) of numbered paragraph 15.1 above provided. Sellers may upon thirty (30) days written notice of intention to accelerate the payment of the entire balance, during which thirty days such default or defaults are not removed, declare the entire balance hereunder immediately due and payable; and thereafter at the option of the Sellers this contract may then be foreclosed in equity and a receiver may be appointed to take charge of said premises and collect the rents and profits thereof to be applied as may be directed by the Court. It is agreed that the periods of redemption after sale on foreclosure may be reduced under the conditions set forth in Sections 628.26 and 628.27, Code of Iowa.

16. ATTORNEY'S FEES. In case of any action, or in any proceedings in any Court to collect any sums payable or secured herein, or to protect the lien or title herein of Sellers, or in any other case permitted by law in which attorney's fees may be collected from Buyers, or imposed upon them, or upon the above described property, Buyers agree to pay reasonable attorneys' fees.

17. INTEREST ON DELINQUENT AMOUNTS. Either party will pay interest at _____ percent per annum to the other on all amounts herein as and after they become delinquent, and/or on cash reasonably advanced by either party pursuant to the terms of this contract as protective disbursements.

18. ASSIGNMENT. In case of the assignment of this Contract by either of the parties, prompt notice shall be given to the other party, who shall at the time of such notice be furnished with a duplicate of such assignment by such assignors. Any such assignment shall not terminate the liability of the assignor to perform, unless a specific release in writing is given and signed by the other party to this Contract.

19. PERSONAL PROPERTY. If this contract includes personalty then Buyer grants Seller a security interest in such personalty. In the case of Buyer's default, Seller may, at his option, proceed in respect to such personalty in accordance with the Uniform Commercial Code of Iowa and treat such personalty in the same manner as the real estate, all as permitted by Section 554.9501(4) Code of Iowa.

20. CONSTRUCTION. Words and phrases herein, including acknowledgments hereof shall be construed as in the singular or plural number, and as masculine, feminine or neuter gender, according to the context. See paragraph 10½, above, for construction of the word "Sellers."

21. SPECIAL PROVISIONS.

Executed duplicate
  triplicate

_____ _____
_____ _____
           SELLERS   BUYERS

_____ _____
        Sellers' Address   Buyers' Address

STATE OF IOWA _____ COUNTY, ss:

On this _____ day of _____ A. D. 19____ before me, the undersigned, a Notary Public in and for said County and State, personally appeared

to me known to be the identical persons named in and who executed the within and foregoing instrument, and acknowledged that they executed the same as their voluntary act and deed.

_____ Notary Public in and for said County and State

Real Estate Contract
Installments

TO

Entered upon transfer books and for taxation this _____ day of _____ 19____
Auditor
By _____ Deputy

Filed for record, indexed and delivered to County Auditor this _____ day of _____ 19____ at _____ o'clock ____M. and recorded in Book _____ of _____ on page _____
County Records.
Recorder's and Auditor's Fee $ _____ PAID
Recorder
By _____ Deputy

WHEN RECORDED RETURN TO

FARMLAND 305

APPENDIX G
SETTLEMENT SHEETS

SELLER COPY

**SALE OF ROBERT R. & BETTY A. BROWN—
HENRY S. & ALICE M. JONES
160 ACRES, STORY COUNTY, IOWA**

Sale Price—5/28/80 Sale—3/2/81 Possession		$528,000.00
Payments Made by Henry S. & Alice M. Jones:		
May 28, 1980—		
Down Payment	$ 30,000.00	
December 15, 1980—		
Additional Payment	100,000.00	
March 2, 1981—		
Final Payment	398,000.00	
		$528,000.00
Disbursements Made:		
Cyclone Abstract Company—abstracting	$ 95.00	
XYZ Law Firm—prepare deed; title work	117.50	
Story County Treasurer—3/81 property tax payment	1,098.24	
ABC Realty Company Escrow Account—funds held for 9/81 property tax payment	1,300.00	
Story County Recorder—revenue stamps	580.25	
ABC Realty Company—real estate commission	26,400.00	
	$ 29,590.99	
Disbursements Made to Robert R. & Betty A. Brown:		
June 15, 1980	$ 29,700.00	
December 15, 1980	100,000.00	
March 2, 1981	368,809.01	
	$498,409.01	$528,000.00

It has been a pleasure to be of service to you.

_____
BROKER

BUYER COPY
SALE OF ROBERT R. & BETTY A. BROWN—
HENRY S. & ALICE M. JONES
160 ACRES, STORY COUNTY, IOWA

Sale Price—5/28/80 Sale—3/2/81 Possession $528,000.00

Payments Made:

  May 28, 1980—
    Down Payment
    (Betty A. Brown) $ 30,000.00
  December 15, 1980—
    Additional Payment
    (Robert R. Brown)   100,000.00
  March 2, 1981—
    Additional Payment
    (Federal Land Bank of
    Omaha)   250,000.00
  March 2, 1981—
    Final Payment
    (Robert R. Brown)   148,000.00
                                         $528,000.00

It has been a pleasure to be of service to you.

_____
BROKER

APPENDIX H
LEASE FORMS

Cash Farm Lease
(with Flexible Provisions)

North Central Regional
Publication No. 76[1]

This CASH FARM LEASE form can provide the landlord and tenant with a guide for devloping an agreement to fit their individual situation. This form is not intended to take the place of legal advice pertaining to contractual relationships between the two parties. Because of the possibility that a farm operating agreement may be legally considered a partnership under certain conditions, seeking proper legal advice is recommended when developing such an agreement.

This lease is entered into this _____ day of _____, 19____, between _____, landlord, of _____
(address)

_____, spouse, of _____
(address)

hereafter known as "the landlord," and

_____, tenant, of _____
(address)

_____, spouse, of _____
(address)

hereafter known as "the tenant."

I. PROPERTY DESCRIPTION

The landlord hereby leases to the tenant, to occupy and use for agricultural and related purposes, the following described property:

_____
_____
_____
_____

consisting of approximately _____ acres situated in _____ County (Counties), _____ (State) with all improvements thereon except as follows:

_____
_____
_____
_____

II. GENERAL TERMS OF LEASE

A. Time period covered. The provisions of this agreement shall be in effect for _____ year(s), commencing on the _____ day of _____ _____, 19____. This lease shall continue in effect from year to year thereafter unless written notice of termination is given by either party to the other at least _____ days prior to expiration of this lease or the end of any year of continuation.

B. Review of lease. A written request is required for a general review of the lease or for consideration of proposed changes by either party, at least _____ days prior to the final date for giving notice to terminate the lease as specified in IIA.

C. Amendments and alterations. Amendments and alterations to this lease shall be in writing and shall be signed by both the landlord and tenant.

D. No partnership intended. It is particularly understood and agreed that this lease shall not be deemed to be nor intended to give rise to a partnership relation.

E. Transfer of property. If the landlord should sell or otherwise transfer title to the farm, he will do so subject to the provisions of this lease.

F. Right of entry. The landlord reserves the right for himself, his agents, his employees, or his assigns to enter the farm at any reasonable time to: a) consult with the tenant; b) make repairs, improvements, and inspections; and c) (after notice of termination of the lease is given) do plowing, seeding, fertilizing, and any other customary seasonal work, none of which is to interfere with the tenant in carrying out regular farm operations.

G. No right to sublease. The landlord does not convey to the tenant the right to lease or sublease any part of the farm or to assign the lease to any person or persons whomsoever.

H. Binding on heirs. The provisions of this lease shall be binding upon the heirs, executors, administrators, and successors of both landlord and tenant in like manner as upon the original parties, except as provided by mutual written agreement.

I. Additional provisions.

_____
_____
_____
_____

III. LAND USE

A. General provisions. The land described in Section I will be used in approximately the following manner. If it is impracticable in any year to follow such a land use plan, appropriate adjustments will be made by mutual agreement between the parties.

1. Cropland
   a) Row crops _____ Acres
   b) Small grains _____ Acres
   c) Legumes _____ Acres
   d) Rotation pasture _____ Acres
2. Permanent pasture _____ Acres

---

1. For cash and flexible rental information see: "Fixed and Flexible Rental Arrangements for Your Farm" NCR publication number 75.

3. Other:

_____ _____ Acres
_____ _____ Acres
_____ _____ Acres
4. Total   _____ Acres

B. Restrictions. The maximum acres harvested as silage shall be _____ acres unless it is mutually decided otherwise.
The pasture stocking rate shall not exceed:
PASTURE IDENTIF.

_____ | _____ acres/animal unit
_____ | _____ acres/animal unit
_____ | _____ acres/animal unit
(1000 pound mature cow is equivalent to one animal unit.)

Other restrictions are:
_____
_____
_____

C. Government programs. The extent of participation in government programs will be discussed and decided on an annual basis. The course of action agreed upon shall be placed in writing and be signed by both parties. A copy of the course of action so agreed upon shall be made available to each party.

IV. AMOUNT AND PAYMENT OF RENT

(If a flexible cash rental arrangement is desired, use material on the last page of this form and omit section A below.)

A. Cash rental rates. The tenant agrees to pay as cash rent the amount as calculated below for each kind of land; or, one total may be entered for ENTIRE FARM UNIT.

Amount of Cash Rent

Kind of Land or Improvements	Acres	Rate/Acre	Amount
Row Crops		$	$
Small Grains		$	$
Legumes		$	$
Permanent Pasture		$	$
Timber		$	$
Waste		$	$
Farm Buildings	XXXXX	XXXXX	$
Dwelling	XXXXX	XXXXX	$
Other			$
ENTIRE FARM UNIT		$XXXX	$

B. Rental payment. The annual cash rent shall be paid as follows:

$_____ on or before _____ day of _____ (month),
$_____ on or before _____ day of _____ (month),
$_____ on or before _____ day of _____ (month),

If rent is not paid when due, the tenant agrees to pay interest on the amount of unpaid rent at the rate of _____ percent per annum from the due date until paid.

C. Rental adjustment—Additional agreements in regard to rental payment:
_____
_____

V. OPERATION AND MAINTENANCE OF FARM

In order to operate this farm efficiently and to maintain it in a high state of productivity, the parties agree as follows:

A. The tenant agrees:

1. General maintenance. To provide the unskilled labor necessary to maintain the farm and its improvements during his tenancy in as good condition as it was at the beginning. Normal wear and depreciation and damage from causes beyond the tenant's control are excepted.

2. Land use. Not to: a) plow permanent pasture or meadowland, b) cut live trees for sale or personal uses, or c) pasture new seedings of legumes and grasses in the year they are seeded without consent of the landlord.

3. Insurance. Not to house automobiles, motortrucks, or tractors in barns, or otherwise violate restrictions in the landlord's insurance policies without written consent from the landlord. Restrictions to be observed are as follows:
_____
_____
_____

4. Noxious weeds. To use diligence to prevent noxious weeds from going to seed on the farm. Treatment of the noxious weed infestation and cost thereof shall be handled as follows:
_____
_____
_____

5. Addition of improvements. Not to: a) erect or permit to be erected on the farm any nonremovable structure or building, b) incur any expense to the landlord for such purposes, or c) add electrical wiring, plumbing, or heating to any building without written consent of the landlord.

6. Conservation. Control soil erosion as completely as practicable; keep in good repair all terraces, open ditches, inlets and outlets of tile drains; preserve all established watercourses or ditches including grassed waterways; and refrain from any operation or practice that will injure such structures.

7. Damages. When he leaves the farm, to pay the landlord reasonable compensation for any damages to the farm for which he, the tenant, is responsible. Any decrease in value due to ordinary wear and depreciation or damages outside the control of the tenant are excepted.

8. Costs of operation. To pay all costs of operation except those specifically referred to in Sections V-A-4 and V-B.

9. Repairs. Not to buy materials for maintenance and repairs in an amount in excess of $_____ within a single year without written consent of the landlord.

APPENDIX H (continued)

B. The landlord agrees:

1. Loss replacement. To replace or repair as promptly as possible the dwelling or any other building regularly used by the tenant that may be destroyed or damaged by fire, flood, or other cause beyond the control of the tenant or to make rental adjustments in lieu of replacements.
2. Materials for repairs. To furnish all material needed for normal maintenance and repairs.
3. Skilled labor. To furnish any skilled labor for tasks which the tenant himself is unable to make satisfactorily. Additional agreements regarding materials and labor are:

_____
_____
_____

4. Reimbursement. To pay for materials purchased by the tenant for purposes of repair and maintenance in an amount not to exceed $_____ in any one year, except as otherwise agreed upon. Reimbursement shall be made within _____ days after the tenant submits the bill.
5. Removable improvements. Let the tenant make minor improvements of a temporary or removable nature, which do not mar the condition or appearance of the farm, at the tenant's expense. He further agrees to let the tenant remove such improvements even though they are legally fixtures at any time this lease is in effect or within _____ days thereafter, provided the tenant leaves in good condition that part of the farm from which such improvements are removed. The tenant shall have no right to compensation for improvements that are not removed except as mutually agreed.
6. Compensation for crop expenses. To reimburse the tenant at the termination of this lease for field work done and for other crop costs incurred for crops to be harvested during the following year. Unless otherwise agreed, current custom rates for operations involved will be used as a basis of settlement.

C. Both agree:

1. Not to obligate other party. Neither party hereto shall pledge the credit of the other party hereto for any purpose whatsoever without the consent of the other party. Neither party shall be responsible for debts or liabilities incurred, or for damages caused by the other party.
2. Capital improvements. That costs of establishing hay or pasture seedings, new conservation structures, improvements (except as provided in Section V-B-5), or of applying lime and other long-lived fertilizers shall be divided between landlord and tenant as set forth in the following table. The tenant will be reimbursed by the landlord either when the improvement is completed, or the tenant will be compensated for his share of the depreciated cost of his contribution when he leaves the farm based on the value of the tenant's contribution and depreciation rate shown in Table I. (Cross out the portion of the preceding sentence which does not apply).

Rates for labor, power, and machinery contributed by the tenant shall be agreed upon before construction is started.

Table I—Compensation for Improvements

Type of Improvement	Date to Be Completed	Estimated Total Cost (dollars)	Proportion to be Contributed by tenant Unskilled			Total Value of Tenant's Contrib. (dollars)*	Rate of Annual Depreciation
			Mat.	Labor	Mach.		
			%	%	%		%
		$				$	
		$				$	
		$				$	
		$				$	
		$				$	
		$				$	
		$				$	
		$				$	
		$				$	
		$				$	
		$				$	
		$				$	
		$				$	
		$				$	
		$				$	

* To be recorded when improvement is completed.

VI. ARBITRATION OF DIFFERENCES

Any differences between the parties as to their several rights or obligations under this lease that are not settled by mutual agreement after thorough discussion, shall be submitted for arbitration to a committee of three disinterested persons, one selected by each party hereto and the third by the two thus selected. The committee's decision shall be accepted by both parties.

AMOUNT OF RENT TO BE PAID WHEN CROPLAND IS RENTED ON A FLEXIBLE BASIS.

A. Cash Rent for Non-Flexible Items (complete at beginning of lease period).
   - a. Pasture .... $_____
   - b. Hayland .... $_____
   - c. Other Non-Flexible Cropland .... $_____
   - d. Timber-Wasteland .... $_____
   - e. Farmstead .... $_____
      TOTAL NON-FLEXIBLE RENT .... $_____

B. Flexible Cropland Rent (From Method I, II, or III below .... $_____

C. TOTAL RENT FOR YEAR .... $_____

D. Flexible Cropland Rent (use Method I, II, or III).

   1. BASIC INFORMATION TO BE USED IN METHODS I AND II.

Crop(s)	Base Cash Rent (per acre)	Base Yield (bu. or ton per acre)	Base Price (per bu. or per ton)	Minimum Cash Rent (per acre)	Maximum Cash Rent (per acre)
_____	$_____	_____	$_____	$_____	$_____
_____	$_____	_____	$_____	$_____	$_____
_____	$_____	_____	$_____	$_____	$_____

   2. THE CURRENT PRICE FOR THE CURRENT YEAR SHALL BE AVERAGE PRICE AT CLOSE OF DAY BASED ON THE FOLLOWING TIME PERIOD(S) AND LOCATION(S):

   Crop(s)   Price Source
   _____ ____ Day ____ Month through ____ Day ____ Month at _____
   _____ ____ Day ____ Month through ____ Day ____ Month at _____
   _____ ____ Day ____ Month through ____ Day ____ Month at _____

FOR EACH YEAR OF THIS LEASE, THE PER ACRE BASE CASH RENT FOR EACH CROP SPECIFIED SHALL BE ADJUSTED AT THE CLOSE OF THE CROPPING SEASON BY ONE OF THE FOLLOWING METHODS:

METHOD I—FLEXING FOR PRICE ONLY.

Crop(s) : Base Rent X $\frac{\text{Current Pr.}}{\text{Base Pr.}}$ = Rent/Ac.[1] X Acres Grown = Adj. Rent for Yr.

_____ : $_____ X ($_____/$_____) = $_____ X _____ = $_____
_____ : $_____ X ($_____/$_____) = $_____ X _____ = $_____
_____ : $_____ X ($_____/$_____) = $_____ X _____ = $_____

                                                    Total all crops = $_____

METHOD II—FLEXING FOR PRICE AND YIELD.

Crop(s) : Base Rent X $\frac{\text{Current Pr.}}{\text{Base Pr.}}$ X $\frac{\text{Cur. Yld.[2]}}{\text{Base Yld.}}$ = Rent/Ac.[1] X Acres Grown = Adj. Rent for Yr.

_____ : $_____ X ($_____/$_____) X _____ = $_____ X _____ = $_____
_____ : $_____ X ($_____/$_____) X _____ = $_____ X _____ = $_____
_____ : $_____ X ($_____/$_____) X _____ = $_____ X _____ = $_____

                                                    Total all crops = $_____

1. If calculated figure is less than Minimum Cash Rent in D-1, use the Minimum. If calculated figure is more than Maximum Cash Rent in D-1, use the Mamixum.
2. The current yield shall be the "farm" yield for the current lease year.

METHOD III—WORK OUT AND RECORD PROCEDURE TO BE USED.

_____
_____
_____
_____

FARMLAND

APPENDIX H (continued)

Executed in duplicate on the date first above written:

_____ _____
(tenant) (landlord)

_____ _____
(tenant spouse) (landlord spouse)

COUNTY OF _____
                                    } SS:
STATE OF _____

On this _____ day of _____ A.D., 19____, before me, the undersigned, a Notary Public in said State, personally appeared _____, and _____, to me known to be the identical persons named in and who executed the foregoing instrument, and acknowledged that they executed the same as their voluntary act and deed.

_____
Notary Public

Issued in furtherance of Cooperative Extension work, Acts of Congress of May 8 and June 30, 1914, in cooperation with the U.S. Department of Agriculture and Cooperative Extension Services of Illinois, Indiana, Kansas, Michigan, Minnesota, Missouri, Nebraska, North Dakota, Ohio, South Dakota and Wisconsin. John O. Dunbar, Director, Cooperative Extension Service, Kansas State University, Manhattan.

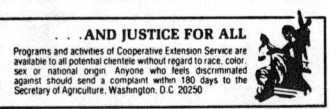

...AND JUSTICE FOR ALL
Programs and activities of Cooperative Extension Service are available to all potential clientele without regard to race, color, sex or national origin. Anyone who feels discriminated against should send a complaint within 180 days to the Secretary of Agriculture, Washington, D.C. 20250

Crop-Share or Crop-Share-Cash Farm Lease

North Central Regional Publication No. 77

This form can provide the landlord and tenant with a guide for developing an agreement to fit their individual situation. This form is not intended to take the place of legal advice pertaining to contractual relationships between the two parties. Because of the possibility that a farm operating agreement may be legally considered a partnership under certain conditions, seeking proper legal advice is recommended when developing such an agreement.

This lease is entered into this _____ day of _____, 19____, between _____, landlord, of _____

(address)

_____, spouse, of _____

(address)

hereafter known as "the landlord," and _____, tenant, of _____

(address)

_____, spouse, of _____

(address)

hereafter known as "the tenant."

I. PROPERTY DESCRIPTION

The landlord hereby leases to the tenant, to occupy and use for agricultural and related purposes, the following described property:

_____
_____

consisting of approximately _____ acres situated in _____ County (Counties), _____ (State) with all improvements thereon except as follows:

_____
_____

II. GENERAL TERMS OF LEASE

A. Time period covered. The provisions of this agreement shall be in effect for _____ year(s), commencing on the _____ day of _____, 19____. This lease shall continue in effect from year to year thereafter unless written notice of termination is given by either party to the other at least _____ days prior to expiration of this lease or the end of any year of continuation.

B. Review of lease. A written request is required for a general review of the lease or for consideration of proposed changes by either party, at least _____ days prior to the final date for giving notice to terminate the lease as specified in IIA.

C. Amendments and alterations. Amendments and alterations to this lease shall be in writing and shall be signed by both the landlord and tenant.

D. No partnership intended. It is particularly understood and agreed that this lease shall not be deemed to be nor intended to give rise to a partnership relation.

E. Transfer of property. If the landlord should sell or otherwise transfer title to the farm, he will do so subject to the provisions of this lease.

F. Right of entry. The landlord reserves the right for himself, his agents, his employees, or his assigns to enter the farm at any reasonable time to: a) consult with the tenant; b) make repairs, improvements, and inspections; and c) (after notice of termination of the lease is given) do plowing, seeding, fertilizing, and any other customary seasonal work, none of which is to interfere with the tenant in carrying out regular farm operations.

G. No right to sublease. The landlord does not convey to the tenant the right to lease or sublet any part of the farm or to assign the lease to any person or persons whomsoever.

H. Binding on heirs. The provisions of this lease shall be binding upon the heirs, executors, administrators, and successors of both landlord and tenant in like manner as upon the original parties, except as provided by mutual written agreement.

I. Landlord's lien for rent and performance. The landlord's lien provided by law on crops grown or growing shall be the security for the rent herein specified and for the faithful performance of the terms of the lease. If the tenant fails to pay the rent due or fails to keep the agreements of this lease, all costs and attorney fees of the landlord in enforcing collection or performance shall be added to and become a part of the obligations payable by the tenant hereunder.

J. Additional provisions.

_____
_____

III. LAND USE

A. General provisions. The land described in Section I will be used in approximately the following manner. If it is impractical in any year to follow such a land-use plan, appropriate adjustments will be made by mutual written agreement between the parties.

APPENDIX H (continued)

1. Cropland
   a) Row crops _____ acres
   b) Small grains _____ acres
   c) Legumes _____ acres
   d) Rotation pasture _____ acres
2. Permanent pasture: _____ acres
3. Other: _____ _____ acres
   _____ _____ acres
4. Total _____ acres

B. **Restrictions.** The maximum acres harvested as silage shall be _____ acres unless it is mutually decided otherwise.
The pasture stocking rate shall not exceed:

PASTURE IDENTIF.
_____ | _____ acres/animal unit
_____ | _____ acres/animal unit
_____ | _____ acres/animal unit

(1000 pound mature cow is equivalent to one animal unit.)

Other restrictions are:
_____
_____
_____
_____
_____

C. **Government programs.** The extent of participation in government programs will be discussed and decided on an annual basis. The course of action agreed upon shall be placed in writing and be signed by both parties. A copy of the course of action so agreed upon shall be made available to each party.

IV. CROP-SHARE-CASH RENT AND RELATED PROVISIONS

A. **General agreement.** The tenant agrees to pay as rent for the use of the land the share of crops shown in Table 1 of this section. The tenant also agrees to furnish all labor, machinery, and cash operating expenses except for landlord's share (percent and/or dollar charge per unit) indicated in Table 1.

Table 1—Landlord's Share (% and/or $) of Crops and Crop Expenses

	Corn example	Corn	Grain sorghum	Small grain	Soybeans	Hay, _____		
SHARE OF CROPS	50%							
SHARE OF CROP EXPENSES:								
Fertilizer:								
Materials	50%							
Application	50%							
Herbicide:								
Materials	50%							
Application								
Insecticide:								
Materials	50%							
Application								
Seed	50%							
Lime, rock phosphate*	100%							
Harvesting (per ac.)	$7.50							
Drying	50%							
Baling								
Delivery to:								
Storage/bu.								
Market/bu.	$ .07							

* Lime, rock phosphate, and other fertilizers having more than one year life paid by the tenant should be recorded in the compensation table in Section V-C-2.

B. Other crop-share-cash agreements.
1. **Operating expenses.** Additional agreements relative to the sharing of expenses are as follows:
_____
_____
_____

2. **Storage, landlord's crop.** At the landlord's request, the tenant agrees to store as much of the landlord's share of the crops as possible, using storage space reserved by the landlord and not to exceed _____ percent of the storage space not specifically reserved.

3. **Delivery of grain.** The tenant agrees to deliver the landlord's share of crops at a place and at a time the landlord shall designate, not over _____ miles distant at the charge shown in Table 1 of this section. Additional agreements are:
_____
_____
_____

4. **Cash rent on non-shared items.** The tenant agrees to pay cash rent annually for the use of the following non-shared items.

Table 2—Amount of Annual Cash Rent
(Complete at beginning of lease)

	Total
Pasture	$_____
Hayland: _____	$_____
_____	$_____
Farmstead: Dwelling	$_____
Service bldgs.	$_____
Timber and waste	$_____
Total cash rent	$_____

Payment of cash rent: The tenant agrees to pay cash rent as follows:

$_____ on or before _____ day of _____ (month)
$_____ on or before _____ day of _____ (month)
$_____ on or before _____ day of _____ (month)
$_____ on or before _____ day of _____ (month)

If rent is not paid when due, the tenant agrees to pay interest on the amount of unpaid rent at the rate of _____ percent per annum from due date until paid.

5. Pasturing. The tenant will prevent damage to cropland and growing crops by livestock.

6. Home use. The tenant and landlord may take for home use the following kinds and quantities of jointly owned crops:

_____
_____
_____
_____

7. Buying and selling. The landlord and tenant will buy and sell jointly owned property according to the following agreement:

_____
_____
_____

8. Division of property. At the termination of this lease, all jointly owned property will be divided or disposed of as follows:

_____
_____
_____

V. OPERATION AND MAINTENANCE OF FARM

In order to operate this farm efficiently and to maintain it in a high state of productivity, the parties agree as follows:

A. The tenant agrees:

1. General maintenance. To provide the unskilled labor necessary to maintain the farm and its improvements during his tenancy in as good condition as it was at the beginning. Normal wear and depreciation and damage from causes beyond the tenant's control are excepted.

2. Land use. Not to: a) plow pasture or meadowland, b) cut live trees for sale or personal use, or c) pasture new seedings of legumes and grasses in the year they are seeded without consent of the landlord.

3. Insurance. Not to house automobiles, motor trucks, or tractors in barns, or otherwise violate restrictions in the landlord's insurance policies without written consent from the landlord. Restrictions to be observed are as follows:

_____
_____
_____

4. Noxious weeds. To use diligence to prevent noxious weeds from going to seed on the farm. Treatment of the noxious weed infestation and cost thereof shall be handled as follows:

_____
_____
_____

5. Addition of improvements. Not to: a) erect or permit to be erected on the farm any nonremovable structure or building, b) incur any expense to the landlord for such purposes, or c) add electrical wiring, plumbing, or heating to any building without written consent of the landlord.

6. Conservation. Control soil erosion as completely as practicable; keep in good repair all terraces, open ditches, inlets and outlets of tile drains; preserve all established watercourses or ditches including grassed waterways; and refrain from any operation or practice that will injure such structures.

7. Damages. When he leaves the farm, to pay the landlord reasonable compensation for any damages to the farm for which he, the tenant, is responsible. Any decrease in value due to ordinary wear and depreciation or damages outside the control of the tenant are excepted.

8. Costs of operation. To pay all costs of operation except those specifically referred to in Sections IV, V-A-4, and V-B.

9. Repairs. Not to buy materials for maintenance and repairs in an amount in excess of $_____ within a single year without written consent of the landlord.

B. The landlord agrees:

1. Loss replacement. To replace or repair as promptly as possible the dwelling or any other building regularly used by the tenant that may be destroyed or damaged by fire, flood, or other cause beyond the control of the tenant or to make rental adjustments in lieu of replacements.

2. Materials for repairs. To furnish all material needed for normal maintenance and repairs.

3. Skilled labor. To furnish any skilled labor tasks which the tenant himself is unable to perform satisfactorily. Additional agreements regarding materials and labor are:

_____
_____
_____

4. Reimbursement. To pay for materials purchased by the tenant for purposes of repair and maintenance in an amount not to exceed $_____ in any one year, except as otherwise agreed upon. Reimbursement shall be made within _____ days after the tenant submits the bill.

APPENDIX H (continued)

5. **Removable improvements.** Let the tenant make minor improvements of a temporary or removable nature, which do not mar the condition or appearance of the farm, at the tenant's expense. He further agrees to let the tenant remove such improvements even though they are legally fixtures at any time this lease is in effect or within _____ days thereafter, provided the tenant leaves in good condition that part of the farm from which such improvements are removed. The tenant shall have no right to compensation for improvements that are not removed except as mutually agreed.

6. **Compensation for crop expenses.** To reimburse the tenant at the termination of this lease for field work done and for other crop costs incurred for crops to be harvested during the following year. Unless otherwise agreed, current custom rates for the operations involved will be used as a basis of settlement.

C. **Both agree:**

1. **Not to obligate other party.** Neither party hereto shall pledge the credit of the other party hereto for any purpose whatsoever without the consent of the other party. Neither party shall be responsible for debts or liabilities incurred, or for damages caused by the other party.

2. **Capital improvements.** Costs of establishing hay or pasture seedings, new conservation structures, improvements (except as provided in Section V-B-5), or of applying lime and other longlived fertilizers shall be divided between landlord and tenant as set forth in the following table. The tenant will be reimbursed by the landlord either when the improvement is completed, or the tenant will be compensated for his share of the depreciated cost of his contribution when he leaves the farm based on the value of the tenant's contribution and depreciation rate shown in the following table. (Cross out the portion of the preceding sentence which does not apply.)

   Rates for labor, power, and machinery contributed by the tenant shall be agreed upon before construction is started.

3. **Mineral rights.** Nothing in this lease shall confer upon the tenant any right to minerals underlying said land, but same are hereby reserved by the landlord together with the full right to enter upon the premises and to bore, search, and excavate for same, to work and remove same, and to deposit excavated rubbish, and with full liberty to pass over said premises with vehicles and lay down and work any railroad track or tracks, tanks, pipelines, power lines, and structures as may be necessary or convenient for the above purpose. The landlord agrees to reimburse the tenant for any actual damage he may suffer for crops destroyed by these activities and to release the tenant from obligation to continue farming this property when development of mineral resources interferes materially with the tenant's opportunity to make a satisfactory return.

VI. ARBITRATION OF DIFFERENCES

Any differences between the parties as to their several rights or obligations under this lease that are not settled by mutual agreement after thorough discussion, shall be submitted for arbitration to a committee of three disinterested persons, one selected by each party hereto and the third by the two thus selected. The committee's decision shall be accepted by both parties.

Compensation for Improvements Table

Type of improvement	Date to be completed	Estimated total cost (dollars)	Proportion to be contributed by tenant			Total value of tenant's contrib. (dollars)*	Rate of annual depreciation
			Material	Unskilled labor	Mach.		
			%	%	%		%
		$				$	
		$				$	
		$				$	
		$				$	

* To be recorded when improvement is completed.

Executed in duplicate on the date first above written:

_____ _____
(tenant)   (landlord)

_____ _____
(tenant spouse)   (landlord spouse)

STATE OF _____ }
COUNTY OF _____ } SS:

On this _____ day of _____ A.D., 19____, before me, the undersigned, a Notary Public in said State, personally appeared _____

_____ and _____, to me known to be the identical persons named in and who executed the foregoing instrument, and acknowledged that they executed the same as their voluntary act and deed.

_____
Notary Public

...AND JUSTICE FOR ALL
Programs and activities of Cooperative Extension Service are available to all potential clientele without regard to race, color, sex or national origin. Anyone who feels discriminated against should send a complaint within 180 days to the Secretary of Agriculture, Washington, D.C. 20250.

Issued in furtherance of Cooperative Extension work, Acts of Congress of May 8 and June 30, 1914, in cooperation with the U.S. Department of Agriculture and Cooperative Extension Services of Illinois, Indiana, Kansas, Michigan, Minnesota, Missouri, Nebraska, North Dakota, Ohio, South Dakota and Wisconsin. John O. Dunbar, Director, Cooperative Extension Service, Kansas State University, Manhattan.

Pasture Lease
North Central Regional

Publication No. 109

This PASTURE LEASE form can provide the landlord and tenant with a guide for developing an agreement to fit their individual situation. This form is not intended to take the place of legal advice pertaining to contractual relationships between the two parties. Because of the possibility that an operating agreement may be legally considered a partnership under certain conditions, seeking proper legal advice is recommended when developing such an agreement.

This lease is entered into this _____ day of _____, 19_____, between

_____, landlord, of _____
(pasture owner)

(address)

_____, spouse, of _____

(address)

hereafter known as "the landlord," and

_____, tenant, of _____
(livestock owner)

(address)

_____, spouse, of _____

(address)

hereafter known as "the tenant."

I. PROPERTY DESCRIPTION

The landlord hereby leases to the tenant, to occupy and use for pasture purposes, the following described property:

_____
_____
_____
_____

consisting of approximately _____ acres situated in _____ County (Counties), _____ (State) and on any other land which the landlord may designate by mutual written agreement.

II. GENERAL TERMS OF LEASE

A. **Term.**—[If a continuing lease is desired, use paragraph (1) and strike out (2).]

(1) Continuing Lease—The term of the lease shall be _____ year(s), commencing on the _____ day of _____, 19_____, and shall continue in effect from year to year thereafter (as an annual lease) unless written notice of termination is given by either party to the other at least _____ days prior to expiration of this lease or the end of any year of continuation. If a definite term is desired, use paragraph (2) and strike out paragraph (1). No notice of termination is necessary if paragraph (2) is used.

(2) Annual Lease—The term of this lease shall be _____ year(s), commencing on the _____ day of _____, 19_____, and ending on the _____ day of _____, 19_____.

B. **Review of lease.**—A request for general review of the lease may be made, by either party at least _____ days prior to the final date for giving notice to terminate the lease.

C. **Amendments.**—Amendments and alterations to this lease shall be in writing and shall be signed by both the landlord and tenant.

D. **No partnerships created.**—This lease shall not be deemed to give rise to a partnership relation, and neither party shall have authority to obligate the other without written consent, except as specifically provided in this lease.

E. **Binding on Heirs.**—The terms of this lease shall be binding upon the heirs, executors, administrators, and successors of both landlord and tenant in like manner as upon the original parties, except as provided by mutual written agreement otherwise.

F. **Transfer of property.**—If the landlord should sell or otherwise transfer title to the farm, he will do so subject to the provisions of this lease.

G. **Right of entry.**—The landlord reserves the right of himself, his agents, his employees, or his assigns to enter the farm at any reasonable time for purposes (a) of consultation with the tenant; (b) of making repairs, improvements, and inspections; and (c) after notice of termination of the lease is given, of performing customary seasonal work, none of which is to interfere with the tenant in carrying out regular operations.

H. Additional agreements regarding term of lease:

_____
_____
_____
_____
_____

FARMLAND

APPENDIX H (continued)

I. Animal units (maximum allowable)—Not more than _____ animal units shall be kept in the pasture at any one time without the express written consent of the landlord. Deliberate violation of this provision shall constitute grounds for termination of this lease. (Each 1,000 pounds of average weight shall be one animal unit. If the pasture owner and the owner of the livestock prefer, they can use the following basis for calculating animal units: 1 bull, 1.25 animal units; one 1,000-pound cow, 1 animal unit; 1 yearling steer or heifer, .75 animal unit; calf, 6 months to 1 year, .5 animal unit; calf, 3 to 6 months, .3 animal unit; sheep, 5 per animal unit; horse, 1.25 animal units.)—

Stocking Rate	Number Head	Number Animal Units
Bulls		
Cows		
Yearling steers		
Yearling heifers		
Calves, 6 mos.-1 year		
Calves, 3-6 mos.		
Other		

III. OPERATION AND MAINTENANCE

A. The livestock owner agrees:

(1) Not to pasture livestock to be breachy. Should any animal be found outside the pasture on at least three occasions, the pasture owner may request its removal.

(2) Not to assign his right and duties under this lease without the written consent of the pasture owner.

(3) Not to put any cattle in pasture without getting specific approval from the pasture owner in advance regarding number, health, sex, breed, and age.

(4) Agrees to furnish health certificate as follows:
_____
_____
_____
_____
_____

B. Both agree:

(1) Not to obligate other party. Neither party hereto shall pledge the credit of the other party hereto for any purpose whatsoever without the consent of the other party. Neither party shall be responsible for debts or liabilities incurred, or for damages caused by the other party.

(2) Responsibilities.—Additional responsibilities for each party shall be divided as follows:

	Landlord	Tenant
Inspect fences not less than once per _____.	___	___
Furnish labor for repair of fences.	___	___
Furnish materials for repair of fences.	___	___
Supervise supply of water to livestock.	___	___
Furnish labor for repair of water system.	___	___
Materials for repair of water system.	___	___
Furnish salt and mineral.	___	___
Count livestock not less than once per _____.	___	___
Return stray animals to pasture.	___	___
Call veterinarian in case of emergency.	___	___
Pay veterinary expenses.	___	___
Provide loading and unloading facilities.	___	___
Furnish supplementary feed, if needed.	___	___
Notify other party of shortage in count.	___	___
Provide facilities for fly control.	___	___
Keep fly control facilities in working order.	___	___
Liability Insurance.	___	___

(3) Additional agreements:
_____
_____
_____
_____
_____

IV. RENTAL CALCULATIONS AND PAYMENT SCHEDULE
(Use Method I, II or III and Strike Out the Two Methods Not Used)

METHOD I—The tenant owner agrees to pay $_____ per acre for use of the property described in paragraph I. Total rent of $_____ shall be paid as follows.

$_____ on or before _____ day of _____ (month),
$_____ on or before _____ day of _____ (month),
$_____ on or before _____ day of _____ (month),
$_____ on or before _____ day of _____ (month),

If rent is not paid when due, the tenant agrees to pay interest on the amount of unpaid rent at the rate of _____ percent per annum from the due date until paid.

Rental adjustment—Additional agreements in regard to rental payment:
_____
_____
_____
_____

METHOD II—The livestock owner agrees to pay the following rates: (The period may be by the month, pasture season or year.)

	Number	X	Rental Rate/Period	=	Total Rent/Period
Bulls	_____	X	$_____	=	$_____
Cows	_____	X	$_____	=	$_____
Yearling steers	_____	X	$_____	=	$_____
Yearling heifers	_____	X	$_____	=	$_____
Calves, 6 mos.-1 year	_____	X	$_____	=	$_____
Calves, 3-6 mos.	_____	X	$_____	=	$_____
Other	_____	X	$_____	=	$_____
Total Rent					$_____

The minimum rent shall be $_____. Such rental shall be required regardless of whether or not livestock are actually being pastured. The Total Rent of $_____ shall be paid as follows.

$_____ on or before _____ day of _____ (month),
$_____ on or before _____ day of _____ (month),
$_____ on or before _____ day of _____ (month),
$_____ on or before _____ day of _____ (month),

If rent is not paid when due, the tenant agrees to pay interest on the amount of unpaid rent at the rate of _____ percent per annum from the due date until paid.

Rental adjustment—Additional agreements in regard to rental payment:

_____
_____
_____
_____
_____
_____
_____
_____

METHOD III—Other Rental Arrangements (Share of gain-etc.)

_____
_____
_____
_____
_____
_____

VI. ARBITRATION OF DIFFERENCES

Any differences between the parties as to their several rights or obligations under this lease that are not settled by mutual agreement after thorough discussion, shall be submitted for arbitration to a committee of three disinterested persons, one selected by each party hereto and the third by the two thus selected. The committee's decision shall be accepted by both parties.

APPENDIX H (continued)

Executed in duplicate on the date first above written:

_____   _____
        tenant (Livestock owner)   landlord (Pasture owner)

_____   _____
           (tenant spouse)   (landlord spouse)

COUNTY OF _____ ⎫
                                   ⎬ SS:
STATE OF _____   ⎭

On this _____ day of _____ A.D., 19____, before me, the undersigned, a Notary Public in said State, personally appeared _____, _____, _____, and _____, to me known to be the identical persons named in and who executed the foregoing instrument, and acknowledged that they executed the same as their voluntary act and deed.

                                                   _____
                                                              Notary Public

Issued in furtherance of Cooperative Extension work, Acts of Congress of May 8 and June 30, 1914, in cooperation with the U.S. Department of Agriculture and Cooperative Extension Services of Illinois, Indiana, Kansas, Michigan, Minnesota, Missouri, Nebraska, North Dakota, Ohio, South Dakota and Wisconsin. John O. Dunbar, Director, Cooperative Extension Service, Kansas State University, Manhattan.

...AND JUSTICE FOR ALL
Programs and activities of Cooperative Extension Service are available to all potential clientele without regard to race, color, sex or national origin. Anyone who feels discriminated against should send a complaint within 180 days to the Secretary of Agriculture, Washington, D.C. 20250.

Irrigation Crop-Share or Crop-Share-Cash Farm Lease

North Central Regional
Publication No. 106

This form can provide the landlord and tenant with a guide for developing an agreement to fit their individual situation. This form is not intended to take the place of legal advice pertaining to contractual relationships between the two parties. Because of the possibility that a farm operating agreement may be legally considered a partnership under certain conditions, seeking proper legal advice is recommended when developing such an agreement.

This lease is entered into this _____ day of _____, 19____, between _____, landlord, of _____
(address)

_____, spouse, of _____
(address)

hereafter known as "the landlord," and

_____, tenant, of _____
(address)

_____, spouse, of _____
(address)

hereafter known as "the tenant."

I. PROPERTY DESCRIPTION

The landlord hereby leases to the tenant, to occupy and use for agricultural and related purposes, the following described property:
_____
_____
_____
consisting of approximately _____ acres situated in _____ County (Counties), _____ (State) with all improvements thereon except as follows:
_____
_____

II. GENERAL TERMS OF LEASE

A. Time period covered. The provisions of this agreement shall be in effect for _____ year(s), commencing on the _____ day of _____, 19____. This lease shall continue in effect from year to year thereafter unless written notice of termination is given by either party to the other at least _____ days prior to expiration of this lease or the end of any year of continuation.

B. Review of lease. A written request is required for a general review of the lease or for consideration of proposed changes by either party, at least _____ days prior to the final date for giving notice to terminate the lease as specified in IIA.

C. Amendments and alterations. Amendments and alterations to this lease shall be in writing and shall be signed by both the landlord and tenant.

D. No partnership intended. It is particularly understood and agreed that this lease shall not be deemed to be nor intended to give rise to a partnership relation.

E. Transfer of property. If the landlord should sell or otherwise transfer title to the farm, he will do so subject to the provisions of this lease.

F. Right of entry. The landlord reserves the right for himself, his agents, his employees, or his assigns to enter the farm at any reasonable time to: a) consult with the tenant; b) make repairs, improvements, and inspections; and c) (after notice of termination of the lease is given) do plowing, seeding, fertilizing, and any other customary seasonal work, none of which is to interfere with the tenant in carrying out regular farm operations.

G. No right to sublease. The landlord does not convey to the tenant the right to lease or sublet any part of the farm or to assign the lease to any person or persons whomsoever.

H. Binding on heirs. The provisions of this lease shall be binding upon the heirs, executors, administrators, and successors of both landlord and tenant in like manner as upon the original parties, except as provided by mutual written agreement.

I. Landlord's lien for rent and performance. The landlord's lien provided by law on crops grown or growing shall be the security for the rent herein specified and for the faithful performance of the terms of the lease. If the tenant fails to pay the rent due or fails to keep the agreements of this lease, all costs and attorney fees of the landlord in enforcing collection or performance shall be added to and become a part of the obligations payable by the tenant hereunder.

J. Additional provisions.
_____
_____

III. LAND USE

A. General provisions. The land described in Section I will be used in approximately the following manner. If it is impractical in any year to follow such a land-use plan, appropriate adjustments will be made by mutual written agreements between the parties.

FARMLAND 321

APPENDIX H (continued)

	Dry	Irrigated	
(1) Cropland			
(a) Corn	_____	_____	Acres
(b) Grain Sorghum	_____	_____	Acres
(c) Wheat	_____	_____	Acres
(d) Sugar Beets	_____	_____	Acres
(e) Silage	_____	_____	Acres
(f) Alfalfa	_____	_____	Acres
(g) Pasture	_____	_____	Acres
(h) Other:	_____	_____	Acres
	_____	_____	Acres
	_____	_____	Acres
	_____	_____	Acres
TOTAL ACRES	_____	_____	Acres

PASTURE IDENTIF.
_____ _____ acres/animal unit
_____ _____ acres/animal unit
_____ _____ acres/animal unit

(1000 pound mature cow is equivalent to one animal unit.)

Other restrictions are:
_____
_____

C. Government programs. The extent of participation in government programs will be discussed and decided on an annual basis. The course of action agreed upon should be placed in writing and be signed by both parties. A copy of the course of action so agreed upon shall be made available to each party.

B. Restrictions. The maximum acres harvested as silage shall be _____ acres unless it is mutually decided otherwise:
The pasture stocking rate shall not exceed:

IV. CROP-SHARE-CASH RENT AND RELATED PROVISIONS

A. General agreement. (1) The tenant agrees to pay as rent for the use of the land the share of crops shown in Table 1 of this section. The tenant also agrees to

Table 1.—Landlord's Share (% and/or $) of Crops and Crop Expenses

	Corn example	Corn	Grain sorghum	Small grain	
SHARE OF CROPS	50%				
SHARE OF CROP EXPENSES:					
Fertilizer:					
Materials	50%				
Application	50%				
Herbicide:					
Materials	50%				
Application					
Insecticide:					
Materials	50%				
Application					
Seed	50%				
Lime, rock phosphate*	100%				
Harvesting (per ac.)	$7.50				
Drying	50%				
Baling					
Delivery to:					
Storage/bu.					
Market/bu.	$ .07				
SHARE OF IRR. EXPENSES					
Well Repairs	100%				
Pump Repairs	100%				
Gear Head Rep.	100%				
Power Unit Rep.	100%				
System Repairs					
Land Maint.					
Irrigation Fuel					
Power Replace.					
System Replace.					
Labor					
Other:					

* Lime, rock phosphate, and other fertilizers having more than one year life paid by the tenant should be recorded in the compensation table in Section V-C-2.

furnish all labor, machinery, and cash operating expenses except for landlord's share (percent and/or dollar charge per unit) indicated in Table 1. (2) Other Provisions relative to Table 1.

_____
_____
_____

3. Other crop-share-cash agreements.

   1. Operating expenses. Additional agreements relative to the sharing of expenses are as follows:

   _____
   _____
   _____

   2. Storage, landlord's crop. At the landlord's request, the tenant agrees to store as much of the landlord's share of the crops as possible, using storage space reserved by the landlord and not to exceed _____ percent of the storage space not specifically reserved.

   3. Delivery of grain. The tenant agrees to deliver the landlord's share of crops at a place and at a time the landlord shall designate, not over _____ miles distance at the charge shown in Table 1 of this section. Additional agreements are:

   _____
   _____
   _____

4. Cash rent on non-shared items. The tenant agrees to pay cash rent annually for the use of the following non-shared items.

   Table 2—Amount of Annual Cash Rent
   (Complete at beginning of lease)

		Total
Pasture	$_____	
Hayland: _____	$_____	
Farmstead: Dwelling	$_____	
Service bldgs.	$_____	
Timber and waste	$_____	
Total cash rent	$_____	

   Payment of cash rent: The tenant agrees to pay cash rent as follows:

   $_____ on or before _____ day of _____ (month)
   $_____ on or before _____ day of _____ (month)
   $_____ on or before _____ day of _____ (month)
   $_____ on or before _____ day of _____ (month)

   If rent is not paid when due, the tenant agrees to pay interest on the amount of unpaid rent at the rate of _____ percent per annum from due date until paid.

5. Pasturing. The tenant will prevent damage to cropland and growing crops by livestock.

6. Home use. Tenant and landlord may take for home use the following kinds and quantities of jointly owned crops:

   _____
   _____
   _____

7. Buying and selling. The landlord and tenant will buy and sell jointly owned property according to the following agreement:

   _____
   _____
   _____

8. Division of property. At the termination of this lease, all jointly owned property will be divided or disposed of as follows:

   _____
   _____
   _____

V. OPERATION AND MAINTENANCE OF FARM

In order to operate this farm efficiently and to maintain it in a high state of productivity, the parties agree as follows:

A. The tenant agrees:

1. General maintenance. To provide the unskilled labor necessary to maintain the farm and its improvements during his tenancy in as good condition as it was at the beginning. Normal wear and depreciation and damage from causes beyond the tenant's control are expected.

2. Land use. Not to: a) plow pasture or meadowland, b) cut live trees for sale or personal use, or c) pasture new seedlings of legumes and grasses in the year they are seeded without consent of the landlord.

3. Insurance. Not to house automobiles, motor trucks, or tractors in barns, or otherwise violate restrictions in the landlord's insurance policies without written consent from the landlord. Restrictions to be observed are as follows:

   _____
   _____
   _____

4. Noxious weeds. To use diligence to prevent noxious weeds from going to seed on the farm. Treatment of the noxious weed infestation and cost thereof shall be handled as follows:

   _____
   _____
   _____

5. Addition of improvements. Not to: 1) erect or permit to be erected on the farm any nonremovable structure or building, b) incur any expense to the landlord for such purposes, or c) add electrical wiring, plumbing, or heating to any building without written consent of the landlord.

6. Conservation. Control soil erosion as completely as practicable; keep in good repair all terraces, open ditches, inlets and outlets of tile drains; preserve all established watercourses or ditches including grassed waterways; and refrain from any operation or practice that will injure such structures.

7. Damages. When he leaves the farm, to pay the landlord reasonable compensation for any damages to the farm for which he, the tenant, is responsible. Any decrease in value due to ordinary wear and depreciation or damages outside the control of the tenant are excepted.

APPENDIX H (continued)

8. Costs of operation. To pay all costs of operation except those specifically referred to in Sections IV, V-A-4, and V-B.
9. Repairs. Not to buy materials for maintenance and repairs in an amount in excess of $_____ within a single year without written consent of the landlord.

B. The landlord agrees:
1. Loss replacement. To replace or repair as promptly as possible the dwelling or any other building or equipment regularly used by the tenant that may be destroyed or damaged by fire, flood, or other cause beyond the control of the tenant or to make rental adjustments in lieu of replacements.
2. Materials for repairs. To furnish all material needed for normal maintenance and repairs.
3. Skilled labor. To furnish any skilled labor tasks which the tenant himself is unable to perform satisfactorily. Additional agreements regarding materials and labor are:

_____
_____
_____

4. Reimbursement. To pay for materials purchased by the tenant for purposes of repair and maintenance in an amount not to exceed $_____ in any one year, except as otherwise agreed upon. Reimbursement shall be made within _____ days after the tenant submits the bill.
5. Removeable improvements. Let the tenant make minor improvements of a temporary or removable nature, which do not mar the condition or appearance of the farm, at the tenant's expense. He further agrees to let the tenant remove such improvements even though they are legally fixtures at any time this lease is in effect or within _____ days thereafter, provided the tenant leaves in good condition that part of the farm from which such improvements are removed. The tenant shall have no right to compensation for improvements that are not removed except as mutually agreed.
6. Compensation for crop expenses. To reimburse the the tenant at the termination of this lease for field work done and for other crop costs incurred for crops to be harvested during the following year. Unless otherwise agreed, current custom rates for the operations involved will be used as a basis of settlement.

C. Both agree:
1. Not to obligate other party. Neither party hereto shall pledge the credit of the other party hereto for any purpose whatsoever without the consent of the other party. Neither party shall be responsible for debts or liabilities incurred, or for damages caused by the other party.
2. Capital improvements. Costs of establishing hay or pasture seedings, new conservation structures, improvements (except as provided in Section V-B-5), or of applying lime and other longlived fertilizers shall be divided between landlord and tenant as set forth in the following table. The tenant will be reimbursed by the landlord either when the improvement is completed, or the tenant will be compensated for his share of the depreciated cost of his contribution when he leaves the farm based on the value of the tenant's contribution and depreciation rate shown in the following table. (Cross out the portion of the preceding sentence which does not apply.)
   Rates for labor, power, and machinery contributed by the tenant shall be agreed upon before construction is started.
3. Mineral rights. Nothing in this lease shall confer upon the tenant any right to minerals underlying said land, but same are hereby reserved by the landlord together with the full right to enter upon the premises and to bore, search, and excavate for same, to work and remove same, and to deposit excavated rubbish, and with full liberty to pass over said premises with vehicles and lay down and work any railroad track or tracks, tanks, pipelines, power lines, and structures as may be necessary or convenient for the above purpose. The landlord agrees to reimburse the tenant for any actual damage he may suffer for crops destroyed by these activities and to release the tenant from obligation to continue farming this property when development of mineral interfers materially with the tenant's opportunity to make a satisfactory return.

VI. ARBITRATION OF DIFFERENCES

Any differences between the parties as to their several rights or obligations under this lease that are not settled by mutual agreement after thorough discussion, shall be submitted for arbitration to a committee of three disinterested persons, one selected by each party hereto and the third by the two thus selected. The committee's decision shall be accepted by both parties.

Compensation for Improvements Table

Type of improvement	Date to be completed	Estimated total cost (dollars)	Proportion to be contributed by tenant – Material	Proportion to be contributed by tenant – Unskilled labor	Proportion to be contributed by tenant – Mach.	Total value of tenant's contrib. (dollars)*	Rate of annual depreciation
Irri. Well		$	%	%	%	$	%
Underground Pipe		$				$	
Land Dev.		$				$	
Tailwater Structures		$				$	
Power Lines		$				$	
Other		$				$	

* To be recorded when improvement is completed.

Executed in duplicate on the date first above written:

_____   _____
          (tenant)   (landlord)
_____   _____
       (tenant spouse)   (landlord spouse)

STATE OF _____ ⎫
COUNTY OF _____ ⎬ S:

On this _____ day of _____ A.D., 19____, before me, the undersigned, a Notary Public in said State, personally appeared _____, _____ and _____, to me known to be the identical persons named in and who executed the foregoing instrument, and acknowledged that they executed the same as their voluntary act and deed.

_____
        (Notary Public)

...AND JUSTICE FOR ALL
Programs and activities of Cooperative Extension Service are available to all potential clientele without regard to race, color, sex or national origin. Anyone who feels discriminated against should send a complaint within 180 days to the Secretary of Agriculture, Washington, D.C. 20250.

Issued in furtherance of Cooperative Extension work, Acts of Congress of May 8 and June 30, 1914, in cooperation with the U.S. Department of Agriculture and Cooperative Extension Services of Illinois, Indiana, Kansas, Michigan, Minnesota, Missouri, Nebraska, North Dakota, Ohio, South Dakota and Wisconsin. John O. Dunbar, Director, Cooperative Extension Service, Kansas State University, Manhattan.

APPENDIX H (continued)

Livestock Share Farm Lease
North Central Regional
Publication No. 108

This form can provide the landlord and tenant with a guide for developing an agreement to fit their individual situation. This form is not intended to take the place of legal advice pertaining to contractual relationships between the two parties. Because of the possibility that a farm operating agreement may be legally considered a partnership under certain conditions, seeking proper legal advice is recommended when developing such an agreement.

This lease is entered into this _____ day of _____, 19____, between _____, landlord, of _____

(address)

_____, spouse, of _____

(address)

hereafter known as "the landlord," and

_____, tenant, of _____

(address)

_____, spouse, of _____

(address)

hereafter known as "the tenant."

I. PROPERTY DESCRIPTION

The landlord hereby leases to the tenant, to occupy and use for agricultural and related purposes, the following described property:

_____
_____
consisting of approximately _____ acres situated in _____ County (Counties), _____ (State) with all improvements thereon except as follows:

_____
_____

II. GENERAL TERMS OF LEASE

A. Time period covered. The provisions of this agreement shall be in effect for _____ year(s), commencing on the _____ day of _____, 19____. This lease shall continue in effect from year to year thereafter unless written notice of termination is given by either party to the other at least _____ days prior to expiration of this lease or the end of any year of continuation.

B. Review of lease. A written request is required for a general review of the lease or for consideration of proposed changes by either party, at least _____ days prior to the final date for giving notice to terminate the lease as specified in IIA.

C. Amendments and alterations. Amendments and alterations to this lease shall be in writing and shall be signed by both the landlord and tenant.

D. No partnership intended. It is particularly understood and agreed that this lease shall not be deemed to be nor intended to give rise to a partnership relation.

E. Transfer of property. If the landlord should sell or otherwise transfer title to the farm, he will do so subject to the provisions of this lease.

F. Right of entry. The landlord reserves the right for himself, his agents, his employees, or his assigns to enter the farm at any reasonable time to: a) consult with the tenant; b) make repairs, improvements, and inspections; and c) (after notice of termination of the lease is given) do plowing, seeding, fertilizing, and any other customary seasonal work, none of which is to interfere with the tenant in carrying out regular farm operations.

G. No right to sublease. The landlord does not convey to the tenant the right to lease or sublet any part of the farm or to assign the lease to any person or persons whomsoever.

H. Binding on heirs. The provisions of this lease shall be binding upon the heirs, executors, administrators, and successors of both landlord and tenant in like manner as upon the original parties, except as provided by mutual written agreement.

I. Landlord's lien for rent and performance. The landlord's lien provided by law on crops grown or growing shall be the security for the rent herein specified and for the faithful performance of the terms of the lease. If the tenant fails to pay the rent due or fails to keep the agreements of this lease, all costs and attorney fees of the landlord in enforcing collection or performance shall be added to and become a part of the obligations payable by the tenant hereunder.

J. Additional provisions.

_____
_____

III. LAND USE

A. General provisions. The land described in Section I will be used in approximately the following manner. If it is impractical in any year to follow such a land-use plan, appropriate adjustments will be made by mutual written agreements between the parties.

For further information see: "Livestock-Share Rental Arrangements for Your Farm" NCR publication number 107.

1. Cropland
   a) Row crops _____ acres
   b) Small grains _____ acres
   c) Legumes _____ acres
   d) Rotation pasture _____ acres
2. Pasture: _____ _____ acres
3. Other: _____ _____ acres
   _____ _____ acres
4. Total _____ acres

B. Restrictions. The maximum acres harvested as silage shall be _____ acres unless it is mutually decided otherwise.

The pasture stocking rate shall not exceed:

PASTURE IDENTIFICATION:

_____ _____ acres/animal unit
_____ _____ acres/animal unit
_____ _____ acres/animal unit

(1,000 pound mature cow is equivalent to one animal unit.)

Other restrictions are:
_____
_____
_____

C. Government programs. The extent of participation in government programs will be discussed and decided on an annual basis. The course of action agreed upon shall be placed in writing and be signed by both parties. A copy of the course of action so agreed upon shall be made available to each party.

IV. **LIVESTOCK PRODUCTION AND SHARING ARRANGEMENTS**

A. It is agreed the tenant and landlord will engage in the production of livestock. Real property including land and the type and number of livestock to be contributed to production by each party are reported in Table 1.

Table 1—Contributions of Property to be Furnished by Each Party

	Approximate number to be kept	Share furnished by Landlord %	Tenant %
1. Land and fixed improvements at beginning of this lease described in Section I		_____	_____
2. Fixed improvements constructed during period of this lease:			
Materials and skilled labor		_____	_____
Hauling materials to farm		_____	_____
Farm labor		_____	_____
_____		_____	_____
_____		_____	_____
3. Livestock:			
Breeding: _____	_____	_____	_____
_____	_____	_____	_____
_____	_____	_____	_____
Replacements: _____	_____	_____	_____
_____	_____	_____	_____
_____	_____	_____	_____
Feeders: _____	_____	_____	_____
_____	_____	_____	_____
_____	_____	_____	_____
4. Machinery and equipment: (crop, livestock, etc.)			
_____		_____	_____
_____		_____	_____
_____		_____	_____
_____		_____	_____
5. Portable farm buildings:			
_____		_____	_____
_____		_____	_____
_____		_____	_____

APPENDIX H (continued)

B. Annual operating expenses shall be supplied by the landlord and tenant as reported in Table 2 except as discussed in Section VI.

Table 2—Percentage Share of Operating Expenses to be Furnished by Each Party

	(L)	(T)
1. Crop Expenses:		
• Fertilizer	___	___
• Lime	___	___
• Seed	___	___
• Herbicide	___	___
• Crop insurance	___	___
• Other supplies	___	___
• Other: _____	___	___
2. Livestock Expenses:		
• Feed purchased	___	___
• Veterinary	___	___
• Breeding fees	___	___
• Medicines and drugs	___	___
• Feed grinding and mixing	___	___
• _____	___	___
• _____	___	___
3. Fuel:		
• Tractor	___	___
• Truck	___	___
• Harvesting	___	___
• Crop drying	___	___
• Feed processing	___	___
• Heating buildings	___	___
4. Electricity	___	___
5. Telephone	___	___
6. General hired labor	___	___
7. Custom:		
• Hauling crops and livestock	___	___
• Harvesting:		
Corn	___	___
Small grain	___	___
Soybeans	___	___
_____	___	___
_____	___	___
8. Insurance:		
• Buildings	___	___
• _____	___	___
9. Taxes	___	___
10. Interest:		
• Operating capital	___	___
• Intermediate term loans	___	___
11. Other	___	___

C. Additional agreements in regard to livestock production.

1. Breeding replacements shall be furnished as follows:
   _____
   _____

2. Sale of breeding stock shall be shared as follows:
   _____

3. Other breeding stock provisions are:
   _____

D. Neither landlord or tenant shall have the authority to bind the other in any contract with third parties. Expenses other than those reported in Tables 1 and 2 shall be shared as follows:
   _____

E. Buying and selling. The tenant shall consult with the landlord regarding time, price, sales agency, and similar matters regarding the purchase and sale of livestock, feed, and crops whenever the transaction exceeds $_____ in value. Additional agreements are as follows:
   _____

F. Livestock restrictions. Neither the tenant nor the landlord shall bring to the farm livestock not included in the agreement without express written permission of the other party.
   Additional agreements relative to livestock are:
   _____

G. Equipment and machinery replacements. The cost of additional and replacement livestock equipment and machinery will be shared as follows:
   _____
   _____

V. DIVISION OF INCOME AND CASH RENT ON NON-SHARED ITEMS

A. Division of income. The tenant shall pay rent to the landlord for the use of the landlord's property described in this lease (Table 1) an amount equal to _____ percent of the gross income. Gross income shall consist of the proceeds from the sale or exchange of all grain, forages, livestock, and other products produced under the provisions of this lease, except for:
   _____
   _____

B. Cash rent on non-shared items. The tenant agrees to pay cash rent annually for the use of the following non-shared items:

Table 3—Amount of Annual Cash Rent
(complete at beginning of lease)

Farmstead: Dwelling	$_____
Service buildings	$_____
Timber and waste	$_____
Other: _____	$_____
_____	$_____
Total cash rent	$_____

Payment of cash rent: The tenant agrees to pay cash rent as follows:

$_____ on or before _____ day of _____ month

$_____ on or before _____ day of _____ month

$_____ on or before _____ day of _____ month

$_____ on or before _____ day of _____ month

If rent is not paid when due, the tenant agrees to pay interest on the amount of unpaid rent at the rate of _____ percent per annum from due date until paid.

VI. OPERATION AND MAINTENANCE OF FARM

In order to operate this farm efficiently and to maintain it in a high state of productivity, the parties agree as follows:

A. The tenant agrees:

1. General maintenance. To provide the unskilled labor necessary to maintain the farm and its improvements during his tenancy in as good condition as it was at the beginning. Normal wear and depreciation and damage from causes beyond the tenant's control are excepted.

2. Land use. Not to: a) plow pasture or meadowland, b) cut live trees for sale or personal use, c) pasture new seedlings of legumes and grasses in the year they are seeded without consent of the landlord.

3. Insurance. Not to house automobiles, motor trucks, or tractors in barns, or otherwise violate restrictions in the landlord's insurance policies without written consent from the landlord. Restrictions to be observed are as follows:

_____
_____
_____

4. Noxious weeds. To use diligence to prevent noxious weeds from going to seed on the farm. Treatment of the noxious weed infestation and cost thereof shall be handled as follows:

_____
_____
_____

5. Addition of improvements. Not to: a) erect or permit to be erected on the farm any nonremovable structure or building, b) incur any expense to the landlord for such purposes, or c) add electrical wiring, plumbing, or heating to any building without written consent of the landlord.

6. Conservation. Control soil erosion as completely as practicable; keep in good repair all terraces, open ditches, inlets and outlets of tile drains; preserve all established watercourses or ditches including grassed waterways; and refrain from any operation or practice that will injure such structures.

7. Damages. When he leaves the farm, to pay the landlord reasonable compensation for any damages to the farm for which he, the tenant, is responsible. Any decrease in value due to ordinary wear and depreciation or damages outside the control of the tenant are excepted.

8. Costs of operation. To pay all costs of operation except those specifically referred to in Sections IV, VI-A-4, and VI-B.

9. Repairs. Not to buy materials for maintenance and repairs in an amount in excess of $_____ within a single year without written consent of the landlord.

B. The landlord agrees:

1. Loss replacement. To replace or repair as promptly as possible the dwelling or any other building regularly used by the tenant that may be destroyed or damaged by fire, flood, or other cause beyond the control of the tenant or to make rental adjustments in lieu of replacements.

2. Materials for repairs. To furnish all material needed for normal maintenance and repair.

3. Skilled labor. To furnish any skilled labor for tasks which the tenant himself is unable to perform satisfactorily. Additional agreements regarding materials and labor are:

4. Reimbursement. To pay for materials purchased by the tenant for purposes of repair and maintenance in an amount not to exceed $_____ in any one year, except as otherwise agreed upon. Reimbursement shall be made within _____ days after the tenant submits the bill.

5. Removable improvements. To let the tenant make minor improvements of a temporary or removable nature, which do not mar the condition or appearance of the farm, at the tenant's expense. He further agrees to let the tenant remove such improvements even though they are legally fixtures at any time this lease is in effect or within _____ days thereafter, provided the tenant leaves in good condition that part of the farm from which such improvements are removed. The tenant shall have no right to compensation for improvements that are not removed except as mutually agreed.

6. Compensation for crop expenses. To reimburse the tenant at the termination of this lease for field work done and for other crop costs incurred for crops to be harvested during the following year, unless otherwise agreed, current custom rates for the operations involved will be used as a basis of settlement.

APPENDIX H (continued)

C. Both agree:
1. Capital improvements. Costs of establishing hay or pasture seedings, new conservation structures, improvements (except as provided in Section V-B-5), or of applying lime and other long-lived fertilizers shall be divided between landlord and tenant as set forth in the following table. The tenant will be re-imbursed by the landlord either when the improvement is completed, or the tenant will be compensated for his share of the depreciated cost of his contribution when he leaves the farm based on the value of the tenant's contribution and depreciation rate shown in the following table. (Cross out the portion of the preceding sentence which does not apply.)

Rates for labor, power, and machinery contributed by the tenant shall be agreed upon before construction is started.

Compensation for Improvements Table

Type of improvement	Date to be completed	Estimated total cost (dollars)	Proportion to be contributed by tenant			Total value of tenant's contrib. (dollars)*	Rate of annual depreciation
			Material	Unskilled labor	Mach.		
			%	%	%		%

* To be recorded when improvement is completed.

2. Mineral rights. Nothing in this lease shall confer upon the tenant any right to minerals underlying said land, but same are hereby reserved by the landlord together with the full right to enter upon the premises and to bore, search, and excavate for same, to work and remove same, and to deposit excavated rubbish, and with full liberty to pass over said premises with vehicles and lay down and work any railroad track or tracks, tanks, pipelines, power lines, and structures as may be necessary or convenient for the above purpose. The landlord agrees to reimburse the tenant for any actual damage he may suffer for crops destroyed by these activities and to release the tenant from obligation to continue farming this property when development of mineral resources interferes materially with the tenant's opportunity to make a satisfactory return.

VII. FARM RECORDS AND FINANCIAL SETTLEMENTS

A. Records of joint interest shall be kept by _____ and shall be made available to the landlord/tenant (cross out one) upon request. Financial and production records shall include a complete inventory of all property used in the farm business. Inventories shall be recorded and financial records summarized by _____ day of _____ (month) or at intervals mutually agreed upon. Specify:

_____

_____

B. The record system to be used shall be:

_____

_____

C. All joint receipts and disbursements shall be handled through _____ bank as follows:

- Receipts: _____
- Disbursements: _____

D. Cash financial settlement shall be made by the _____ day of each month or at intervals mutually-agreed upon. Specify:

_____

_____

_____

VIII. ARBITRATION OF DIFFERENCES AND DIVISION OF PROPERTY

A. Arbitration of differences. Any differences between the parties as to their several rights or obligations under this lease that are not settled by mutual agreement after thorough discussion, shall be submitted for arbitration to a committee of three disinterested persons, one selected by each party hereto and the third by the two thus selected. The committee's decision shall be accepted by both parties.

B. Division of property. Upon termination of this lease, unused production shall be divided as follows:

1. Feed grain and supplies. All grain, silage, other feeds, and all co-owned supplies including straw and other bedding materials shall be divided by measure or value, whichever is more equitable, with the landlord and tenant each receiving title to his respective share as reported in Section V-A.

2. Livestock. If the livestock are owned equally (50-50), the tenant shall divide each class of livestock, as cows, steers, calves, hogs, etc., into two groups and the landlord shall take his choice of the two groups of each. In case the groupings cannot be made equal, a difference in monetary value shall be assigned before the choice is made and added to the choice.

3. Undivided interest of co-owned property. If both parties mutually agree not to accept the above described plan for dividing co-owned classes of property including livestock, it is agreed the tenant shall set a value on the entire amount of the respective co-owned classes of property on the basis of which he will either sell his undivided interest or buy that of the landlord, at the option of the landlord; or the co-owned property may be disposed of by private or public sale arranged for that purpose at a reasonable time and place.

4. Home use. The tenant and landlord may take annually for home use the following kinds and quantities of jointly owned crops, livestock, and/or livestock products:

_____

_____

_____

Executed in duplicate on the date first above written:

_____   _____
           (tenant)   (landlord)

_____   _____
        (tenant spouse)   (landlord spouse)

STATE OF _____ ⎫
                                ⎬ SS:
COUNTY OF _____ ⎭

On this _____ day of _____ A.D., 19____, before me, the undersigned, a Notary Public in said State, personally appeared _____

_____, _____, _____,

and _____, to me known to be the identical persons named in and who executed the foregoing instrument, and acknowledged that they executed the same as their voluntary act and deed.

_____
         (Notary Public)

Issued in furtherance of Cooperative Extension work, Acts of Congress of May 8 and June 30, 1914, in cooperation with the U.S. Department of Agriculture and Cooperative Extension Services of Illinois, Indiana, Kansas, Michigan, Minnesota, Missouri, Nebraska, North Dakota, Ohio, South Dakota and Wisconsin. John O. Dunbar, Director, Cooperative Extension Service, Kansas State University, Manhattan.

. . . AND JUSTICE FOR ALL
Programs and activities of Cooperative Extension Service are available to all potential clientele without regard to race, color, sex or national origin. Anyone who feels discriminated against should send a complaint within 180 days to the Secretary of Agriculture, Washington, D.C. 20250.

INDEX

A

Abstract of title, 172
Accelerated cost recovery system, 225-227
Adjusted gross estate
    charitable deduction, 274-275
    marital deduction, 273-274
    orphan's deduction, 274
Adjustments for comparable sales
    improvements, 61
    location, 61
    size, 61
    time, 60-61
Advertising land for sale
    papers and magazines, 154
    showing a property, 154
After-tax interest rate, 112
Alternative minimum tax; 248-250
Amortization of loans, 75
Annual exclusion
    for gifts, 278-279
Annuity
    nontax considerations, 255-256
    tax considerations
        estate tax, 256
        gift tax, 256-257
        income tax, 257-258
Appraisers, 52
APR (annual percentage rate of interest)
    computation of, 129-131
    defined, 111-112
    effect of service charge on, 113-114
    effect of stock purchase requirement on, 113-114
ASCS, 37, 45, 52-53, 206
Auction sales
    involuntary, 149
    sealed bid, 148-149
    voluntary, 150

B

Balloon payment, 92-93, 133
Basis
    acquisition by gift, 222-223
    acquisition by inheritance, 223-224
    acquisition by purchase, 221-222
Bona fide sales, 59
Boundaries, 48-49
Buildings
    depreciation, 67-68
    gathering information on, 46-47

Buyers of land
    absentee, 15-16
    owner-operators, 14-15
    price interest rate tradeoffs, 116
    tenants, 14-15
Buying land
    economic reasons
        expected capital gains, 26-27
        expected income, 26-27
    personal reasons for, 23-26

C

Capital gains on land
    economic factors, 42
    historical rates, 26-28
    justification of gains, 30-32
Capital gains taxes
    corporations, 249
    effects on land values, 81-83
    individuals, 248-249
    method of calculating, 243-244
Capitalization rate, 65-66
Capitalizing cash rents
    for use valuation, 262-263
Carryover basis, 224
Cash deficiency
    defined, 94
    measures of, 94-96
Cash flow projections, 98-100
Cash flows
    associated with financing, 89-93
    feasibility of purchase, 89
Cash lease
    problems, 187-188
    risks of, 187
Cash rent for use valuation, 262-263
Cash surplus
    defined, 94
    measures of, 94-96
Charitable deductions
    effect on gross estate, 274-275
    for gifts, 280-281
Commercial banks
    lending limits per customer, 144
    sources of loanable funds, 143-144
    structure, 143
Commodity Credit Corporation (CCC)
    nonrecourse loans, 207-208
    recourse loans, 207-208
Comparable sales, 60-61

Compounding
    equal payments, 73
    single amounts, 72–73
    sinking funds, 73
Consideration furnished rule, 267
Contractual interest rate, 111
Corporations
    amount of farmland owned, 13–14
    regular, 260
    reporting of capital gains, 249–250
    restrictions on land ownership, 212–213
    Subchapter S, 260–261
Cost approach
    building valuation, 67–68
    inventory of land, 68
Cost sharing, 247
Courthouse records, 58–59
Credit for services rule, 270–271
Cropland
    acres, 12
    amount by region, 13
    how used, 13
Crop share leases
    for use valuation, 263–264
    problems, 190
    sharing arrangements, 188–189
Custom hiring, 184–185

D

Debt servicing requirements, 102–108
Deeds
    general warranty, 175
    quit-claim, 175
    recording, 179–180
    restrictions, 50–51
    special warranty, 175
Depreciation
    accelerated cost recovery, 235–238
    amount depreciable, 234
    buildings, 86
    expense method, 234–235
    limits for real property, 238
    machinery, 85–86
    recapture, 245–246
Determinable fee estate, 5
Discounting
    capital recovery, 74–75
    equal payments, 74
    single amounts, 74
Dower interest, 164–165
Downpayment, 89–90, 100–102
    effect on prices paid for land, 121–123
    tradeoffs between price and downpayment, 123–124
Drainage, 47

E

Easements, 49, 160, 173–174
Eminent domain, 4, 33

Equity capital
    outside sources, 145–146
    corporations, 146
    limited partnerships, 146
Escheat, 4
Escrow, 178–180, 253–254
Estate tax
    consideration furnished, rule, 269
    credit for services rule, 270–271
    effects of method of ownership, 268–273
    fractional share rule, 269
Excess deductions account, 247–248

F

Fair market value, 57, 68
Farmers Home Administration, 100
    emergency loans, 140
    farm ownership loans, 139
    focus of loan programs, 138
    loan supervision and graduation, 140–141
    soil and water loans, 140
    sources of loanable funds, 138–139
Farmland sales
    amount sold by region, 18
    financing, 19–20
    price per acre, 19
    size of tract sold, 19
    type of seller, 18
Farm managers
    farm visits, 196
    liability considerations, 198–199
    plans developed by, 195–196
    selection of, 195
Federal estate tax returns, 275
Federal gift tax
    annual exclusion, 278–279
    charitable deductions, 280–281
    marital deduction, 279–280
    unified credit, 281–281 ???
    valuation of assets, 277–278
Federal Land Banks, 98–100
    description, 135–136
    effect of stock purchase requirement, 113–114
    effective interest rates, 264–265
    methods of capitalization, 136
    sources of loanable funds, 136–137
    terms of loans, 137–138
    types of loans, 136
Federal programs
    agricultural conservation program, 205–206
    commodity programs, 206–208
    emergency conservation measures, 208
    emergency livestock feed program, 208–209
    loan programs, 209
    water bank program, 209
Fee simple estate, 5

FARMLAND   **333**

Fences, 47
First mortgage, 134
Fixed interest rate, 134
Foreclosure, 34
Foreign ownership of land, 17, 212-213
Fractional share rule, 269-270

G

Gifts
    income tax basis, 222-223
    of future interest, 279
    part gift/part sales, 285-286
    within three years of death, 272-273
Government farm programs, 42, 46

I

Income approach to valuation
    capitalization rate, 65-66
    estimating expenses, 64-65
    estimating gross returns, 63-64
Income tax basis
    under installment sales, 254
Income tax rates
    effect on land values, 83-84
Increasing payment plan
    method of computation, 108-110
    structure, 103-107
Individuals
    prepayment of loans, 145
    seller financing
        first mortgage, 145
        second mortgage, 145
    tax treatment for ownership of land, 259
Inflation, 40
    effect on land values, 80-81
    farmland as a hedge against, 27-28
    historical levels, 27
Information sources, 196
Inheritance
    income tax basis, 223-224
Installment land contract, 177-178
    disposition of contract, 254-255
    income tax aspects, 250-254
Insurance
    against liability, 204
    specification in land contract, 170-171
Interest rates
    effect on net present value of land, 114, 115
    tradeoff with price, 115-123
    types of rates
        after tax interest rate, 112
        annual percentage rate (APR), 111-112
        contractual rate, 111
        real rate, 112
Investment tax credit
    amount eligible, 232
    claiming as a lessor, 229-231
    how calculated, 226-227
    limits on used property, 228
    passthrough to lessee, 230-231
    property eligible, 225-226
    property not eligible, 228-229
    recapture, 232-234, 244-245
    recovery period property, 227-228
    when claimable, 231-232
Irrigation, 48

J

Joint tenancy, 165-166
    inadvertent severance, 284-285
    severance, 284-285

L

Land
    acres in U.S., 11
    concepts
        legal, 2
        economic, 2-3
    definitions
        real estate, 2-3
    real property, 3
    uses, 12
Land Contract (see installment land contract)
Landlord
    claiming investment tax credit, 229-230
Landowners
    custom operator, 184-185
    liability considerations, 198-199
    owner-lessor, 185-194
    owner-operator, 183-184
Land use
    cropping history, 45
    fertilizer history, 45-46
    yields and carrying capacity, 46
Land use restrictions, 159-160
Leasehold estate
    estate from year to year, 7
    estate for years, 7
    tenancy at sufferance, 7
    tenancy at will, 7
Leases
    adaptations to particular farm, 193-194
    cash leases, 187-188
    combination leases, 193
    crop-share leases, 188-190
    labor share leases, 191
    lease agreement, 186
    livestock-share leases, 190-191
    standing rent lease, 191-193
Legal descriptions of land
    congressional survey, 167-168
    courses and distances, 167
    metes and bounds, 166-167
    plats, 168
Length of loan
    effect on price of land, 124-129
    tradeoffs with price and downpayment, 129

Life estate, 6
Life insurance companies, 98–100
    contacts for loans, 142–143
    investment objectives, 141
    terms of loans, 142–143
    types of farm loans, 141
Like-kind exchanges, 21–22
Lineal descendants, 263
Listing agreements
    exclusive agency listings, 153
    exclusive rights to sell, 153
    multiple listing, 153
    net listing, 153–154
    open listing, 152–153
Litigation
    reasons for, 197–198
    status, 198–199
Livestock-share leases
    problems, 191
    sharing arrangements, 190–191
Loan payment plans
    defined, 134
    equal principal, 93
    full amortization, equal payments, 90–92
    partial amortization, equal payments, 92–93
    variable amortization, 93–94

M

Marital deduction
    effect on gross estate, 273–274
    for gifts, 279–280
Market data approach to valuing land
    comparisons to subject property, 60–62
    sales data, 58–59
    verification of sales, 59–60
Material participation, 187, 264
Millage, 219–220
Mineral rights, 49
Mortgage
    assuming subject to, 176
    foreclosure, 176
    novation, 176

N

Negligence
    conditions to be liable for, 200–201
    invitee, 201–202
    special situations, 202–204
    trespass, 201–202
Negotiating
    to improve cash flows, 100–108
    debt servicing, 102–108
    downpayment, 100–102
    role of broker, 154–155
Net present value
    components of calculation
        discount rate, 78
        expected growth rates, 77–78
        expected net after-tax returns, 76–77, 87
        planning horizon, 79
        tax rates, 78–79, 86
        terms of financing, 78
    defined, 76
    for different levels of after-tax returns, 94–96
    for discount rates, 94–96

O

Occupational Safety and Health Act, 211
Offer to buy (see also land contract)
    general considerations, 163–164
    legal description, 166–168
    names of parties, 164–166
    property tax payment, 169–170
    purchase price, 168–169
    risk of loss, 170–171
Owners of land
    by occupation, 15
    by type, 14
    corporations, 13–14
    government, 13
    partnerships, 14
Ownership patterns and gift taxes, 282–285

P

Partnerships
    amount of land owned, 13–14
    income tax aspects of owning land, 259
Payout period, 134
Preliminary considerations for purchase
    availability of utilities, 160–161
    loan committment, 160
    selecting an attorney, 161–162
    trend of development, 160
    visual inspection, 162–163
    zoning and land use restrictions, 159–160
Prepayment penalty, 134
Prepayment privilege, 134
    determined by net present value, 76–84
    tables, 80–82
    tradeoffs with interest rates, 115–123
Private agreements
    defined, 147
    problems, 148
Productivity index for soil, 263
Property assessment
    deductions and credits, 218
    objections, 218
    valuation methods, 217–218
Property taxes
    ad valorem tax, 215–216
    assessments, 217–219
    millage or tax rate, 219–220
    recording taxes, 217
    severance taxes, 217
    special assessment taxes, 216–217
    specification of payment upon sale, 169–170

R

Real estate brokers
    choice of, 155–156
    functions performed, 152–155
    licensing requirements, 150–151
    relationship to seller, 151–152
Real rate of interest, 112
Recapture
    depreciation, 245–246
    investment tax credit, 244–245
    land clearing expenses, 246–247
    soil and water conservation expenses, 246–247
Refinancing, 133
Regulations
    environmental, 210–211
    labor, 211
    marketing, 211–212
Rent with option to buy, 101
Rental terms, 63
Returns to land, 40–41
    capital gains, 28
    compared to other investments, 28–30
    income returns, 28
Rights of ownership
    limitations of rights, 4
    right of disposition, 3–4
    right of exclusion, 3
    right of use, 3

S

Sole proprietors, 14
Second mortgage, 101–102, 134
Sellers of land
    absentee, 16–17
    estate, 16–18
    owner-operator, 16–18
    price/interest rate tradeoffs, 119–121
Selling land
    economic reasons, 33–34
    personal reasons, 32–33
Settlement sheet, 179
Soils
    depth and subsoil, 43
    productivity, 43
    texture, 43
    topography, 43–44
Soil classification
    land capability classes, 44
    soil series, 44
    ratings, 44–45
Soil conservation service, 44–45, 52–53
State programs
    cost-sharing, 209–210
    loan programs, 210

T

Tax Reform Act of 1976, 261
Tax sales of land, 33
Taxable estate, 275
Term of loan, 103
Tenancy in common, 165–166
Tenant liability considerations, 198–199
Time value of money, 71–76
Title defects
    clearing up, 174
    death of landowner, 173
    easements, 173–174
    judgments, 173
    mechanics' liens, 173
    spouses' interests, 173
    unreleased mortgages, 173
    variance in names, 172–173
Title insurance, 171–172
Title registration, 171
Transfers with retained powers, 271
Trespass, 201–202

U

Unified credit, 281–282
Unstated interest rule, 253
Use valuation
    capitalization of cash rents, 262–263
    crop share rents, 263–264
    effective interest rates, 264–266
    formula for determining, 262
    incentives for, 261–262
    in relation to gifts, 277–278
    post-death requirements, 267–268
    pre-death requirements, 264–267

V

Value of real estate
    capitalization, 62–63
    climatic factors affecting, 39–40
    economic factors affecting, 40–42
    elements of value, 8
    factors creating value, 9–10
    general factors affecting, 37–39
    market value, 8
    objective value, 7–8
    subjective value, 7
Valuing land
    cost approach, 66–68
    income approach, 62–65
    market data approach, 57–61
Variable interest rate, 135

W

Water rights
    appropriation doctrine, 5
    permits, 5
    riparian rights, 4–5
Wells, 48
Wrap-around mortgage, 134–135

Z

Zoning, 49, 159, 213

ABOUT THE AUTHORS

David A. Lins holds a joint appointment as associate professor of agricultural finance at the University of Illinois and as an agricultural economist with USDA.

Lins received his B.S. and M.S. degrees in agricultural economics from the University of Wisconsin. He then spent two years in Washington, D.C. working as a research economist with USDA. He then transferred to the University of Illinois where he completed a Ph.D. degree in agricultural economics in 1972.

As an employee of USDA Lins has been heavily involved in forecasting future financial structure and in providing economic analysis of policy options that influence financial outcomes in agriculture. He also serves as editor of Agricultural Finance Review and secretary of the National Agricultural Credit Committee.

Dr. Lins teaches both graduate and undergraduate courses in agricultural finance at Illinois. He has also spent a year teaching and doing research at Texas A&M University. He is co-author of AGRICULTURAL FINANCE: AN INTRODUCTION TO MICRO AND MACRO CONCEPTS, an undergraduate textbook. He has authored or co-authored over 60 professional articles and publications. Dr. Lins is a frequent speaker at conferences for lenders, agribusiness firms, and farmers.

Neil E. Harl is Charles F. Curtiss Distinguished Professor in Agriculture and professor of economics at Iowa State University. He received a B.S. degree from Iowa State in 1955, a Juris Doctor (law) from University of Iowa in 1961, and a Ph.D. in economics from Iowa State University in 1965. He is a member of the Iowa State Bar Association, American Bar Association and the American Agricultural Economics Association and is a member of several honorary fraternities.

Dr. Harl joined the Iowa State staff in late 1964 working in research, graduate and undergraduate teaching and extension. He was elected to the Executive Board of the American Agricultural Economics Association in 1979. Also, in 1979, Dr. Harl was appointed to the Advisory Group of the Commissioner of Internal Revenue, and in 1980 he was elected President of American Agricultural Law Association.

Harl is a recognized authority on farm estate and tax planning. He is author or co-author of more than 125 publications in legal and economic journals and bulletins and more than 350 in various farm and financial publications. More than 130,000 copies of 6 editions of Farm Estate and Business Planning are in use. He is author of the 15 volume technical treatise Agricultural Law, published by Matthew Bender. He has spoken widely on estate planning and organization of the farm business with speaking appearances in 36 states.

Thomas L. Frey is professor of agricultural finance at the University of Illinois as well as a Certified Public Accountant and Accredited Rural Appraiser. He has a B.S., M.S., and in 1970 received a Ph.D. in agricultural economics from the University of Illinois.

He spent eight years as a loan officer, two years as a fieldman for a Production Credit Association, and six years as manager of a Federal Land Bank Association. Dr. Frey regularly teaches at the Graduate School of Banking, University of Wisconsin, Madison, the Illinois Bankers Association School at Southern Illinois University, and in 1978 helped organize and teach a new advanced Agricultural Lending School at Illinois State University, sponsored by the Illinois Bankers Association.

Dr. Frey is a frequent lecturer at schools and conferences held by both banking and academic associations. He has authored or co-authored over 50 professional articles and publications on agricultural finance including Coordinated Financial Statements for Agriculture, You and Your Balance Sheet, and Covering Ag's Loan Needs. He is also co-author of a just published book, Lending to Agricultural Enterprises and he is a member of many honorary and professional organizations, most recently receiving the Paul A. Funk Award.

He has served a nine month sabbatical with Arthur Anderson and Company, the second largest accounting firm in the country, working to bridge the gap between current agricultural accounting practices and generally accepted accounting principles.

ABOUT OUR OTHER PUBLICATIONS

FARM ESTATE AND BUSINESS PLANNING
The Economic Recovery Tax Act of 1981 is the most significant piece of legislation to be passed by Congress since the historic Tax Reform Act of 1976. The seventh edition of the best-seller Farm Estate and Business Planning contains all the latest legislation and information regarding these changes affecting agriculture — from gift tax provisions to special use valuation of land.

Highlights include family estate trusts; looks at "carryover basis," joint tenancy rules; "use" valuation of farmland; chapter on father/son agreements; discussion of low or no interest loans to family members and update on the use of "flower bonds".

The author is Neil E. Harl, lawyer and Distinguished Professor in agriculture and economics at Iowa State University.

Available in soft cover.

COORDINATED FINANCIAL STATEMENTS FOR AGRICULTURE
This system of financial analysis specifically designed for agriculture, includes an 84-page manual and 16 statements and schedules. It was developed to aid farmers in organizing uniform accounting data, as it utilizes both a cost basis and current market basis valuation of assets.

Also available are slide/tape sets "Describing Your Financial Position with a Balance Sheet", and "Using Coordinated Financial Statements to Manage Your Farm Business Dollars". The authors are Thomas L. Frey and Danny Klinefelter.

COVERING AGRICULTURE'S LENDING NEEDS
This 32-page booklet explores and identifies the sources of financial market funds and traces the process of getting those funds into agriculture.

The authors are Thomas L. Frey and David A. Lins.

AGRI FINANCE MAGAZINE
Agri Finance Magazine is the magazine for agricultural lenders and professional farm managers designed to help them understand the financial needs and trends of their clients and the agricultural finance industry.

For price information and ordering, write Agri Business Publications, 5520-G W. Touhy Ave., Skokie, IL 60077, or call 312-676-4060.